Contents

List of Figures and Tables

FIGURES

TABLES

Modernising Hunger

FAMINE, FOOD SURPLUS
& FARM POLICY
IN THE EEC & AFRICA

Philip Raikes
Research Fellow
Centre for Development Research, Copenhagen

CATHOLIC INSTITUTE FOR
INTERNATIONAL RELATIONS
in collaboration with
JAMES CURREY · LONDON
HEINEMANN · PORTSMOUTH (N.H.)

Catholic Institute for International Relations (CIIR)
22 Coleman Fields, Islington, London N1 7AF

James Currey Ltd.
54b Thornhill Square, Islington, London N1 1BE

Heinemann Educational Books Inc
70 Court Street, Portsmouth
New Hampshire 03801

British Library Cataloguing in Publication Data
Raikes, Philip
  Modernising hunger: famine, food surplus & farm policy in the EEC and Africa.
  1. Food supply.   International economic aspects
  I. Title   II. Catholic Institute for International Relations
  338.1'9

  ISBN 0–85255–111–8 (James Currey) ✓
  ISBN 0–85255–112–6 Pbk (James Currey)

  ISBN 0–435–08030–X (Heinemann)

Cover photograph by Mike Goldwater/NETWORK

Typeset in 9/11 pt Mallard
by Colset Pte Ltd, Singapore
Printed in England by Villiers Publications, London N6

Preface and Acknowledgements

This book started life as the final report from a research project, studying how EEC policies affected agriculture (later revised to 'the food situation') in tropical Africa. It was originally intended as a study of the direct effect of aid policies (someone else had already 'covered' trade) on food production, which meant finding out how many EEC aid projects had anything to do with production of foodstuffs and considering how well they did it. Many research projects are composed by those proposing to do the research. This one had been devised by or for the body giving the funds.

From the start, this seemed to me an excessively narrow approach to the question, and as I proceeded, it seemed steadily more so. The 'food situation' clearly encompassed very much more than simply the aggregate tons of food produced (let alone food produced by aid projects), there being far more important determinants of the incidence of hunger. 'EEC policies' turned out to be less easily divisible between aid and trade than envisaged, or distinguishable from those of other powers. How, for example can one discuss food aid without looking at the trade policies which generate the surpluses which compose it?

This process of broadening out, of finding new and relevant areas of study, continued throughout the research and writing of the book. In a sense it still continues. Given the time over again, I would pay more attention to environmental issues, to the dynamics of peasant agriculture, and above all to trying to bring the material together in a more coherent shape. Whether I would succeed in the latter endeavour is quite another matter, as I have been trying to do just that for the past several years without success. One reason for this (apart from personal disorganization) is the way in which different organizing principles turn out to be partial, misleading or outright wrong.

One (pragmatic) way to organize such a book is to work towards a series of 'policy conclusions'. Indeed this is widely expected from the author of a research report. One has been granted funds to consider a specific aspect of development policy and come up with practical suggestions as to how things might be improved. But it becomes increasingly difficult to say what are practical suggestions, when one's research tends to show that what is politically feasible is usually too minor to make any difference, while changes significant enough to be worthwhile are often unthinkable in practical political terms. In any case, genuine practicality in making policy suggestions requires detailed knowledge of a particular country or area; its history, culture, vegetation, existing situation and much more besides. Lists of general 'policy conclusions' make it all too easy for the rigid-minded to apply them as general recipes, without thought, criticism or adjustment for circumstances. And yet, even with all these excuses at the ready, I still have this sneaking feeling that, as an employed (at least most of the time) 'development specialist', I should be drawing policy conclusions.

The book could also be ordered in disciplinary terms. But here a difficulty for one trained as an economist has been the increasing realisation that economics invariably leads one off at a tangent to any problem connected to the distribution of incomes and access to material resources. Economic analysis is so crucial to the understanding of what is going on in tropical Africa, and elsewhere, that it feels wrong not to be able to use it to draw conclusions. And yet to a very large extent one cannot. If there is one message I would like to come through clearly from this book, it is that a discipline which can only handle peoples' requirements when backed with purchasing power to become 'effective demand', can produce few helpful conclusions about issues like poverty and hunger, where needs are specifically not effective demand.

Much of this will already be obvious to those who are not economists, though many will harbour the delusion that their own discipline or profession provides the answer. For others, including to some extent myself, one or other brand of Marxism serves as an organizing principle and one with more serious pretensions that the separate social sciences. Here the problem is not that of 'the light that failed'. Along with many others of my generation and situation, Marxism never was that sort of shining religious light, so there was never any reason to suppose it perfect. At the same time it remains, in my opinion, by far the most useful source of insights about socio-economic structure and organization and much else besides. But considered as an overall system within which to locate a study of hunger, its weaknesses in areas like gender, environment and the analysis of peasantries raise major problems, as does a tendency to uncritical evolutionism. The more formalistic versions also share with western economics a quantitative approach to value and valuation.

One of the central problems in finding a clear positive message for this book, is that of value; of how one sets values on, and so chooses between different actions, outcomes, work-processes, goods, material resources and services. Most disciplines which seek to give answers to such questions do so by assigning quantitative values to things either, as with western economics, in the forms of prices generated by the market, or as with most 'official' Marxisms, as assigned by the state (often rhetorically but little more, on the basis of the 'labour theory of value'). Where large numbers of choices have to be made in a complex depersonalized society, this is to some extent inevitable, but considered as the underlying basis for valuation and choice, it misses a lot. Some things are easier to value than others (those easier to set a price on) and thus tend to be relatively over-valued. Valuation on the basis of things shifts the focus from people, reducing them from full humans to mere consumers. It also assumes away the problem that what is best for some may well be far from best for others. An extreme form of this is western welfare economics, for which the statement that a crust of bread is worth more to a starving beggar than to a millionaire is an inadmissible 'interpersonal comparison of utility'. But economistic Marxism uses the claimed 'primacy' of the economic level and of production over circulation to achieve a not dissimilar result – that criticized by Marx as commodity fetishism.

Other versions of Marxism do treat this issue seriously, through the notion of use-value, the value which people assign to concrete goods, services, relationships, activities (and qualities of all of these), in terms of their usefulness, beauty, humour, interest or simply congeniality. It becomes clear in any such discussion that the values assigned to different things and relationships vary enormously between different societies and positions within them, so that any discussion of the values to be assigned by the society as a whole is necessarily a discussion of how power and wealth are organized and distributed within them.

Crucial as such discussion is for the development of values upon which to found a worthwhile politics, it is much harder to use for 'policy-making' than exchange-value and its close relation price. A further problem is that such a discussion can only too easily drift off into a vague utopianism – and still more easily be dismissed as such by the hard-nosed. This is particularly the case with the general shift to the right of the past decade which has brough to many circles such an uncritical veneration for the market and its values that to propose any principle other than individual self-interest is to invite derision as a half-baked utopian. 'That's just the way the cookie crumbles baby, take it or leave it.' There is indeed increasing evidence that the world cookie is crumbling, and rather more rapidly than thought a few years ago, under the abrasion of market individualism by states and corporations. Nor is there much chance for most of us to leave it other than feet first. Sheer survival would seem to point towards broader decision criteria than market optimization.

But self-interested market-oriented behaviour is deeply socialized into people through the market itself and the whole political-ideological structure of capitalism, while at the same time appearing on the surface as the most 'direct', apolitical, asocial behaviour. By the same token, reversing this process will involve a major re-socialization of values to form an alternative basis for politics. How this can be achieved is quite another matter and one on which I would probably not reach tidy conclusions even if I had the whole time over again. In the meantime, as I have told myself a thousand times, the book may still be useful even without firm conclusions, so long as it provokes thought and discussion among others.

The study on which this book has been based was funded by the Danish Council for Development Research, via the Centre for Development Research, in Copenhagen, to which thanks. The book itself was mostly written while unemployed.

Since the book includes things which I have picked up from and discussed with colleagues and friends over a number of years, it is very hard to acknowledge even the more important sources of help, inspiration and encouragement. At the very least such a list would include Ole Moelgaard Andersen, Ann-Lisbet Arn, Henry Bernstein, Fred Bienefeld, Jannik Boesen, Debbie Bryceson, Lene Büchert, Sofus Christiansen, Lionel Cliffe, Sara Crowley, Len and Leslie Doyal, Poul Engberg-Petersen, The Food Aid Group at Copenhagen's Agricultural University, Barbara Grandin, Colin Hines, IPRA Food Policy Study Group, Tony Jackson, Cora Kaplan, Gavin Kitching, Mette Kjaer, Finn Kjaerby, Joergen Sams Knudsen, Gerda Landgreen, Roger Leys, Jesper Linell, Lipumba, Vicky Meynen, Britha Mikkelsen, Sam Moyo, Susanne Mueller, Benno Ndulu, Rie Odgaard, Chresten Petersen, Alanagh, Ben, Bobbie, Cilla, Jane, Jessica, Pam, Rob and Toby Raikes, Salva Rugumisa, Hans-Otto Sano, Sheila Smith, Mike Speirs, Ole Therkildsen, Mike Watts, Gavin Williams, Fiona Wilson, Ben Wisner, Ann Whitehead, Kate Young and Orla Zinck. This excludes authors of works I have found useful, since these are included in the bibliography, but Sen's *Poverty and Famines* perhaps did more than any other book to get me started. Kevin Watkins of CIIR has been particularly supportive and helpful since reading a rough first draft, while David Seddon and Margaret Cornell have read more recent versions and given useful comments. None of the above are, of course, responsible for my opinions and errors.

1
Introduction

Two years ago, the newspapers were full of news about famine in Africa. Famine persists, as do far more widespread 'normal' food shortage and hunger, though they are no longer news. But in a few years' time, if not sooner, there is every likelihood of another major and newsworthy famine, and when that happens, the same official spokesmen or their successors will intone the same platitudes about population growth and the need to increase food production as rapidly as possible through technical innovation. One major purpose of this book is to show that, while there is some truth in this argument, it is a very partial truth, and one with exceedingly dangerous policy consequences. Focusing solely on such technical factors is more likely to aggravate than to relieve the underlying problem, leading to policies which distribute income, food and control over resources *away* from the people most vulnerable to famine and food shortage.

In writing this book I have tried to get behind the bald facts of the food crisis in Africa: to put together the more important of the very many different factors and processes involved, to give some idea of what is going on, why it is going on and what might possibly be done to improve things. Here, I am concerned specifically with the level and reliability of food supply of those currently or potentially short of food. The 'other food crisis' – the steadily increasing import of basic foodstuffs into tropical Africa – is considered primarily in relation to how it affects hunger, which it does in a number of ways. I am also concerned to show that, while the import and hunger crises are related, they are separate. They are often wrongly collapsed into one crisis in which increasing imports stem directly from stagnating local food production. This exaggerates the per capita decline in food production and generates technicist policies, whose main, if not sole, aim is to increase aggregate production as rapidly as possible. One major difficulty with this approach is that today's short-term 'technical fix' can only too easily add to the problems of tomorrow. Another is that it ignores the vital issue of distribution.

People die of starvation, or go hungry, not because there is no food in their country (or region), but because they cannot afford it and have no other means of access. Food may indeed be physically absent from famine areas. But this reflects the fact that aggregate 'effective demand' (demand with the money to back it) is insufficient to draw it there from richer areas (or to prevent its outflow to such areas). And while famines have affected huge numbers of people in recent years, much larger numbers go hungry seasonally or most of the time, without any physical food shortage at all. To assure the food supply of poor people in Africa, as elsewhere, is both a matter of increasing aggregate food supply and one of providing the means for them to produce their own food or earn the incomes with which to purchase it.

1

Since those who focus on the distribution of resources are invariably accused of ignoring the importance of increasing production, it is worth stating from the outset that the book is very much concerned with production. I am only too well aware of the worrying implications of the existing combination of rapid population growth, environmental degradation and inadequate growth of production. But focusing on production does not (or should not) mean a concern solely with the quantity produced and the technical means for achieving this. Every bit as important are the socio-economic relations under which production is increased – who gets to control the process and product, how labour is allocated, by whom and to what ends. Ill-conceived projects to accelerate the growth of production only too often fail to achieve their stated aims. But even when they succeed in increasing aggregate production, they can fail to increase food availability to those most in need – or destroy the environment and long-term basis for future production. Increasing food production on the large farms of the wealthy usually does nothing to improve the food situation of the poor, and can make them worse off by diverting land and other resources away from them. Increasing next year's food production by monocropping maize in areas of fragile soils can all too easily reduce production potential in the long run. Many, if not most, current policies for Africa's agricultural growth, and the social processes with which they are inextricably entwined, tend to concentrate control over resources upon the wealthier farmers (and non-farmers) in the most climatically favoured areas of the continent and often do so in ways peculiarly inimical to the sustained and sustainable growth of production, in both ecological and social terms.

Gone are the days when famines were attributed simply to 'natural causes'. Almost everyone now recognizes the importance of environmental degradation – deforestation, erosion and desertification. Nor can there be much doubt that rapid growth of population has something to do with this, together with failure to adjust production methods and social systems to the changing environment and resource base. But this also misses out some crucially important factors. It is not just population growth which causes environmental problems; development policies and the class and other factors underlying them are also very significant. Neither is it simply a matter of failure to adjust production and social structures. On the contrary, the way they have been adjusted has been every bit as important. Nor does every harvest shortage lead to famine or major hunger; indeed it is not always the greatest harvest failures which lead to the worst famines. At every point, social, political and economic processes intervene, for better or (more commonly) for worse.

There are two simple explanations of the dominant food problems of Africa today, which appear to be diametrically opposed. On the one side are those, mostly representing tropical African states and their supporters, who place the blame on imperialism and the international capitalist system, or more specifically on the policies of the International Monetary Fund and the World Bank. The other argument, espoused most particularly by the World Bank and IMF, is that high population growth and low productivity, compounded by the policies of African governments themselves, are primarily to blame. There is some truth in both these viewpoints, and this fact itself points one towards those apparently non-controversial areas where they agree. One important example of this is the agreement among almost all official decision-makers, from IMF bankers to radical African leaders, on the need to modernize agriculture through technologies, inputs and equipment derived, if not imported, from the advanced Western (or socialist) countries. The specific means for achieving this vary widely, from private farms, both large and small, to state farms and producer co-operatives. But one thing they share is the assumption of the necessarily, beneficial and socially neutral impact of 'modern agriculture'. A major purpose of this book to demonstrate the falsity of this assumption.

Of course, modern agricultural methods are generally superior in terms of yield and output, under the conditions for which they were developed, but this does not mean that they will necessarily always be beneficial. The infrastructure and back-up services upon which they depend may be missing, or may cost so much to provide that they divert resources from other, more important areas. Techniques developed for large capitalist farms may not be suitable for small peasant holdings, and techniques which yield profits to individual farmers (especially where agriculture is subsidized) may not be the most socially beneficial. Even without subsidies, basing decisions on individual profitability can lead to practices like monocropping, which yield short-run profits at the expense of long-run damage to the soil. And this sort of problem can be especially serious when interest rates are high.[1]

There is abundant evidence that modern crops and techniques tend to concentrate resources both geographically and socially. Hybrid maize requires a higher and more stable rainfall than most local varieties of maize, let alone sorghum or millet, and its adoption shifts the focus of production to more favoured areas. The more money is invested in inputs and equipment, the more important it is in economic terms that investment should not be lost through climatic vagaries. This is reflected not only in decisions over investment but in research priorities, as in the recent decision by the International Livestock Centre for Africa (ILCA) to downgrade its previous emphasis on pastoralists in semi-arid and arid areas, in favour of a focus on mixed farming in areas of higher rainfall.[2]. The reasons given for this decision are that the latter can achieve greater values of increased production for given funds invested. The result is to shift resources away from one of the most drought- and famine-prone sections of the African rural population towards the development of dairy production for urban consumers by 'progressive farmers' in areas of higher potential and rainfall.

There is also a clear tendency for programmes of agricultural modernization to focus on the better-off farmers in any given area. In part this results from deliberate policy, since they are seen as more 'progressive' and open to advance – which is usually true since they have the means to purchase the requirements and can more easily sustain the risks involved. But it is often intensified by the provision of subsidies in various forms, including cheap credit, the lion's share of this being usually snapped up by the wealthier, who are also characteristically those with the most political influence.

But a third factor is perhaps most important in the long run and least observed. Increasing agricultural productivity through purchased inputs and equipment reduces the labour requirements for a given level of production, while mechanization also reduces the labour requirements per hectare. In an ideal society this would reduce the, often very heavy, labour burdens on those involved, to the benefit of all. Under capitalism its characteristic form is the expulsion of labour. Even the much lower rate of technological advance in European agriculture during the nineteenth century had such effects. In addition to the huge transfer of population to the towns, over 60 million people were pressed into migration to the 'new world' between 1860 and 1920. In present-day Africa, advances in industrial technology have vastly reduced the capacity of industry to absorb labour from agriculture, population growth is far higher, the pace of agricultural advance is much greater, and there is no new world to migrate to.

The above is not a plea for the retention of manual labour, low-yielding crop varieties and shifting cultivation. Of course, advance is needed to keep production increasing in line with population, while intensive forms of smallholder cultivation require more labour than do less intensive ones. Nor is it an argument that 'peasant wisdom' provides a sufficient basis for policies. For one thing, the subordinate position of peasants makes any such proposal utopian. For another, the wisdom applicable to subsistence production with abundant land is not necessarily applicable with land shortage and market integration. For a third, market forces have led to internal differentiation within peasantries, not only raising questions of 'which peasants?', but leading significant numbers of them to adopt harmful beliefs and practices. But I most emphatically do believe that far more account should be taken of peasant views and practices, in adjusting policies to take account of the long-term survival needs of the environment and the most vulnerable social strata. This book tries to analyse what this implies and why it will be, to put it mildly, an uphill struggle.

The EEC and Africa

While the primary focus of the book is Africa, it has another: the policies of the European Community and how they affect events in Africa. There are some fairly obvious reasons why this should be of interest. The EEC, considered as a bloc (the Community as such and its member states) is by far the major external trading partner of sub-Saharan Africa and overwhelmingly the largest project or programme aid donor (the USA is a more important food aid donor). This is so because the countries which now make up the EEC were colonial masters in tropical Africa for the best part of a century, not to mention the primary external actors in previous centuries of slavery, trading and colonization.[3] But it is impossible to select out the policies of the EEC and study their effects in isolation, since this ignores their interaction with other factors. One of the points to have emerged most strongly in researching for this book is the strong degree of inter-connection between different factors – the way they work through one another and alter one another's effects.

The combined members of the European Community certainly form the most important external economic group interacting with tropical Africa. But the term 'EEC policies' applies only to the policies of the EEC as such – the Commission and its various sub-agencies. And the impact of these is far more limited and recent. EEC policies are thus only one small part of the whole complex of external factors bearing upon the food situation in Africa and are in any case invariably filtered through the international system and the local tropical African state, rather than having a direct impact. Each individual aid project, food aid transfer or price effect feeds into a variety of existing local processes and affects, and is affected by, them.

There are also questions about the degree to which EEC policies can be considered *independent* factors. One may be able to measure the effects of EEC protection (or may not, see Chapter 7), but how much of that can be attributed to the Community, when the member states would be protecting their agriculture even in its absence? Specific EEC policies are often responses to international commodity price changes and general economic trends or to the policies of other blocs or countries, notably the USA.

Thus, although the absurdities of the Common Agricultural Policy make it an easy target, I have avoided the temptation to make this book a diatribe against the European Community, or to blame it for the situation in Africa. Not only would this be inaccurate, but it would exclude many interesting and relevant connections. What I have tried to do is rather to show how the EEC fits in with the international agricultural system and exemplifies some of its characteristics. I have then tried to indicate some of the more important relationships between processes occurring in Africa and those at the international level.

The hunger problem and tangents to it

The focus on adequacy and reliability of food supply to those most threatened has proved quite hard to maintain. In part this is because much of the book is concerned with national policies and international factors, in discussion of which the focus on food needs easily gets lost. But another reason is that in order to understand the processes by which the current situation came into being, one must often use methods of analysis which proceed at a tangent to the problem, replacing it with one or other technical or economic 'proxy', such as the level of agricultural production or national income per capita.

For example, economics, the major discipline at hand for the study of resource allocation, resolutely excludes matters relating to income distribution from the agenda. This is not simply a matter of the choices of individual economists. The methodology and theory underlying the optimization around which the discipline is built require abstraction from questions of distribution.[4] Moreover, economics works only with 'effective demand', that is, demand backed by purchasing power. It thus has no means of considering needs which are not so backed – like food shortage and hunger. Of course, distribution and needs can be re-inserted into economic calculations, but always in such a way that they become a 'constraint upon' the optimization process rather than part of the process itself.

Nor is this simply a matter of theoretical or mathematical convenience. That economics ignores problems of distribution and needs not amounting to 'effective demand' is in part a reflection of the reality which it attempts to model (though also of its acceptance of the values involved). The conflict between growth and equity is not a mere figment of the economist's imagination. In some ways the concentration of resources makes them easier to deploy for investment purposes. So long as national income or wealth is defined as the sum total of goods and services in a given society, weighted by the prices generated by a given income distribution, it is quite likely that its growth will be accelerated by a high degree of inequality. In such a situation, to assure food security, income, employment or other basic needs to all can be seen as a 'constraint' or drag on development, most especially as this will often require state activities, involving control over and revenue extraction from the private sector and thus 'market distortion'.

All this makes economics a murky and distorting lens through which to view the hunger problem. The latter continually disappears from focus, to be replaced by the functioning of markets and the value of aggregate production. But economic analysis is unavoidable as a tool for understanding what has happened (one of the most useful analyses of how famines occur is largely an exercise in microeconomics) and what is feasible. One can scarcely ignore the power of market trends in studying agriculture in tropical Africa, or at the world level, come to that. One can, however, seriously question whether their 'unrestricted' operation is always beneficial to those at greatest risk from hunger.[5] I shall try to demonstrate this is in more detail in the book. For the present, the point is that economic analysis will never be sufficient on its own to yield conclusions appropriate to solving problems of hunger, food shortage and insecurity. On the other hand, conclusions which ignore economic forces and processes are unlikely to be realistic.

More generally, I believe it to be no longer tenable, if it ever was, to treat the hunger problem as something which will be (or can only be) solved by 'development' or 'the market' or 'socialism'. The record of the past several decades of 'development' in tropical Africa inspires little confidence on this score and, as I hope to demonstrate, recent policy changes have been insufficient to tackle the problem or have been

misdirected. Some of this has to do with an exaggerated respect for the benign powers
of free markets, coupled with an unrealistic optimism as to the likelihood of their
occurrence. But then socialism, as practised in Africa, has shown itself still less
successful at tackling food problems. Some of this can be attributed to hostile external
action, but by no means all.[6] The hunger problem cannot be usefully understood or
tackled in the context of one or other general 'paths' of development. Nor can it can be
reduced to narrow technical and economic grant analysis. There is, in my opinion, no
one model, be it neo-classical, Keynesian or Leninist, which can provide a full theoretical underpinning for a study of food and hunger.

Theoretical and methodological implications

This belief has had a number of implications for the way in which the present study
was performed and the book written. I have found Western economics indispensable
as a means of analysis but inadequate as a source of conclusions. The same is true of
Marxism, which I find by far the most useful source of theoretical insights, while being
unable to espouse any school of Marxist doctrine and preferring to avoid the heavier
end of its terminology. I am not in doubt that class formation and the operation of the
capitalist system lie behind most of the problems to be discussed in this book. But the
aspects of capitalism considered here are not always those central to standard
Marxist analysis. Most forms of Marxism, for example, consider the issue of distribution to be in some sense 'secondary' to production, where I consider the two so intimately related that it makes no sense to make such a distinction. Again, I make
virtually no reference to concepts like 'surplus-value', since I regard them as largely
unusable in discussing real world phenomena. Nor is class formation in Africa easily
fitted into the traditional Marxist model, though the general approach is crucial to
understanding the process.[7] This relates in part to the purpose of the book. While it is
concerned to make a variety of different processes intelligible, which inevitably
involves analysis and theory, it is not intended as a theoretical work. But this itself
derives in part from the opinion that (at least for the purposes of this book) an attempt
to reduce all the processes described to one abstract theoretical structure would
involve losses of accuracy outweighing any gains in theoretical rigour.[8]

Nor is the problem confined to choice of theoretical school. Any study of the African
food problem has to span topics as widely separated as peasant household decision-
making, the operation of international markets, the development of agricultural technology and the political factors lying behind and linking these. Clearly this requires
that one select from different types of analysis and risks a degree of inconsistency
between the parts. It also poses problems about the order in which to present the
material, which I cannot pretend to have solved satisfactorily.

There is another problem, perhaps less important for this book than for the research
report from which it is adapted, but worth mentioning. Academic research within the
social sciences, including development problems, is much concerned with 'scientific
method', which, by analogy with what is taken to be the method of the physical
sciences, is much concerned with statistical verification. I have given a number of
examples in the book of cases where it seems to me that a heavy artillery of statistical
method is brought to bear to reach conclusions which are either trivial or meaningless.
But, in general, such a method – and its corollary of picking out one or a few limited
hypotheses for testing – seems inappropriate to the subject and purpose of this book.
By this I do not mean to imply that questions about the validity and generalizability
of statements made are not important. Of course they are, and I have attempted

wherever possible to indicate the strength of the evidence for arguments put forward. What I mean is rather that to spend whole chapters, or even a whole book, applying sophisticated models and statistical techniques to specific questions (like the impact of EEC protection on world prices of specific commodities) cannot be justified either on grounds of relevance or on the quality of the data available.

This relates back to the purpose of the book, which is to try to illuminate issues and provide some basis upon which the reader can develop his or her own conclusions. Given this aim, the sort of analysis which I find most useful interweaves concrete description with interpretation and analysis, often employing extended chains of reasoning in which statistical results can at most only be referred to if the flow is not to get bogged down. Examples are intended to illustrate points made and to provide insights into the sorts of things which can happen. They do not, and are not intended to, 'prove' anything.

Policy and politics

The further I have proceeded with the research for this book, the more difficult I have found it to present 'policy conclusions'.[9] There are, of course, plenty of obvious conclusions to be drawn. One could, for example conclude (for the n-hundredth time), that more resources should be devoted to solving the hunger problem. Of course they should. But there is no sign that this will happen and little evidence (other than rhetoric) that solving the hunger problem is an overriding concern of the world's political leaders, most of whom are little enough concerned with the conditions of the poor in their own countries, let alone those of black Africa. Nor is there any sign that if increased resources were forthcoming, they would be applied so as to tackle the central problems. When one turns to the European Community, it is again not hard to think of changes to its policies which could be helpful, but little evidence that the situation of poor people in Africa or the rest of the Third World is a significant determinant of EEC policy. Nor is there much evidence that this is a major priority of those in control of most African states, except to the extent that certain sections of their populations (primarily urban and/or middle-class) can bring down governments if their food needs are not met.

The problem is thus not so much to think of policies which could help, but to think of policies which could help but which are not simply utopian as things stand at present. The book is full of implicit policy conclusions, though they are often not worth formulating as 'policy conclusions' because there is little chance that they could be implemented, at least without such distortions as to nullify their effects. 'Policy' does not exist in a vacuum. The particular mix of policies, politics, patronage and entrepreneurial activity which has emerged in many African states is not merely accidental, nor is it the fault of corrupt or incapable persons. It is closely bound up with the emergence of a complex and unstable class structure in African countries, and its interaction with a far more developed and powerful international one. Certain tendencies in policy are part and parcel of that process and of the concurrent emergence of dominant ideologies. But there are also problems to which there seems to be no correct 'policy solution'.

A significant example of this is state monopoly control over agricultural marketing – a widespread phenomenon in tropical Africa, virtually regardless of formal political ideology. There is a super-abundance of evidence that the operations of such monopolies reduce producer prices to peasants and thus lower the amounts of agricultural produce sold. It is not just that these monopolies are an important source of state revenue (indeed in many cases costs of operation are so high that they are not). More importantly, they are a significant means of accumulation for members of the emerging ruling classes. Thus one criticism of those (like the World Bank) who propose de-monopolization and privatization is not that this is anti-socialist but that it is unrealistic. Where privatization does occur, it is likely to take forms which preserve much of the monopoly power and thus the capacity of the well-placed to continue to extract a revenue surplus from the process.[10] But another problem with the 'market solution' is that, since it operates solely in terms of 'effective demand', it is often a most efficient means to channel food *away* from the most needy in times of stress, and especially during famines. While it is true that most existing state monopoly agencies provide for the needs of the poor ineffectively, if at all, there definitely is a need for some agency which can collect and distribute food in times of need, on terms other than those of the market.

Another significant example is that of aid policies. There has been increasing recognition, including by aid-agency officials, that large-scale capital-intensive irrigation, ranching, state-farm and similar projects provide answers to neither the economic problems of African states nor those related to the well-being of the mass of their inhabitants. Yet they continue to be funded. The reasons for this are not simply the interests of the donor countries and firms involved, significant though these are.[11] Nor are they just the material interests of ruling classes and powerful individuals in recipient countries. They are closely bound up with a whole statist ideology of development through modernization which, though closely related to the above interests, cannot simply be derived from them, as propounded by the more economistic streams within Marxism. Since state-to-state aid is subject to such pressures, one solution proposed is the routing of more aid programmes through voluntary agencies and non-governmental organizations (NGOs) – which has indeed happened increasingly in recent years. While, in general, this seems a positive trend, it should not be seen as a panacea. Such agencies themselves vary widely in their commitment and capacity. Perhaps more importantly, the routing of state or international agency funds through voluntary agencies submits them to many of the same pressures operating on official aid agencies – as the more aware NGOs willingly admit. Apart from bureaucratization, arising from the necessity to comply with state-agency accounting systems, this can lead to the adjustment of policies or personnel to make them more acceptable to sources of funds.

EXPORT CROPS VERSUS FOOD CROPS

Another issue is perhaps worth considering at this point, since it figures widely in discussions of food needs and hunger. A debate[12] has raged for many years between those who claim that emphasis on export-crop production leads to deterioration in food availability and those who argue, to the contrary, that increased earnings from export-crop production allow higher levels of food consumption. The argument is that the foreign exchange earned from export-crop production would buy more (imported) food than could be produced locally with the same resources. Alternatively the argument can take the conditional form that it is preferable to concentrate on export crops *when* prices are such that more food could be imported from their export proceeds than could be grown with the same land, labour and inputs. So far as it goes, this is a perfectly reasonable economic argument.[13] Other things being equal, it seems obviously true.

But other things, and notably the distribution of income, are not necessarily equal, and the question cannot be answered in solely economic terms, since there are a number of distributional issues involved, upon which most criticisms of an export-crop strategy focus. Encouraging export crops may imply concentrating on the already richer regions of a country. It may precipitate a land-grab in which the rich and politically influential dispossess the powerless. Men may increase their cash income and expenditure, while women and children lose a part of their previous food supply, as seems to be demonstrated by a number of the nutritional studies cited in Chapter 4. Mechanization may reduce labour requirements and thus wage income. Ill-advised or high-risk innovations can lead to losses or debt. But all this could happen with the development of food crops for commercial sale, or non-food crops for local sale, for that matter. These are processes related to the commercialization of agriculture rather than to any particular end-use of the product. One cannot answer the question about food crops versus cash crops in the terms in which it is set. One needs to know *what* crops and their specific implications in a particular situation.

There are strong reasons, however, for doubting whether poor people's access to food is likely to be improved by increased emphasis on export-crop production. Just because the foreign exchange from export crops could pay for food imports, does not necessarily mean that it will do so. There is no reason to suppose any relationship between the two. Nor is there any assurance that if food is imported, it will be destined for, or suitable to, those most in need – or priced and distributed to allow them access to it. If local producers switch from maize to cotton production for export, and this allows an increase in imports of canned meat or dairy products, this will not do much for poor consumers. There is, moreover, a general tendency for a high proportion of imported foodstuffs to remain in the towns, especially where, as is not uncommon in tropical Africa, capital city and main port are the same.

This does not necessarily mean that it is best to concentrate on food crops. One cannot decide what is the best pattern of production without information on how relative incomes and access to resources are affected by the choice, which will be as much a matter of *how* the crops are developed as of whether they are destined for export, the home market or subsistence. This will also relate to the quite separate discussion about trends in international prices of export commodities.

HOW THE BOOK IS SET OUT

The book is divided into two Parts. In Part I the following three chapters concentrate on the situation in Africa. Part II introduces international factors, considers the European Community in that context, and ends with an attempt to integrate conclusions from both parts.

Chapter 2 looks at the 'conventional wisdom' on the food and economic crisis in Africa, and demonstrates that in several important respects it is either wrong or dangerous oversimplified. Much of it is based on the misinterpretation of unreliable and biased data. The chapter goes on to discuss some of the processes within the food sector which have given rise to rapid increases in imports, and their interpretation as aggregate food shortfalls.

Chapter 3 starts with an outline and discussion of the more general African crisis, relating it to processes of class, state and policy formation, and showing how these underlie some of the problems discussed in Chapter 2. It then goes on to look at agricultural policies and a number of their specific effects.

Chapter 4 is concerned with how hunger and food shortages emerge, and who suffers from them, when and under what circumstances. It starts by looking briefly at different types of food shortage, and considers the impact of seasonality and seasonal variability. It then looks at the long-term factors lying behind food shortage, before discussing the more immediate causes of famine and A.K. Sen's analysis of 'entitlement to food'. Examples are then presented to illustrate both some of the points made in the chapter and some of their limitations.

Opening the second part of the book, Chapter 5 examines the development of the world and African food systems, indicating some of the important factors which affect and are affected by EEC policy. The chapter starts with a background sketch of the development of production and international trade in foodstuffs (and more recently feeding stuffs), before proceeding to the events of the post-1945 period and their implications.

Chapter 6 then introduces EEC policy in general, and is followed by three chapters which treat specific aspects of that policy in more detail. Chapter 7 looks at agricultural protection and the Common Agricultural Policy (CAP). The first section indicates some of the effects of the CAP *within* the Community, notably its impact on the structure of production. This is followed by consideration of its external effects, starting from some econometric studies, and proceeding to a more general argument. The final section considers recent changes in EEC thinking and what changes in strategy seem likely to emerge from them.

Chapter 8 looks at food aid and continues a discussion started in the previous chapter on the likelihood or otherwise of 'disincentive effects' on local production and/or the development of dependence on external sources of supply. The chapter also considers the history and institutional structure of food aid over its relatively short lifetime, noting the dominance throughout of the USA. While most of the chapter concentrates on cereal food aid, a section looks briefly at dairy food aid, the experience of various attempts at a 'white revolution' and their implications.

Chapter 9 focuses on project aid, starting with a discussion of aid to Africa under the Lomé Conventions, both generally and by separate country programmes. The chapter ends with a discussion of the main trends in aid policy and of whether there are any significant differences between the European Development Fund and other agencies.

Chapter 10 draws the material from the different chapters together, to show the interrelationships. What emerges clearly (to me at any rate) is that current trends in policy seem unlikely to solve either the hunger problem or the food-import problem. A general alternative direction is sketched very briefly, since it departs far too radically from current assumptions and visions to be even a starter as 'policy'. It may, however, help to provide a focus for small-scale or local-level activities, which seem to me to offer the best hope.

NOTES

1. Economic calculations of viability almost invariably involve some means of 'discounting the future' and normally use the prevailing real rate of interest for this purpose. To give some idea of the effect of this, when the rate of interest rises above, say, 10%, anything occurring more than ten years from the date of calculation is reduced almost to irrelevance. Under almost any rate of interest, factors occurring forty or fifty years hence have no impact at all on decisions based on such criteria.
2. Article in *New Scientist*, 25 June 1987 by Deborah MacKenzie subtitled: 'Scientists are learning how cattle rearing works in the drought-ridden lands of Africa. . . just as Western aid agencies are losing interest'.

3. More prosaically, this book derives from a research project whose purpose was to consider the impact of EC policies on the food situation in Africa.
4. For an explanation of why this is so, see the first chapter of virtually any economics textbook.
5. The quotation marks indicate that there are few, if any, markets which are in reality unrestricted in one or other sense.
6. Some of the examples in the book discuss these issues, cf. Raikes 1984 (Mozambique) and 1985 (Tanzania). One of the less often discussed problems with socialism in Africa (and in agriculture over a wider scope) is that it tends to imbue its advocates with the self-confidence and self-righteousness to undertake exaggerated modernization policies, with great disruptive effect. However, this is certainly not the only source of unjustified self-confidence. The ill-advised policies of (say) the World Bank or the Nigerian Government can scarcely be ascribed to socialism.
7. For one thing, as discussed in the final chapter of this book, it tends to assign such large proportions of the population to the residual category, the petty bourgeoisie, as to lose much of its analytical bite.
8. I am well aware that the above is highly over-simplified. Its purpose is to indicate roughly where I stand rather than to justify that position.
9. This was, again, perhaps a greater problem for the research report from which the book is adapted. The conditions of funding for such reports often require that one should offer 'practical' solutions to problems, primarily for the use of aid agencies. Much of the book is concerned with showing that this is not really the most practical approach.
10. See Chapter 3 for examples and discussion, including the views of Robert Bates, who recognizes this fact but proposes a solution which is arguably more harmful than the disease – the concentration of land and other resources in the hands of a powerful agrarian bourgeoisie. Not only would this imply a further loss of control over such resources by the most vulnerable sections of African populations but, as Bates himself admits, there is no guarantee that it would have the desired effect. To the extent that an 'agrarian bourgeoisie' has emerged in Africa, it has been heavily dependent on accumulation through state resources and there is no reason to believe that its further dominance would result in altruistic self-denial on this score.
11. See Birch 1987 for a detailed analysis of such factors in relation to tractor mechanization.
12. Polemic might be a more accurate term since most concerned simply talk past one another.
13. See, for example, Pinstrup-Andersen (1985). This is to be distinguished from a much more dubious argument put forward in the World Bank's Berg Report (World Bank 1981), that there is a complementarity between food and export-crop production such that encouraging export-crop production tends to increase food-crop production. Since different crops, whatever their final use, compete for land, labour and inputs, this is an odd argument resting largely on the thin reed of residual fertilizer effects (fertilizer applied to one crop in one year has some effect on whatever is grown there the following year) which is valid only if export and food crops are rotated, which is seldom the case. Slightly more convincing is that money earned from export-crop production may be used to purchase inputs for food-crops, though evidence for this is at best patchy. The empirical evidence rests on the greater success of certain market-oriented countries in attracting sales of food products on to official markets. As shown in Chapter 2, this is not a valid indicator of total food production, since these countries have often been successful in attracting smuggled produce from their neighbours.

PART ONE

Africa

2

Africa's Food Gap:

Why Has it Emerged?

Introduction

Over the past decade and a half, tropical Africa has become a significant net importer of basic foodstuffs, notably cereals. Since the continent is among the world's most 'land-rich' areas and since the vast majority of the population is composed of agricultural producers, this clearly points to serious problems, relating in some way to hunger and famine. But the purpose of this chapter is to show that the relationship is much less close and direct than is commonly thought. It is thus more of a ground-clearing exercise than a direct approach to the hunger problem, though it does have something to say about the latter.

While opinions vary as to the underlying causes, most observers attribute growing food imports to an increasing imbalance between population and food supply. One of the most commonly accepted explanations of the crisis runs somewhat as follows:

Population is growing more rapidly in Africa than anywhere else in the world. Given low levels of technology and low rates of development, growth of food and other agricultural production cannot keep pace with that of population. Therefore imports have increased to fill the gap. This is made worse by inept and wasteful government policies, which have favoured the urban areas and the industrial sector at the expense of agriculture. As a result, export production has fallen and imports of capital goods have increased. Together with increased food imports, this has led to increasing indebtedness and economic crisis.

Thus what is needed can be fairly simply stated (if not quite so easily achieved). The terms of trade to the agricultural sector must be improved (better prices to farmers); government expenditures (especially in the urban areas and for social services) must be cut, to reduce the tax burden on commodity prices; and all efforts must be focussed upon increasing productivity within the agricultural sector, to 'get production moving'. The major means for this is seen as technological development – if possible, a 'Green Revolution' for Africa.

It is not my purpose to deny that rapid population growth and misjudged government policies have led to major problems. But there are serious difficulties with parts of this explanation. In the first part of the chapter I shall try to demonstrate that:

(i) The growth of food imports into sub-Saharan African has been exaggerated.

(ii) The direct evidence for stagnation in food production is both very thin and subject to significant bias.

The net effect of these biases has been to exaggerate the aggregate production shortfall and thus to place excessive emphasis on policies aimed at increasing production rapidly through technological change. This shifts emphasis away from the need to generate incomes for poor people and the need to base production increases on technologies which are environmentally and socially sustainable.

In the second half of the chapter, evidence is presented to show that food imports have increased primarily, though not solely, as a result of rapidly increasing urban populations, incomes and political clout, combined with state pricing and marketing policies, significantly assisted by donor and developed country policies. The purpose of this is in no sense to deny the existence of a major food problem. It is simply to elaborate the rather obvious point that it is not countries which suffer hunger (and certainly not their political leaders), but specific disadvantaged sections of the population, often situated in particular hunger-prone areas.

Food imports into sub-Saharan Africa

> Cereal imports into sub-Saharan Africa . . . have been growing fast, doubling every seven years approximately . . . the food situation begins to look really alarming when recent trends are projected into the future . . . The food deficit in 2020 would correspond to the entire present-day agricultural production of India. (FAO, 1983:12)

Anyone who has looked at FAO reports in recent years will have seen dozens of dire predictions like the above. To put this in perspective, Indian cereal production in 1983 was approximately 166 million tons (FAO, 1985c, Table 7), roughly three times that of sub-Saharan Africa, sixteen times the 1984 peak level of cereal imports into tropical Africa and more than three-quarters of the level of total world cereal trade.

Such predictions are less common today than they were two or three years ago, but a glance at Figure 2.1 shows that the prediction was never very likely. Tropical

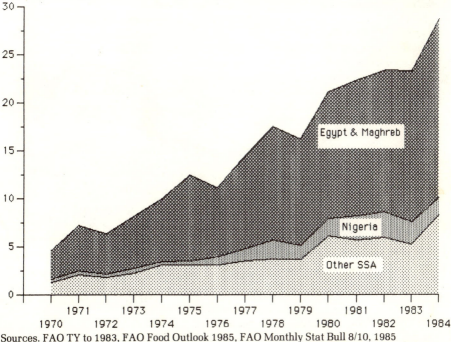

Sources. FAO TY to 1983, FAO Food Outlook 1985, FAO Monthly Stat Bull 8/10, 1985

Figure 2.1: *African cereal imports 1970–82 (million tons)*

African food imports *have* grown very rapidly – to five times their 1970 level in fourteen years. But even a straight-line projection, based on the maximum growth period 1979–84, would give a figure for 2020 of not much more than 40 million tons which, large as it is, is about one quarter of the Indian figure quoted above. But Figure 2.1 also shows that by far the largest proportion of cereal imports into Africa does not go to sub-Saharan Africa at all, but to Egypt and the Maghreb, whose imports have also grown far more rapidly over the period since 1970. One seldom sees the argument about disparity between population growth and food production capacity put forward in regard to the North African countries, however. This is because it is clear that the reasons lie elsewhere.

Egypt, the major African cereal importer, which alone imports about as much as the whole of sub-Saharan Africa combined, is also the world's largest cereal food aid recipient. It is US policy to allow the Egyptian Government to operate a 'cheap bread' policy, so long as it maintains the approved stance with respect to Israel. But since Egypt is also a traditional market for EEC (French) grain, there is a significant element of competition for markets (see Chapter 6). Of the other Maghreb countries, Algeria, Libya and Tunisia are 'Upper Middle-Income Countries' (UMICs), with per capita income levels from 5 to 15 times the sub-Saharan average (OECD, 1984: Table II.1.13). Morocco and Tunisia are also major food aid recipients. Only Egypt falls into the OECD's second lowest income category for Third World countries (LICs) and there is some reason to suppose that the upper boundary of that category is defined as being 'above Egypt'.[1] However this may be, food expenditure levels are on average far higher than in tropical Africa. North African grain imports are increased by oil incomes, food aid and deliberate state policies.

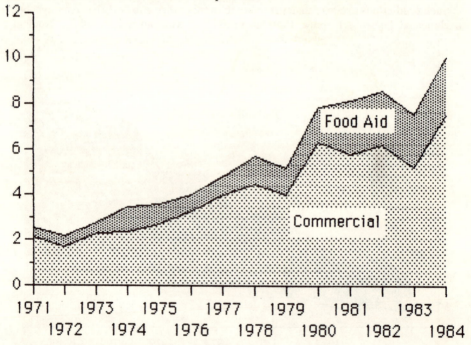

Source. FAO, Food Aid in Figures & as for Fig. 2.1

Figure 2.2: *Cereal Imports into sub-Saharan Africa, commercial and food aid 1971–84*
(million tons)

Food aid has also been an important factor in tropical Africa, as is shown in Figure 2.2. Over the period 1971–84, commercial cereal imports increased by three and a half times, while food aid transfers increased by over 6 times. It is not just that increases in food aid account for an increasing proportion of total cereal imports; they can also act as 'loss leaders' in encouraging increased commercial imports. It is no coincidence that the food commodities which showed the most rapid growth in imports into Africa were wheat and dairy products, the two most important food aid products (see Chapter 8).

Another thing which emerges from comparison of food aid and food import figures is that an increase in food aid shipments tends to be followed one year later by a reduction in commercial imports and two years later by a considerable increase. One explanation for this (and one which fits with other evidence) is that food aid (which takes considerably longer to ship than commercial imports) arrives in time to pre-empt some of the following year's commercial trade, but has itself an impact on subsequent levels of demand and imports. Each major crisis leads to an increase in food ship-ments, some of which subsequently is transferred to commercial trade, funds permit-ting. This effect has been less noticeable in the aftermath of the 1984–5 shortage, because of increasingly severe foreign-exchange constraints.

To summarize, food imports into tropical Africa certainly have increased rapidly since 1970 and do weigh heavily on the balance of payments of many countries. But the trend has been much exaggerated in the more alarmist reports and the impact of food aid in this respect has been largely ignored.

The evidence on production

FAO and other agencies have issued a flood of reports during recent years, indicating that food production in sub-Saharan Africa has been growing more slowly than popu-lation and food requirements. Thus the FAO *World Food Report*:

> Africa South of the Sahara is losing the race to keep food production ahead of population growth. New technology, new attitudes toward agriculture, and new forms of assistance from the international community are needed. (FAO 1983:9).

> Of 41 sub-Saharan countries with a significant agricultural sector, only five (Cameroon, Central African Republic, the Ivory Coast, Rwanda and the Sudan) have kept food production consistently ahead of population growth. (ibid.12).[2]

Two well-known and respected researchers have used FAO data as the basis for a study in which they reach similar conclusions:

> It is particularly notable that land-rich Africa had a growth-rate of land-area in food-crop over 2.5 times that of Asia in the 1960s, while in the 1970s the growth-rate of crop area was even less than that in Asia. Food production in Africa during the 1970s increased at less than half the rate in Asia. The rapid increases in food imports into Sub-Saharan Africa are readily understood in view of this abysmal production record. (Paulino and Mellor, 1984:298).

One could choose another twenty or thirty references from respected experts, making similar claims with similar self-assurance. One might thus assume that the data upon which these conclusions were based were reasonably well-established and accurate. Nothing could be further from the truth.

The data upon which most of these predictions are based come from FAO *Produc-tion Yearbooks*, which present estimates of area cultivated, yield per hectare and tons produced for about 75 different crops or groups thereof and a variety of different livestock categories. Up to 40 sub-Saharan African countries are covered – since not all of them produce all of the products included. The data come from two main sources. Where they are considered reliable, data provided by African governments are used. Where such estimates do not exist (or are not considered sufficiently reliable), the FAO makes its own estimates. In some cases, there are FAO experts who can assist in this process, though they do not always seem to be used.[3] Otherwise they seem to reflect the best estimates or guesses of local FAO administrative personnel. FAO estimates are normally heavily rounded (i.e. to the nearest hundred thousand or half million tons).

In 1982, no less than 75 per cent of all cereal production figures for tropical African countries were based, wholly or in part, upon 'FAO estimates'. The 1984 *Production Yearbook* would appear at first sight to have been more successful in getting African governments to provide figures. Slightly over half of all estimates of 'total cereal production' have no 'F' or 'X' mark denoting either 'FAO estimates' or 'unofficial estimates'. But if one looks a bit closer, the estimates of production of different cereals on which they are (presumably) based have rather more 'FAO estimates' – 70 per cent, for example, in the case of maize, the most important cereal grown on the continent. The conclusion can only be that someone forgot to type in a large number of the 'F's and 'X's.

In point of fact, however, this makes little difference, for whether FAO or the government makes the estimates, there are few countries in sub-Saharan African where the level of total food production is known to within plus or minus 20 per cent. Even Zimbabwe, where a high proportion of production comes from a small number of well-covered large commercial farms, has failed in recent years to predict commercial deliveries of maize. The reason for this is the poor basis for estimating peasant production. And this is the country which (at least by reputation) has the best extension and predictive service in tropical Africa. By no means entirely at the other end of the scale is Tanzania, where different estimates of food-crop production from the Ministry of agriculture vary by a factor of up to three. In this case, FAO also varies its estimates considerably. Thus the 1982 *Production Yearbook* shows Tanzanian maize production in 1982 to have been 0.8 million tons. The 1984 *Production Yearbook* shows 1.55 million tons for 1982. In May 1985, the FAO expert most qualified to estimate maize production in Tanzania was of the opinion that total maize production was well over 2 million tons and had been for some years.[4]

If one compares estimates of maize production for 1982 in the 1982 and 1984 *Production Yearbooks*, for the twenty largest sub-Saharan African countries, only five remain unchanged. Eight are revised between 1 and 10 per cent, another five by between 11 per cent and 50 per cent and two by over 50 per cent. But while this gives some idea of the degree of uncertainty, there is no reason to suppose that because figures have not been revised, they must necessarily be accurate. When, as is often the case, they are heavily rounded, this could as easily simply reflect that no new information has emerged. This seems, for example, to be the case with unchanged estimates for Angola (250,000 tons of maize). On the other hand, the apparently more detailed estimate for Uganda (293,000 tons, changed to 393,000 tons) hardly inspires more confidence. Apart from wondering which contains the typing error, the situation in Uganda in 1982 (or 1984 for that matter) was hardly such as to allow accurate estimation of maize production. Similar considerations apply to the many other countries in which armed incursion or civil war preclude large areas from virtually all aspects of civil administration.

Even in the developed countries, agricultural production statistics are probably less accurate than is normally supposed, but the opportunities for accurate counting are far superior. Major cereals are almost entirely grown on large farms, which not only

declare the area planted and harvested to government statistical agencies, but produce almost solely for the market.[5] With area cultivated known and the quantity sold
close to the amount produced, the two provide an independent check on each other.
Moreover, with larger more homogeneous farms, yield measurement and prediction
are far easier, providing a further independent check on the figures.

In Africa, most staple food production comes from small peasant farms, the vast
majority of which are covered by no system of registration or crop reporting. Where
these do exist, they refer almost without exception to large and medium-scale farms,
whose contribution to total production tends, in consequence, to be exaggerated.[6] It is
almost impossible to produce even reasonably accurate estimates of the area cultivated. Aerial photography and satellite imagery are less easy to use in Africa than in
the developed countries. Aerial photography is too expensive to be carried out more
than once every ten or twenty years, while the area planted can vary widely between
years. In any case funds and personnel are seldom available to analyse the data.
Satellite imagery should be a help, but the minimum resolution of currently available
satellites is 80 sq. m., which is considerably larger than the average African peasant
field.[7] There are serious issues of definition when crops are interplanted and when
fields are being semi-cultivated while on their way back to bush-fallow, to name but a
few of the problems.[8]

Ground-surveys are limited by funds and personnel, by transport and the road network, and yet more by the cost and time required to survey areas off the roads, since
there is often reason to expect different patterns of cultivation in the vicinity of roads.
Detailed surveys, even on a sample basis, are an expensive rarity and most estimates
depend on reports from local Ministry of Agriculture staff, among their various other
duties.

Even where serious efforts are made to collect production data directly, 'agricultural censuses and sample surveys, no matter how well-organized and administered,
are subject to numerous sources of error and uncertainty, ranging from farmers'
unwillingness to disclose information to potential tax-assessors, to problems of
designing a representative sample in countries where there is no reliable census from
which to (work)' (Berry, 1984: 61). Even where there is a reliable census, problems can
arise, as exemplified by an FAO sample agricultural census in Tanzania in 1970. In
spite of the generally accepted accuracy of the 1967 population census, upon which
sampling was based, the FAO 'agricultural census' got the *direction* of population
change wrong for all four districts of the region where I was working. Moreover it
entirely missed the major cash crop for one of the districts.[9]

Thus although FAO *Production Yearbooks* present estimates for area cultivated,
yield per hectare and total production, this provides no means for independent
checking. There is simply no way of getting accurate estimates of either area cultivated or yield. What normally happens is that the estimator (usually an extension
officer) takes the previous year's figure as basis and 'estimates' the proportion by
which it has increased or decreased. A common practice is simply to copy last year's
figure unchanged.

If area cultivated and yield are virtually impossible to estimate with any accuracy,
there exists no basis for estimating production, except what is marketed. But a large
proportion of total production is destined for subsistence consumption and can only be
estimated by guessing at average consumption and multiplying by population. Where
marketing is not state-controlled, primary purchasing tends to be dominated by small
to medium operators, many of whom cannot write well and very few of whom are likely
to make a voluntary gift of tax to the government by declaring their level of sales. Even
where marketing is officially monopolized by a state agency, this will seldom handle

more than a minority of total produce in reality (see below), and the same private traders will still handle the greater proportion unofficially, and have even less incentive to report their activities to the government. In short, cereal production in most African countries is estimated through a combination of the best guesses of agricultural officers who have many other things to occupy their time and estimates based on that part of production which is marketed through official agencies. There is, moreover, reason to suppose that the budgetary crisis of the past several years has led to a worsening in the quality of crop-production statistics.

But in addition to the low level and downward trend in the general level of accuracy, there is every reason to expect a *significant downward bias* in the reporting, and one increasing over time, for a variety of reasons. One of these is what might be called 'automatic' downward bias. This arises when, in the absence of new data, the previous year's figure are simply reproduced, sometimes for several years on end, although with populations growing at 3 per cent or more per annum this is not the most logical of assumptions.

More important is the bias which arises from estimating food production from the portion of it which is officially marketed, or letting the latter figure influence the estimation. This is very common because it is often the only figure available and is also the figure of greatest interest to government officers (and to their superiors). The quantity officially marketed indicates how much grain the state has at its disposal for supplying urban and institutional markets, and also forms the basis for estimates of import and food aid requirements (see below).

There is good reason to believe that the proportion of total food marketed through official agencies has declined since the early 1970s, as a result of state policies to hold producer and consumer prices down, and the consequent diversion of produce on to black markets. This process is to a significant degree self-perpetuating, if not self-expanding. Holding official producer prices down leads to diversion of produce on to unofficial markets, leading to urban shortages on official markets. This forces consumers to buy on black markets, leading to price increases on those markets and increased diversion of produce on to them, leaving still less for the official market, and so on. As a basis for estimating total production, this will give not only a downward bias, but one which increases over time.

There are also a number of reasons why most people concerned would prefer to see this as production decline rather than as a decline in the proportion officially marketed. Firstly, it is a simpler thesis and fits well with the assumption held by many officials and experts that peasant conservatism and resistance to innovation are the real roots of the problem. Secondly, to admit that produce is being diverted on to black markets is an indication of the ineffectiveness of government policy, especially if, as is not infrequently the case, civil servants are involved.[10] But most importantly, it points to conclusions which most policy-makers would like to draw in the first place: that more food needs to be imported and that rapid technical change is needed to increase aggregate production. Even authorities which refer to black markets, when arguing about the necessity for private trading, tend to ignore the implications of their arguments for flows of produce, assuming that all produce not sold through official markets is 'subsistence'. Finally, as discussed below, there is some reason to suspect dramatization and exaggeration of 'food gaps' by governments anxious to increase or maintain levels of food aid.

There is thus every reason for the compilers of food production statistics to attribute declining official deliveries by domestic producers to declining overall production. Once the assumption of declining production has been made, and built into explanatory models, it imparts further bias. One finds FAO retrospectively revising ten-year-old figures upwards, to ensure that current increases in estimates of production do not risk showing a rising trend. I am not suggesting deliberate dishonesty. The figures are estimated on such a weak basis that one can easily argue for revision. That the direction of revision tends to reflect standard assumptions is nothing new. But it does further bias the estimates, some of which are simply ridiculous.

For example, the total cereal area in Africa (other than South Africa) was estimated by FAO to have increased between 1974–6 and 1984, by less than 5 per cent, and this in a 'land-rich' continent, with a rate of population increase estimated at 3 per cent or more per annum. But the full absurdity of this does not emerge until one looks at some of the specific instances on which the average is based, like the estimate that Zambia's maize area in 1984 was only 43 per cent of the amount cultivated in 1974–6. Maize is Zambia's major staple grain, the population must have incresed by around 35 per cent over the period, land is not short in Zambia *and* there was a major maize production programme for the northern part of the country in the intervening period, which is said to have increased production very significantly.[11] In this case the estimation seems to have been strongly affected both by problems with maize marketing and by the decline in maize production from a group of large-scale, mostly white, farmers. Similarly, in Mozambique the area cultivated with maize is claimed to have remained unchanged at 600,000 ha for ten years, during which time the population has probably grown by over 30 per cent.

Again, in Kenya, with a population growth-rate of over 4 per cent per annum, the area cultivated with maize is estimated to have fallen by 28 per cent between 1974–6 and 1984, and that of total cereals almost as rapidly. Still more remarkably, total cereal production is estimated to have fallen by 44 per cent over the same period, and that of maize by 48 per cent. With the population having increased (according to standard estimates) by over half, this would imply a reduction in per capita grain availability by between two-thirds and three-quarters! Obviously no such thing has happened. Looking further in the *Production Yearbook*, one finds a possible explanation. The estimate for total cereal area in Table 15 for 1974–6 is 1.95 million ha. But in Table 1, total arable area for the same years is only 1.77 million ha, making the part some 10 per cent greater than the whole. By 1984, the cereal area was safely below the total, though a minor embarrassment had been covered at the cost of a ridiculous estimate of the trend in production (FAOa, 1985). Here, as in the Zambian case, one encounters another source of bias – the tendency to overestimate the significance of large-scale commercial farms and underestimate that of small peasants.

An example from Tanzania may illustrate how this comes about. Prior to independence, wheat was grown in northern Tanzania largely by expatriate settlers, who cultivated not only their own large farms but the farms of African peasants on a share-cropping basis. The latter was forbidden by the colonial government in 1959, and the demise of wheat production was widely predicted in colonial agricultural circles. However, there were also, during the 1950s, a small number of African tractor-farmers and from 1958 to 1967 they expanded production from some 800 tons to around 20,000 tons, far more than compensating for the decline in settler production and taking total production to almost three times the level of the 1950s (Raikes, 1975: Appendix 1).

In 1969, I started a study of wheat production in northern Tanzania, and went, as a first step, to interview officers from the agricultural department. This rather significant development seemed simply not to have been noticed by them. Almost without exception, they spoke of wheat production as being in decline from an earlier heyday. Questioned about African production, they referred to it as of minor importance and hastened to tell stories about non-economic farming, low yields and peasant tractor-owners who were primarily concerned with 'prestige'. Closer study showed that not one of these observations was correct. In the event, wheat was grown almost solely for sale to official agencies, so this made no difference to production estimates. Had this not been the case, one could have expected very low estimates from the local extension personnel.

From this, and from many other examples, it seems likely that government officials will tend to underestimate both area cultivated and production from peasant farms, except those on government-organized projects and schemes. In the latter case one sometimes finds overestimation as officers strive to give the impression that targets have been achieved, but this is seldom on a scale sufficient to affect national figures.

Nor does this end the list of sources of bias, for there is also evidence of deliberate reduction of production figures. In a conversation with an official from FAO, he freely admitted the uselessness of Production Yearbook data. He also mentioned (and justi-fied) deliberate reduction in aggregate production figures in certain cases. Famines and food emergencies may occur in particular regions of countries, even though the country as a whole may not be seriously (or even at all) in deficit. Aid donors, however, are not often willing to accept that an emergency exists unless it can be shown that there is an overall deficit. If one is to make the case for rapid deliveries of supplies to cover an emergency, it may be 'necessary' to report deficits but omit to mention surpluses in other parts of the same country.

It is, of course, perfectly true that famine conditions may exist in one part of a country, while there are surpluses elsewhere. This is a major part of Sen's (1981) argument *against* the FAO view of famine as resulting from overall 'food availability decline' (FAD), and in favour of the alternative view that famine results primarily from the disappearance of *entitlements to food*. An example of this is that the market will fail to bring surplus food to deficit areas, if the demand is not 'effective' (i.e. backed by money). It is thus ironic to hear of the FAO misrepresenting local shortages as aggre-gate food availability decline, in order to keep donors happy. If the donors are wedded to the view that only aggregate FAD can produce famine, then aggregate FAD must be shown to exist. One can hardly imagine a more perfectly closed circle.

I should make two things clear in this respect, however. Firstly, the intention with this 'adjustment' was quite specifically to increase the availability of food transfers in a situation in which it was felt that these were desperately needed. Secondly, this is probably only a minor source of the overall bias in the FAO production figures, far exceeded by 'normal' aspects of the estimation procedure.[12]

To summarize, the statistical basis for estimating food production in Africa is extremely weak, has probably deteriorated with budgetary strain in recent years and is biased downwards, with the degree of bias increasing. One should thus be extremely cautious about accepting trends which appear to emerge from FAO *Production Yearbook* figures, unless there is good independent reason to support them.

In the case of the supposed stagnation in area cultivated, both common sense and all available evidence indicate almost precisely the reverse, at least in all areas where there is spare land. In point of fact, the evidence relates much of the crisis itself to increased area cultivated, and specifically to the *new cultivation of marginal lands*. The form which this often takes is movement on to uncultivated land, very often previously grazed by pastoralists and quite often with unreliable rainfall and unstable soils. Its ill-effects, often made worse by destructive short-term soil-mining techniques aimed at extracting a fast profit, are discussed in the following chapter. Here the point to be made is that this is totally incompatible with stagnation in the area cultivated – unless other land is being taken out of cultivation, for which there is no evidence.

Where overall production is concerned, it is impossible to generalize. There are certainly areas where yields are declining from over-intensive cultivation without corresponding soil maintenance, or from movement into unsuitable marginal areas. But there are also others where yields are increasing. The important point is that the trends and processes are not national, let alone continental, but specific to particular areas. But one thing can be said with a fair degree of certainty; given the low level of accuracy and the multiple sources of bias, the aggregate growth of per capita food production in tropical Africa is higher than as presented in FAO *Production Year-books*. Indeed, there are hints of some recognition of this from the FAO itself, since recent *Production Yearbooks* have significantly increased their estimates of production.

Food imports: an alternative analysis

If increased food imports and famine do not arise from decline in aggregate national food production, what do they derive from? This question has to be anwered at different levels. The remainder of the present chapter looks at some specific reasons within the food sphere, focussing on demand. The following chapter looks first at the genesis of the more general crisis and continues with policies, programmes and problems in the production and supply of food.

The short answer to the question above is that increasing food imports derive primarily from an imbalance between rapidly increasing urban food demand and the capacity of official food purchasing agencies to meet this demand, exacerbated in a number of cases by factors making food imports (especially in the form of food aid) more attractive to African governments than local procurement. These three factors will be considered in turn.

URBAN DEMAND FOR FOOD

While aggregate population in Africa is estimated to be growing at between 3 and 4 per cent, rates of urban growth are often over twice that high, ranging between 5 and 9 per cent per annum. Since these growth rates have remained more or less constant, while the size of urban populations have increased, the additions to the urban population have been increasing every year. Some of the cities of tropical Africa now have ten to fifteen times the population they had in the early 1960s, while the urban population as a whole has more than doubled as a proportion of the total. Since urban populations typically produce only a small fraction of the food they consume, this alone would be sufficient to create problems of food supply, since it generates a demand for surplus food from the agricultural sector which itself is often growing by up to 3 per cent per annum.

But this is by no means all, for average incomes are also significantly higher in the urban areas, despite the presence of large numbers of poor and sometimes destitute people. Not only are the rapidly-growing state sectors and professional or business middle classes largely concentrated in the towns, but the growth of industry and services implies the rapid growth of a larger stratum ranging from clerical personnel to skilled and unskilled workers in formal employment whose incomes, though low by Western standards, are still well above the rural average.

As incomes rise the proportion of them spent on food declines, according to what is known as 'Engels' Law'. But at very low income levels, a high proportion of total income is spent on food, and the income elasticity of demand for food (the proportion of any increase in income spent on food) is very high and at first falls relatively slowly. Increasing incomes under African conditions will thus normally lead to very considerable increases in the demand for food and urban populations will consume significantly more per capita than rural.

The composition of the demand for food also changes as income rises, from basic staples and non-preferred goods towards more convenient, palatable, sightly or prestigious types of food. Of course, what is considered palatable, convenient or prestigious will vary between different societies, being affected by a combination of cultural and practical factors, but it will also tend to change over time and with changes in circumstances. Wheat products, for example, require a longer and more fuel-intensive cooking process than most alternative staples. They are thus not well suited to rural areas where people do their own cooking. But in urban areas, or where there are bakers and chapatti makers, the disadvantage turns to an advantage since wheat products can be purchased ready made and transport or keep better than products of other staples. Other important factors affecting food preferences are social contacts and advertising, which once again tend to increase the demand for wheat products. Bread is known as a high-income and 'modern' foodstuff from its dominance in the developed countries and among colonial elites. It is also characteristically produced by larger companies which pay more attention to advertising.

A number of general tendencies are observable. Wheat and rice tend to increase their importance in the diet as incomes increase, at the expense of 'lower value' (though not necessarily less nutritious) staples like millets, sorghum and root crops. Maize consumption gains from non-preferred staples, while losing to wheat and rice. Consumption of fresh vegetables increases with incomes as does that of meat and dairy products. The latter are particularly significant since production of meat or dairy products for regular urban markets almost always depends on purchased feed – the transformation of crops to livestock products, a process in which a calorie loss of 80–90 per cent is common.[13] Where livestock are pastured on land which could otherwise be cropped, a smaller, but nonetheless substantial, calorie loss occurs. Increasing consumption of livestock products thus implies not only a greater money expenditure per calorie, but considerably increased use of land and other agricultural resources per calorie consumed.

It is thus not hard to see why urban food demand has increased explosively since independence in most African countries. Under most colonial regimes, a very small white population not only monopolized virtually all high- and middle-income jobs but took considerable pains to avoid social contact with the African population. Wages were held to the lowest levels possible, to the extent that many regimes had to rely on institutionalized force to get hold of African labour. City populations were kept to a minimum (often including controls on movement and compulsory repatriations of the unemployed) and normally strictly segregated. Africans not only had insufficient money to purchase items like fresh milk; at least in settler colonies, they would often not have been served in the shops where it was sold unless they carried a 'chit from Memsahib' to show that they were servants buying for a European employer.

With independence, this form of sumptuary legislation or custom was usually relaxed (though not always completely). At the same time, most newly independent governments found it politic to increase wages (as indeed had most colonial governments in the last years before independence). Africans were promoted into positions previously held by whites and in some cases at their wage-levels. In most cases employment in the state, parastatal and urban private sectors began to expand rapidly. Moreover, since most of these changes occurred much more rapidly in the towns than in the country, this attracted a flood of rural-urban migrants, including the families of previous migrants who could afford (or who were allowed) to have them there for the first time. Many post-independence governments tried to continue controlling the influx into the towns (and periodically to expel those without papers), but with decreasing success.

Of course, this was not always a sudden process happening at independence. Where colonial regimes were more liberal, Africans had been drawn into better jobs and colour-bars lowered before independence. In other cases, it took some time after independence for barriers to be lowered and for bureaucracies and urban populations to grow. In general, however, urban populations doubled as a proportion of populations growing at around 3 per cent per annum, between 1960 and 1980 (World Bank, 1982: Tables 17, 20).

In most cases, the same ingredients were present: rapid growth of the urban population, increases in income levels, greater freedom of choice in what to buy and increased contact with 'European' styles in living and eating. The latter was further emphasized by expansion of education. African students were sent abroad to Europe and the USA, while new foreign teachers and experts arrived, often (though by no means always) less racialist and stand-offish than their colonial predecessors and usually more numerous. This in itself would have been enough to lead to an expansion in urban food demand well above the 5–9 per cent per annum at which the towns were growing, but government and aid donor policies further accelerated the increase in urban food demand.

Many of the governments of post-colonial Africa have tried to keep the urban population satisfied and wage demands low by subsidizing the prices of basic foodstuffs. Others have focussed rather on keeping the middle class (and their own employees) satisfied, by subsidizing prices of items like meat and dairy products. In both cases, this has often been forced upon governments, as indicated by a number of urban riots and even coups resulting from increased food prices.[14] But whatever the reason for food subsidies, they have an obvious effect in increasing demand for food since, apart from any income effect, they reduce the prices of the affected foodstuffs in relation to those of non-subsidized commodities. This has a direct impact on imports when the commodities subsidized (wheat, for example, and often milk) are not well-suited to domestic production.

Donors, on the other hand, provided subsidized loans and sometimes even grants for the building of wheat mills, bakeries, dairy plants and abbattoirs, on the grounds that these were 'developmental' and required for hygienic purposes. Where wheat and dairy products have been concerned, the effect has been almost exclusively to increase imports, though the rhetoric surrounding such 'development projects' has nearly always stressed increased local production. In almost all cases, however, such projects have had the effect of further increasing demand for the products in question – and often led to pressure for further subsidies.

From all this, it can be seen that the task of feeding the towns in Africa would have been a heavy one under any circumstances, and one which inevitably required increased imports for those many countries where wheat cannot be produced or where dairy products for the towns must either be imported or depend on imported feedstuffs.[15] But the other side of this 'scissors crisis' has been problems in procuring foodstuffs from the rural areas.

State monopoly food marketing

For reasons to be discussed in the following chapter, many if not most of the states of tropical Africa have attempted to control, if not perform, the primary marketing of agricultural products (both food and export), through marketing boards or other monopoly agencies. Reference was made above, in the context of estimating total food production, to the fact that these have tended to command a decreasing proportion of the total food supply over time and to the impact of low producer prices in this respect. Some of the more specific reasons for these low prices (and often long delays in making payment) are again discussed in Chapter 3, but two of them, one 'demand-oriented', the other structural, are relevant to the present discussion.

The first is a rather obvious corollary of controlled food prices for consumers. If prices of basic foodstuffs are to be kept low for urban consumers without massive costs to overloaded government budgets, then producer prices must also be kept low – though this has not always, or even often, prevented significant budgetary costs. While low consumer prices stimulate demand, low producer prices reduce supply.

In the event, this supply reduction is further intensified by structural characteristics of the market and the pricing policies of most marketing boards. Most of the basic foodstuffs produced in tropical Africa come from small peasant producers, who produce both for family subsistence needs and for sale. While there is some evidence that significant numbers sell 'too much' of their total produce in poor years, thus leaving themselves short of food at the end of the season, it is generally the case that the amount sold varies more than total production. To take a simple example, if the family produces twenty bags of maize in a normal year, and has a normal consumption of ten bags, a 10 per cent reduction in production to 18 bags will reduce sales by 20 per cent, if consumption is maintained at the same level. Similarly, a 10 per cent increase in production may generate a 20 per cent increase in sales. Thus, under any circumstances, marketed production will fluctuate more than production.

But a further element of variation derives from price factors. While it is often illegal for peasants to sell other than small amounts locally outside the official marketing channels, this provision is widely evaded and extremely hard to control. In Tanzania it is estimated (guessed) that something like 50 per cent of the maize produced in a normal year is sold, while the official marketing board gets somewhere between 10 and 40 per cent of this (5–20 per cent of total production). This is probably below, but not enormously so, the proportion which passes through official channels in many other countries. But the amount passing through official channels also varies enormously from year to year, depending on weather and other price-affecting conditions. Marketing boards characteristically set prices before the beginning of the season and do not change them during it. Private traders, on the other hand, offer the going market rate, which may shift from day to day and will certainly vary widely between years. The result is that in years of shortage, prices on unofficial markets can rise up to several times the level of official prices, diverting produce from every peasant who can possibly evade the official channels and leaving the marketing board to meet urban demand with a much reduced fraction of an already much reduced amount available for sale. To take another simple hypothetical example, total maize production from a given area in a normal year is 200,000 tons, of which about 100,000 tons is consumed by the producers. If production falls by 10 per cent, total sales will fall by 20 per cent. But if at the same time the proportion sold to the marketing board falls from 20 per cent of this to 5 per cent (by no means impossible), then marketing board purchases will fall from 20,000 tons to 4,000 tons, i.e. by 80 per cent. By contrast, in a year of bumper harvest, prices on official markets may fall below the official price, leading to a flood of deliveries to the marketing board, overtaxing its storage facilities and leading to spoilage, while probably generating huge delays in collecting and paying for produce.[16]

The above concerns fluctuations deriving from variations in production but, as indicated above, gaps between urban demand and available official supplies can have not dissimilar effects. If a monopoly agency sets producer prices below those on private markets even in a normal year, this will lead to diversion of produce on to the latter, increasing the gap between urban demand and available (official) supplies. This in turn is often aggravated by hoarding, as consumers rush to assure their supplies in the face of expected shortage and speculators do so to profit from it. This in turn increases demand and prices on unofficial markets, resulting in further diversion of produce on to them. This is some evidence that this has been a significant factor in the declining availability of produce to state purchasing agencies since the early 1970s.

Given these sorts of problems, it is hardly surprising that marketing board and government officials who are responsible for urban food supplies are inclined to exaggerate declines in total production and to call for imports to make up a shortfall which may not exist in aggregate terms. It is not of interest to them that supplies may be available on the private market. These supplies are not available to them, and do not significantly reduce the demand for foodstuffs from them, since all consumers would rather have official supplies at the lower fixed price than pay more on the black market. But even this does not exhaust the incentives to import.

Food import decisions

In Chapters 7 and 8, the vexed question of the impact of cheap food imports and food aid on local production will be considered in more general terms. Critics of food aid claim that it predisposes countries to import rather than produce locally, that it leads to lower producer prices and reduces incentives to produce locally. Opponents claim that this cannot be proved. Proof is taken to imply measurement of some difference and poses the problem that the available economic models by which the hypothesis might be tested rely on highly unrealistic assumptions about the nature of the markets and decision-procedures. For purposes of measurement, that is, it has to be assumed that African countries can be likened to textbook 'rational consumers' operating in a free market and making decisions on the basis of relative prices. Without this assumption one cannot use or measure elasticity coefficients, the purpose of which is to predict the response of consumers to changes in price.

While there are clearly some ways in which states are affected or constrained by prices, it is grossly unrealistic to suppose that decisions on food imports are made in this way. If one considers the standard case in which the state either controls imports directly or licenses foreign exchange for specific items, then a process something like the following seems more likely.

Due to a combination of urban growth and problems of procurement, an urban 'food gap' emerges and expands. If prices are not controlled, this shows itself in price increases. If they are controlled, it shows in shortages. In either case it generates hardship and political dissatisfaction. The government is thus obliged to act – that is, to import food to fill the gap.

Where prices are controlled and shortages emerge, the size of the gap can be estimated and then defined as the country's 'import requirement'. This practice has become widely prevalent in recent years, since FAO, the World Food Programme and other donors are interested in knowing the 'import requirement' as a means of planning their allocations of food aid. A recipient country has an interest in estimating on the high side, since food aid saves foreign exchange, as compared with commercial imports, and since there is some tendency to provide less than requested.

The decision over how much of what cereals to import commercially will thus depend in part upon the size of the defined 'food gap' (itself a function of consumer and producer prices), and in part upon the amount of food aid which the country can get hold of. Commercial imports will, for many countries, be a last resort and often made only when problems are becoming desperate. Decision-makers thus often face complex sets of choices. The gap which they have to fill is often time-specific (i.e. the period until the next main harvest). They must also make decisions for several months in advance, to allow for delivery times. While the price of commercial imports can often be reduced by timely purchase, the chance of getting food aid is increased by putting off commercial purchases until the last moment. In many cases the agency concerned takes some time to estimate or admit to shortages, so that decisions have to be taken rapidly to avoid serious suffering or political upheaval. In such a situation, one can scarcely imagine state decision-makers behaving like 'rational consumers' shopping around for the best value. Doubtless they will be affected by prices, but only as one among a number of competing factors.

But there are also a number of reasons why a state agency might prefer imports to local produce, so long as price relations are favourable. In the first place, it is simply less trouble. Local purchases normally require collection from thousands of small peasants, scattered and mostly at the other end of appalling roads. To organize this collection is a major operation, and one which only too easily goes wrong.[17] This is particularly the case if there is already a foreign-exchange crisis, since there will then be shortages of fuel, spart parts, tyres, sacks and various other necessary items – not to mention money to pay for the produce. By contrast, importing food from commercial firms can be arranged by telephone, so long as credit (and the telephone system) is in order. Food aid takes far longer, but requires no major labour input once a request has been made. In either case, the produce arrives in bulk, ready to be unloaded into the country's dockside storage (often the most extensive there is), and is then concentrated for distribution to the urban population.

Another reason for preferring imports, and especially food aid, is that they can contribute to the budget. If, for one or other reason, the local currency price of imported cereals is below that of local supplies (as is very often the case)[18], the state has a number of attractive options open to it. It can use this to subsidize consumer food prices below those generated by local supplies, thus satisfying one or more political constituency. It can hold the consumer price to the level generated by local supplies (plus transport and marketing costs) and keep the difference. In some cases, this can amount to a significant proportion of government revenue, though food-aid donors sometimes place limits on what can be done with the money. The third alternative is a combination of the two, and this is probably the most common. Whatever the choice, cheap imports have an advantage over local products in this respect.

The question of what effect this is likely to have on prices is taken up again later in this book and depends largely on how the state distributes food aid or imported food. There are three main effects which it might have. Cheap food expands urban demand and increases the future size of the 'food gap' for given price and harvest conditions. It competes with local production and reduces unofficial prices (where it is available). It also probably affects the setting of official prices by the state. One can well imagine that governments might take the occasion to hold producer prices and so marketing agency deficits down, especially since they gain in budgetary terms thereby. But since this refers to the decisions from closed meetings, in which a variety of things may (or may not) be taken into account, it is not possible to say with certainty. A whole series of further complications attend efforts to assess effects on production, one obviously being the timing of imports and their distribution. Clearly if official producer prices are held down as an effect of the availability of food imports, this reduces deliveries to the official market. But what is the effect on total production is much harder to say, except that it will generally be negative.

These different processes have been combined in different ways. But that they have had a significant effect on the food economies of tropical African countries can be demonstrated with some examples.

Mozambique became independent in 1975, after a century of backward and repressive Portuguese colonialism. Wages were held so low under the Portuguese that forced labour was still in operation up to the year before independence. Rapid in-migration of Portuguese in the years before independence had led to expansion of both consumption and production of livestock and wheat products (the latter mostly imported), but this was kept entirely within the white community, except for a small minority of educated Africans. If Africans were not forbidden to enter 'white' bakeries and milk shops, it was certainly not made easy for them to do so.

With independence, many of the controls were lifted, prices of bread, meat and other 'basic staples' were fixed and the FRELIMO Government made it a plank of its policy that there should be enough for all. But while this led to an enormous increase in demand, intensified by massive in-migration into Maputo, an equally massive exodus of Portuguese farmers, businessmen and artisans brought the economy, and notably commercial agriculture, grinding to a halt. Given an extremely difficult position to start from, the Frelimo Government made further problems for itself by defining a food policy in terms of wheat, irrigated vegetables and livestock products (which had previously been grown only on white commercial farms) rather than maize and beans which are grown by peasants. While this was certainly not the only or even the main reason for a disastrous decision to concentrate resources for agriculture almost entirely upon state farms, it may well have been among the reasons for this decision. (Raikes, 1984b). This combination had started to generate serious urban food-shortages, and the whole process of peasants withholding produce from official markets, even before the situation was made catastrophically worse by the activities of the South African-backed MNR terrorists. Mozambique's major coastal cities are currently virtually under siege, while disruption to transport, supplies and practically all aspects of commerce and civil administration have further disrupted agricultural production, exacting an enormous toll in human life, both directly and through starvation.

Commentators on Mozambique, myself included, have been highly critical of the agricultural policies of the Frelimo Government. So it is worth remembering that the extreme repressiveness and very low income-levels of the colonial period made an explosive increase in urban food demand almost inevitable, thus increasing vulnerability to the subsequent terrorist attack.

Zimbabwe has been much more successful in maintaining food production, but has also experienced major increases in urban demand for food. Here the major reasons were wage increases and price controls (with subsidy) for bread, dairy products, meat and maize.[19] Prior to independence, low wage levels, combined with the concentration of cereal production on large-scale white-owned commercial farms, had allowed the export of surplus maize in most years. Since independence, there have been maize exports in some years, but most expectations are that they will become progressively rarer. This has more to do with rapid growth in local demand than with lagging production, in spite of the most serious and protracted drought of the century. Maize production by peasants in Zimbabwe is growing rapidly, though this increase comes primarily from the richer peasants in the more fertile areas.[20]

Zimbabwe's major food problem lies elsewhere, in the enormously skewed distribution of land, under which a few thousand white farmers still control over 40 per cent of arable land and most of the top-quality land. This implies serious over-crowding in the 'communal areas', including some appallingly dry and desolate places where people are forced, for lack of alternative, to try to cultivate.

Tanzania exemplifies a number of the points made above, though it took some time after independence for the problems to show. One reason for this was the very low level of development and urbanization at independence. Very few Tanzanians had been trained beyond secondary school level and even secondary schooling had only just begun to expand from minimal levels. The country was self-sufficient in basic staples, though it had been a net importer until the mid-1950s, when a system of state-controlled cereal marketing was dismantled by the colonial government (Bryceson, 1978). This was re-imposed a few years after independence as part of a process of institutional development aimed at producing the conditions for later more radical policy ventures.

Urban food demand did rise quite rapidly, but the urban sector was so small that this increase proved well within the capacity of local production. There was little change in producer prices for foodstuffs from the mid-1950s until the late 1960s, but correspondingly little inflation, so that prices were sufficient to maintain the flow of cereals on to official markets. As Bryceson (1982: Fig. 2) shows, after increased imports due to drought, immediately after independence, average grain imports fell and were lower during the late 1960s than in the early 1950s. There was already a significant black market (to avoid co-operative marketing charges and local taxes) but at this time planners were at least as concerned to avoid losses from exports of cereals as to increase deliveries.[21] Wheat was imported, but the country was more or less self-sufficient in maize and rice (Raikes, 1985b).

The beginnings of urban food shortages in the later 1960s concerned meat and were the result of nationalization of butcheries and price-control rather than any change in the underlying supply situation. After the nationalization of import-export trade in 1967, shortages also began to appear of certain imported items, as was the case with cooking oil after the transfer of wholesale trade to co-operatives and regional trading corporations.

A combination of drought and the disruption caused by forced villagization of the rural population led to widespread food shortages and to disastrous reductions in deliveries of cereals to official agencies, during the period 1973–5.[22] Food imports and food aid took a quantum leap in this period and have hardly fallen since. An almost continuous succession of food production programmes failed to solve the problem of supply to official channels (see Chapters 3), though recent relaxations in controls on private marketing (of all commodities) have somewhat eased the market problem.

Problems with the official food marketing system (which became even worse after co-operatives were replaced by a parastatal corporation in the mid-1970s) were important reasons for this change, but growth in urban and institutional markets was also important. Not only has Dar es Salaam grown at over 8 per cent per annum to a city of some 1.6 million inhabitants, but the expansion of the state sector, education, the army and rationing systems has vastly increased the institutional market which has to be fed through official channels. Food subsidies have also played a significant part. Maize prices were controlled from the early 1970s until 1983, at levels which involved a very considerable element of subsidy. Supplies of cereals were probably better in Dar es Salaam than in the countryside for much of this period, and the rationing system more effective (not that it provided adequately for the poorest residents, some of whom lacked papers and fell outside the system). Not only are these both likely to have increased demand for cereals, but they seem likely to have contributed to the high urban growth rate. With an urban population some eight times the level at independence, it is scarcely surprising that food deliveries (let alone those to official markets) have been unable to keep up with urban demand. But this says nothing about the relationship between production and population growth (Raikes, 1985 a and b).

   Foreign aid donors have also played their part. Food aid for the 1973–4 crisis mostly arrived in 1975 when local supplies were improving, and helped to make feasible the subsidization of urban maize prices. Food aid has accounted for almost the total import of wheat (though saving foreign exchange in this case by replacing commercial imports). During the 1970s, there were also significant food aid imports of powdered milk, mainly to supply a number of donor-funded dairies (see Chapters 3 and 9).[23]

Kenya developed a large-scale cereal and livestock producing sector during the colonial period, which generated export surpluses in most years, assisted by low wages and a wealth of restrictions on activities by the African population. Like Tanzania, Kenya has experienced a very rapid rate of urbanization since independence, while the growth of an urban upper and middle class has been still more rapid. In this process, the food export surpluses have largely disappeared. Kenya has had to import cereals officially in some recent years, while there is good reason to suppose that it imports unofficially from Tanzania in many years.[24] Plans for the development of 'feedlot' beef production for the tourist trade had to be abandoned because, with increasing human demand, the price of maize rose enough to make them uneconomic. Kenya's international trade in cereals is characterized by wide variations, but an apparently increasing deficit. During the period 1969–76, there were net exports in seven years and net imports in one. During the following eight-year period 1977–84, there were net imports in six years and net exports in only two (FAO c,e,f, various), though the past two years have seen net exports of maize.[25] It also appears that the inter-annual variability of production has increased over time. This could be an inappropriate reading from variations in officially recorded procurements, though these are in fact unusually stable between years for tropical Africa, reflecting the important of large producers. But there are independent reasons to expect increased variation because of the settlement of new and marginal areas (see below).

   While it seems likely that Kenya's cereal production has failed to keep pace with its (record-high) population growth, it seems highly unlikely to have fallen to the extent indicated by FAO *Production Yearbooks*. There have been serious food shortages in recent years, most especially 1984. But the impact of these has been local and strongly connected to loss of entitlements (see Chapter 4).

Nigeria is perhaps the classic case of food imports growing rapidly, under the combined stimulus of commercial promotion and government policy. In 1970, prior to the rapid development of oil exports, Nigeria was more or less self-sufficient in foodstuffs and exported food items like groundnuts. This was the case in spite of rapid urbanization during the previous decade. Some quarter of a million tons of wheat was imported but, for the most part, local producers were increasing grain production rapidly enough to keep up with urban demand (Andrae and Beckmann, 1985).

But oil production was just beginning to take off in 1970, and Andrae and Beckmann show that this has much to do with the explosive growth of wheat consumption and imports which they refer to as the 'wheat trap'. Bread is a convenient form of food, especially for urban residents, but this alone would not have ensured its rapid spread.

Firstly, oil-induced inflation within the Nigerian economy meant that the price of imported wheat fell in relation to that of local foodstuffs. Secondly, government price controls ensured that this was also true of bread and flour, with no taxation on wheat imports even during the austerity budget of 1984 when most other imports were hit. The state seems to have participated enthusiastically in the spread of bread and its achievement of a position as 'the cheapest staple food of our people' (cited in *ibid*: 2). Of equal importance, commercial firms, ranging from multinational giants to local small-scale bakers, have rapidly built up a comprehensive network of mills, bakeries and delivery systems, taking bread to a large proportion of the population. The combined effect of all this was a growth in wheat imports, during the 1970s, of well over 10 per cent per annum, continuing into the 1980s. By 1984, in spite of a sharp decline in oil revenues and serious balance-of-payments problems, wheat imports had grown to 1.7 million tons, six and a half times the level of 1970. Nigeria had clearly fallen into the wheat trap.

While the above examples each consider only a part of the food situation in their respective countries, one thing which emerges clearly is that a host of reasons other than a simple 'food gap' can account for increased food imports into tropical Africa. Another is that growth of imports may itself act to obstruct the growth of local food production.

NOTES

1. The US Congress specifies that at least 75% of US food aid should go to countries in the LIC (Lower Income Country) category or below, which would give rise to problems if Egypt (which normally receives about 30% of the total) did not fall in that category. A number of revisions to the category have had the effect of reinstating Egypt as a LIC, after income increase or inflation had moved it above the limit.
2. It is worth remarking on the absurdity of some of these examples. By 1983, the disastrous policies which would intensify famine in the Sudan in the following years, had already been in operation, with donor approval and support, for several years (Chapter 3 and O'Brien, 1985). Rwanda may possibly have increased per capita food production, but is one of the countries where difficulties in maintaining food production among small peasants, and consequent serious malnutrition, have been long-term problems. If the Central African Empire/Republic had increased agricultural production, this seems unlikely to have been the result of state policy. Cameroon and Ivory Coast are relatively wealthy countries with market-oriented policies, which probably did help to keep the flow of food products on to official markets going (including from neighbouring countries). the arbitrary nature of the list is indicated by the fact that Grigg (1985), using similar data for a slightly different period, produces a list of six 'success-stories', of which only Sudan and Ivory Coast overlap with that cited here.
3. It seems, for example, unlikely that the manager of the FAO crop early warning project in Tanzania could have been consulted in drawing up the figures cited in the following note. He told me in May 1985 that, in his opinion, maize production in Tanzania had been over 2 million tons for a number of years.
4. See previous note. In revising Tanzania's cereal production figures upwards between 1982 and 1984, FAO was careful to revise those for 1974–6 upwards as well, in order to maintain the purported declining trend in per capita production. For maize the average production figures for 1974–6 were revised as follows:

Yearbook for:	1982	1984
Area ('000 ha)	1,167	1,167
Yield (kg/ha)	716	1,037
Production ('000 tons)	835	1,210

It is hard to make any sense of this, since 1974 and 1975 were both poor harvest years in which there were severe food shortages and yields were almost certainly below average. Presumably those involved in the revision thought the yield figures the least likely to be noticed and assumed that no-one would look through decade-old figures to check.

5. Since producer prices in many developed countries are heavily subsidized, there is some possibility of overestimating production, as it may be presented more than once to collect further subsidies. This is likely to be small in comparison with the errors likely to arise in developing countries.

6. This would appear to be the case for Zimbabwe and to account for anomalies in the production figures for Zambia and Kenya. In all three cases a significant proportion of the farms in question are or were operated by white settlers, who reported (and sometimes over-reported) their production for purposes of subsidies or guarantees. In each case, moreover, since they produced solely for the market they were regarded by the colonial and post-colonial states as the most secure source of cereals.

7. This means that satellite pictures are made up of dots, each of which represents the 'average' colour (and so vegetation) of an 80 sq. m. piece of land, though a new satellite, introduced in the past two years has a 10 sq. m. dot. The satellite pictures of which one reads in spy stories, which can show individual blades of Russian wheat, are photographs and far too expensive for use in anything so dully useful as crop prediction.

8. Even in Europe, the calibration of satellite imagery has not yet been refined to the point where it can provide independent estimates of crop production. S. Christiansen, personal communication, relating to the very homogeneous farming of Fyn in Demark.

9. This was Kagera Region, where I was working as a regional planner. Population was estimated to be in decline in three of the then four Districts (Bukoba, Karagwe and Biharamulo), while the 1967 census and all available knowledge indicated an increase, especially in Karagwe and Biharamulo, both areas of significant in-migration. It was only estimated to be increasing in Ngara District, the only district in Tanzania where population had declined through the past two intercensal periods (largely as a result of tsetse infestation – and government policies to halt its advance!). Cotton, the major and most rapidly growing cash-crop in Biharamulo, was estimated to be unimportant. All this arose from the fact that a small random sample missed the areas of population and production growth. When I mentioned this to the FAO statistician in charge, he seemed of the opinion that reality had got itself out of line with the figures rather than the other way round.

10. Until quite recently, it was considered rather poor taste in African government circles, and in the consultancy reports written for them, to refer to things like black markets and smuggling. I have been requested on several occasions to excise or tone down such references, (even once for an academic paper).

11. There is some doubt how great this increase actually was, since it seems likely to have been augmented by smuggling from Tanzania (see Chapter 3).

12. This point has been cited by others from an earlier version of this chapter to read that deliberate falsification of the figures was the major source of bias. I do not believe this to be the case.

13. In the intensive production of pigs and poultry, the loss may be reduced by 65%, but this does not account for a significant proportion of African meat consumption. In any case, the resource-loss is still considerable.

14. These often being the result of IMF conditionality or other donor pressure.

15. Producing a steady flow of fresh milk for urban consumers involves far greater use of purchased feedstuffs than the previous pattern of allowing production to vary with the season and availability of grazing and water.

16. The figures above may be hypothetical, but such violent fluctuations in deliveries are certainly not impossible. In 1973/4 Tanzania's National Milling Corporation procured some 23,000 tons of maize locally. In 1977/8 the amount delivered was some 220,000 tons, while a lot more probably could have been purchased if the money and physical facilities had been available.
17. For a perhaps extreme, but most revealing, example, see Good (1986).
18. Reasons for this include subsidies by developed countries, notably the EC, exchange-rate overvaluation by African countries and, of course, food aid.
19. For the first few years after independence, maize consumer prices were kept so low that peasants travelled to town to buy their requirements, having sold their own produce.
20. On this see Weiner et al. (1985).
21. There was some conflict here between the political authorities, who defined self-sufficiency in maximal terms of zero maize imports, implying exports in favourable years, and economic planners who preferred a balanced and minimized trade. This latter would have minimized the losses of the parastatal grain purchasing agency since, with maize prices between import and export parity, both imports and exports implied losses. In reality, through disorganization, the NAPB often both imported and exported in the same year.
22. Food shortages were also probably aggravated by the very localized notion of self-sufficiency held by many local administrators, who forbade 'exports' from their districts even when there were surpluses. For example, producers in Karagwe District were forbidden to sell their substantial (and perishable) surpluses of bananas to other parts of the country. They did so, of course, but with greater inconvenience, reflected in higher prices.
23. Although the Tanzanian study is much longer than the other case-studies here, it still omits important aspects of the food-supply situation. Further details on this can be found in Raikes (1985 a and b, 1986), and in Bryceson (1978, 1982, 1985) and others.
24. Disparities in exchange rates, easier availability of many consumer goods in Kenya and a long uncontrollable border have been the main reasons behind this.
25. Some 150,000 tonnes were exported between March and July 1987, though the article from which this information comes indicates that this may involve running down stocks excessively for private profit. 'Grain Drain?', *Financial Review* (Kenya) 17 August 1987.

3
The African Crisis

Outline & Implications for Food Supply

Food imports are part of a more general crisis in Africa, involving indebtedness, falling incomes and a significant degree of breakdown in the state administration and its provision of basic services. The same could be said of hunger, though, given the direction of development in most African states, this would be present and growing even if there was no crisis of the state – though it would be more effectively hidden from view.

This chapter argues that to no small extent the crisis derives from policies aimed at development and from the definitions of development which emerge from local dominant classes and from international forces, notably aid donor agencies. In recent years, this has been much aggravated by the rigid conditionality imposed by the IMF and World Bank upon debtor nations. Part of the problem can be seen as an almost inevitable contradiction between virtually any reasonable definition of development and the limited resource-base with which to achieve it. But another significant part derives from misallocation of resources, waste and corruption, related to and fuelled by the way in which dominant forces define development. Having outlined the contours of the crisis and of the above argument, the chapter turns to agriculture and food production. Here the purpose is to show how many of the policies, programmes and projects intended to develop agriculture have contributed as much to the hunger problem as to its solution. Once again, there are two parts to the argument. One is that agricultural development in the context of the need for general development and the growing crisis almost inevitably allocates resources away from those most in need. The other is that, as above, class ideologies and interests considerably aggravate this tendency.

It is perhaps as well to start with a disclaimer. I am only too well aware of the problems inherent in presenting an 'identikit' picture of 'tropical Africa', given the wide variety in history, culture, economy, climate, resource-base and much more besides among the different countries, especially as my own experience has been limited to eastern Africa. All the same, I think there is enough in common in the experience of post-independence tropical Africa to make the effort worthwhile.

The economic crisis in tropical Africa

Central to the current crisis in tropical Africa is huge and growing indebtedness. According to a 1985 World Bank Report, Third World debt in 1984 amounted to some 34 per cent of GNP, with debt service accounting for 20 per cent of exports. For Africa, the corresponding figures were 52 per cent of GNP and 20 per cent of exports.[1] This

35

represents an almost threefold increase in African debt as a proportion of GNP since 1970. Debt service has grown more than threefold as a proportion of exports, from 6 per cent in 1970 to 20 per cent in 1984.[2] As regards the aggregate amount owed, an OECD survey showed that the debts of 43 sub-Saharan African countries increased by about three and a half times between 1975 and 1982, with almost half of the total in 1982 accounted for by the top five borrowers.[3] Since that time, the situation has grown worse, with the average debt-service ratio climbing to over 40 per cent by 1986[4] and in some cases over 100 per cent, implying that, even if total export proceeds were applied to debt and interest repayment, they would fail to prevent the debt increasing further, even without further borrowing.[5] Figures of this sort, combined with a further slump in export commodity prices, make repayment all but impossible for many countries. As debt-service and other payments abroad have soared, financial flows to Africa have plummeted during the 1980s, leaving net financial inflows negligible, if not negative.[6]

While the aggregate foreign debt of tropical African countries is small in comparison with that of the major Latin American debtors, the much smaller size of the countries, their much greater poverty and incapacity to pay make it every bit as serious. In addition, this very factor reduces their bargaining strength. Brazil, Argentina and Mexico, simply because of the huge size of their debts, have a certain leverage in bargaining with international creditors, since default by them could precipitate a general crisis in the international financial system. By contrast, the debts of, say, Tanzania or Ghana are minor on the international scene, though no less crippling to those two countries. Even if totally repudiated, this would cause no more than a ripple on the international financial ocean – unless, of course, the ripple spread to other debtor countries. This latter possibility is no help to small debtor nations, since it means that the terms applied by international creditors become yet more stringent, to ensure that this does not happen.

With falling oil, mineral and agricultural commodity prices, almost all sub-Saharan African countries now run significant balance-of-payments deficits and are thus increasing their debts, quite apart from current repayments. There is much talk of new initiatives to improve the situation, but even the most generous proposals (which are invariably reduced) come too little and too late to prevent further deterioration.

If foreign indebtedness is central to much of the crisis, its effects have spread to every corner of economies and societies. Real incomes have fallen, and Ravenhill (1986: 2) quotes the World Bank to the effect that they may well soon fall below the level of the early 1960s when most countries became independent. This implies that real levels of living have fallen *below* that level on average and especially for the poor.[7] Foreign-exchange shortages and budgetary deficits lead to the breakdown and closing of industries for lack of fuel and spares, and threaten such advances as have been made in the provision of social services. A further threat to both is the conditionality imposed by the IMF and World Bank on the large number of countries whose debts force them to apply for structural adjustment loans and other credit to tide them over. These increase unemployment and urban price levels, leading to general deterioration in income levels, which have been reflected in a large number of cases by riots and even coups. In many cases, economic life is much disrupted by shortages and delays.

Reasons for the debt crisis

A trade deficit means that more is being imported than exported in value terms and prompts the question whether the underlying cause is slow growth of exports or rapid

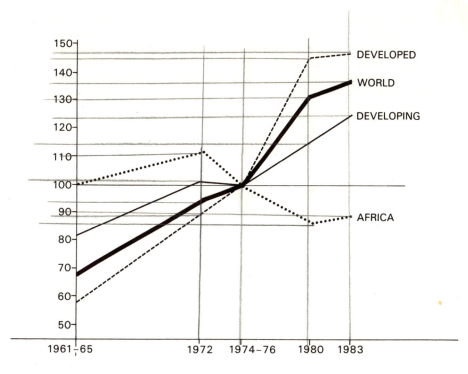

Figure 3.1: *Volume index of African agricultural exports* (1974–76 = 100)

growth of imports. Both have occurred, though growth of imports has been slowed in recent years by difficulties in getting credit. Generally speaking, more attention has been paid to the trend in exports, the rise in imports being considered a necessary accompaniment to development, especially as an increasing proportion of the total import bill goes to paying for intermediate goods, fuels, spares and services for local production facilities.[8]

While the current dimensions and the growth of the crisis are clearly heavily determined by trends in the international economy, it is worth considering how this situation came about. Exports from tropical Africa appear to have taken a fairly marked downturn from about 1973, according to FAO indices, as shown in Figure 3.1. Since then growth has lagged significantly behind that of other parts of the world. For the first part of this period, this arose mainly from stagnant or declining volumes of exports. Since the late 1970s, rapidly deteriorating terms of trade for agricultural and mineral export products have reduced export proceeds directly and contributed to continuing stagnation in volume. But the downturn in export volume dates from before the major deterioration in terms of trade, and its onset cannot thus be attributed to it. Moreover, the effects of indebtedness and shortage of foreign exchange were beginning to affect export production negatively already by 1970 in some cases, so it makes little sense to concentrate on exports alone.

Körner *et al.* (1986) list seven 'typical causes of debt crisis', of which five are primarily concerned with expanding imports, while the other two concern the size and stability of export production. Revising the order to reflect this distinction, they are as follows:

Indebted industrialization. i.e. industrialization financed through incurring foreign debt, which fails to yield a return sufficient to repay that debt. Körner *et al.* assert, and with some justification, that a large proportion of import-substitution industrialization in the Third World, from Latin America in the 1930s to Africa from the 1960s, has been of this nature. It has increased imports instead of reducing them – replacing consumer goods imports with sophisticated equipment, fuel and spare parts. At the same time, it has affected export proceeds, by diverting resources from agriculture and through subsidies which shift the terms of trade against agriculture. There is nothing here that the World Bank or IMF would disagree with, and it does seem generally true, as a plethora of examples from most of the countries in Africa indicates. It does, however, need contextualization.

Indebted militarization. Military expansion in Africa is by definition a debt-incurring mechanism, since none of the countries produces its own military hardware – and even if it did, it would be an extreme case of indebted industrialization. Körner *et al.* cite eight tropical African countries (excluding oil-exporters, and thus Angola) whose military expenditure exceeded 20 per cent of the total government budget in 1980, headed by Ethiopia 43 per cent, Chad and Mozambique 29 per cent and Zimbabwe 26 per cent.[9] Since then it seems highly likely that the number and proportions have increased. While military expenditure (and the use of the equipment in military action) is always harmful to economic and social welfare, one's judgement as to its necessity will invariably be coloured by one's sympathy for the government or the opposing forces and the degree to which military build-up is aimed at protection against external attack as opposed to repression of the civilian population. Thus it seems easier to sympathize with Mozambique, the victim of savage attacks from a South African-backed terrorist group which has made scant attempts to build political institutions, concentrating almost entirely on destruction and the infliction of hideous mutilations on its victims, than with Ethiopia, where secessionist movements appear to enjoy a very considerable measure of popular support. Least of all could one sympathize with Zimbabwe in 1980, when the expenditure was that of a white minority, grimly clinging on to power by savage repression of the African population. But these cases of outright civil war are only the tip of the iceberg and in many of the countries of tropical Africa military expenditure places a heavy burden on economy and society, either as a result of dictatorship by the military or as pay-offs to them, in the hope of avoiding this.

Developmental gigantomania. By this is meant huge, expensive capital-intensive projects, which absorb enormous quantities of foreign exchange for limited or negligible returns. Körner *et al.* cite the Sudan and six projects with a combined cost of over $4,300 billion in this respect.[10] But Nigeria (see below) and another score of countries or hundred projects would also serve to exemplify this important source of indebtedness. Since many of these huge projects are infrastructural or agricultural (usually irrigation), this serves to indicate that industry is not the only sector where wasteful investment occurs. Moreover, as shown below and in Chapter 9, the problem is by no means confined to huge projects. A very significant proportion of development projects, including those funded by the World Bank, increase indebtedness far more than they increase production or exports.

Debt-inducing social reforms. Improved provision of education, water and health services inevitably increase debt, so long as they are considered within a narrow economic framework[11], and so do food subsidies to urban populations. The effect on indebtedness is twofold. In the first place, they do tend to increase domestic consumption and imports, but secondly, since they are seen by the IMF and World Bank as major problems (far more important than corruption, though their effect is arguably less), countries which follow such policies invariably encounter major problems in getting stand-by credit once they have become indebted. The best-known example from Africa is Tanzania, which had to wait four years for IMF and World Bank tiding-over and structural adjustment loans, because of its refusal to give up various social programmes and devalue the currency (because it would increase prices to low-income, notably urban, consumers).

At the same time, it is worth stressing that not everything which appears as 'social expenditure' on government budgets is necessarily beneficial to the population. Apart from problems of waste and corruption, items like Tanzania's forced villagization are often assigned to this category, where many (myself included) would argue that their negative effects on the rural population far outweighed any benefits. That the programme was enforced is itself sufficient testimony to its lack of popular appeal.[12] Urban food subsidies may also fail to improve consumption standards where (as described in Chapter 2) they increase official food gaps, forcing a large proportion of the urban population (and typically the poorest) to purchase their food requirements at high black-market prices.

Kleptocracy. By this is meant the rule of a class based on theft from the public purse, a major example being Mobutu's Zaire, where the President, his personal coterie and clan have misappropriated literally billions of dollars of public money, not only increasing foreign indebtedness but running the economy into the ground. One may also note the unusually favourable treatment of Zaire by the IMF and other donors, because of its strategic importance to the USA and the enormous value to the West of its mineral wealth. Time after time, IMF credits have been granted on the basis of patently worthless promises to abide by conditionality. An IMF official, appointed as what Körner *et al.* call '*de facto* director of the Banque du Zaire' (though *de imaginatio* might have been a better term), wrote in his report on leaving that 'Mobutu and his government regard the repaying of their debts as a joke' (Körner et al. 1986, p. 97). This continues up to the present, with Zaire recently having 'won a landmark agreement from the Paris Club of Western creditor governments to put off paying part of its debt for much longer than ever agreed before'.[13] It is instructive to contrast this with the several years delay experienced by Tanzania in getting a much smaller credit. In part this reflects disapproval of Tanzania's (only moderately) anti-Western statements and socialist policy; in part the fact that a worthless promise is of more value to bankers than none.[14]

While Zaire, Uganda and the Central African 'Empire' under Bokassa are pre-eminent examples, the phenomenon is by no means limited to these cases, however, or to the personal coteries of grossly corrupt leaders. Corruption, misappropriation of public funds and manipulation of subsidies and other state controls are prominent among the means of accumulation of what Körner *et al.* call the 'state class' (see below) in many if not most of the countries of tropical Africa. While the individual amounts extracted may be less dramatic, their total size is almost invariably significant. Moreover, development projects, especially but not only the gigantic ones, are a fruitful source for such exactions.

The two export-decreasing processes listed by Körner *et al.* are *neglect of agricultural development* and *failure to diversify exports*. In the first case, they cite low agricultural prices and urban bias in investment patterns, both clearly related to the sorts of high expenditure considered above. These are dealt with in more detail below, where the point is made that the problem is as often misapplied and wasteful investment as lack of it. It is quite true that most African countries have failed to diversify their export bases, but it is often hard to see how they could have done so. Indeed, at least some of the wasteful expenditures have had precisely that as their purpose. As considered at greater length in Chapter 5, the scope for and returns to primary export production have deteriorated significantly in recent decades and seem likely to continue in this direction. The option for export-oriented industrialization was never the panacea it has been portrayed as; has largely been pre-empted by 'Newly Industrializing Countries' with far superior economic and infrastructural bases, and is tightly constrained by growing protection in the developed West.

While this distinction into seven different causes is useful, there are certain problems with this method of presentation, notably that it can give the impression that these are *separate* causes of indebtedness. In reality, many of them are intimately bound up with one another, as the following thumbnail sketch of developments since the early 1960s shows.

Post-Independence development in tropical Africa

Budgetary deficits of any magnitude were virtually unknown in colonial Africa. Monetary systems were tied closely into those of the colonizing power, making deficit financing virtually impossible. But in any case, colonial regimes were fiscally parsimonious in the extreme. Budgetary imbalance was avoided at the cost of grossly inadequate investment in economic infrastructure and even less adequate provision for education, health, water-supplies and production for the African population. At independence, most tropical African countries thus attempted to break with the stagnation of the colonial period, by instituting ambitious plans for national development and integration, with correspondingly large budget and import requirements.

In recent years it has become common to assert that large proportions of this expenditure went into 'unproductive' social services (education, health services and water supply), and, by implication, that this was a major cause of waste. But firstly, in nearly every case it was agreed at the time, by local and external observers alike, that the countries of tropical Africa at independence were disastrously lacking in the minimum economic and social infrastructure for national integration and development. Apart from this, most of them lacked the trained manpower simply to run the existing very low level of services. Secondly, the development goals of most countries included the eradication of 'poverty, ignorance and disease'[15], while improvements were certainly expected by their populations. It was thus not only reasonable but politically necessary to raise social expenditures from their abysmally low level. Thirdly, the problem is at least in part one of measurement. If an agricultural project leads to increased production, this is attributable to that project, but if increased production arises from better health or education, or from a reduction in the time spent collecting water, there is no way to attach that benefit to the social service. Finally, when all is said and done, social expenditures have accounted for the smaller proportion of state expenditure. The major portion went to 'directly productive' purposes and to economic infrastructure (especially in transport) having as purpose to develop production. Given the state of the transport networks and the prohibitively high level of transport costs both then and now, this was an entirely comprehensive focus.

Attempts to develop industrial production rapidly have often led to the emergence of uneconomic and inefficient industrial sectors, whose protection has imposed major costs on the remainder of the economy, and notably agriculture. But it is worth citing Robert Bates on this score. He notes that common sense, the evidence of history, and economic doctrine, all communicate a single message: that (development) can best be assured by shifting from economies based on the production of agricultural commodities to economies based on industry and manufacturing (1981: p. 11).

It is worth noting that the 'economic doctrine' referred to covers that of Rostow and the World Bank as well as that of Marx and Samir Amin. What is more, there are senses in which this proposition is clearly true; there are aspects of development (virtually however defined) which require the growth of towns, industry, and specialization.[16]

The problem is how much emphasis to give to industrial growth, given that in most cases it will require protection to survive competition from much larger and more developed economies, and often further forms of subsidy into the bargain. This, at any rate, is true of import-substituting industry, which has predominated in the investment patterns of most tropical African countries. It is thus usually *not* true, as tends to be claimed in development plans and industrialization policies, that import substitution will improve and cheapen supplies of the product in question – and certainly not in the short run. All too often, local investment has led to higher costs, lower quality and limited foreign-exchange saving, or even loss, often at the same time as receiving significant direct state subsidies.[17]

While one can argue with hindsight that this emphasis was excessive, it can scarcely be denied, however, that it was understandable and widely shared among recipient governments and donor agencies alike. Nor should it be forgotten that, while the latter are now highly critical, they provided most of the funding loans. My own opinion is that the problem is less one of an excessive emphasis on industry than of a misapplied emphasis on the wrong investments and capital-intensive techniques, coupled with a tendency to keep misbegotten projects alive whatever the cost.

Be that as it may, the first years after independence saw a major expansion of investment in all sectors of most tropical African countries and a corresponding expansion in the state sector. In those cases where minerals or a large colonial settler economy provided conditions attractive to investors, foreign capital accounted for a large proportion of industrial investment. In others, lack of interest by foreign investors and the absence of a significant local capitalist class encouraged states themselves to become involved in, or underwrite, industrial investments.

Thus growth of the state sector was often supplemented by the rapid growth of a parastatal sector, composed of state corporations operating in industrial production, provision of essential services and agricultural marketing. Pressures for expansion came from two different sources. On the one hand, there were development policies to be implemented, and institutions and other accoutrements of a modern nation-state to be built. On the other hand, the educational institutions were turning out rapidly increasing numbers of educated people, looking for jobs and most unwilling to follow their parents into peasant farming. It is not hard to see how a statist development ideology, building on their education, would appeal to such strata.

Thus the development activities of the first years after independence led to major increases in state expenditure (and thus revenue needs), together with a rapid growth in import requirements, especially of capital and intermediate goods for development activities. With imports increasing rapidly, trade balance deficits began to emerge and expand. At the same time, development and national integration-related activities, including the growth of the state sector, were generating budgetary deficits, notably in the recurrent expenditures which, by tradition, are paid for out of taxation rather than borrowing. Development projects tend to expand the activities of governments and notably wage and fuel costs. In the first instance (while the activity is being run by an aid donor or its agents), such costs are attributed to the development budget (financed by borrowing and aid grants). As each project is 'completed', the new activities must either be transferred to the host country's (recurrent) budget or be terminated.[18] Apart from this, there is a certain self-expanding tendency to many of these processes, for the more energies are expended on development planning, the more problems are turned up, to be overcome with more projects and personnel, more transport, administrative structures and imports. This tendency is further strengthened by its close compatibility with the interests and ideology of an emerging ruling class.

'THE STATE-CLASS' AS A REVENUE EXTRACTOR

In the absence of any significant landlord or capitalist class in most Africa countries at independence, the state became more than usually central, not only as 'engine of development' but as the basis of an emerging local ruling class, often referred to, for want of a better term, as the 'state-class'. The defining characteristic of this class is that its means of accumulation lies in control of the levers of political power and flows of funds through the state budget.[19] In some cases, this is achieved through outright theft and misappropriation, in others through favourable access to subsidized credit, grants and scarce resources distributed through state agencies; in others it may consist of no more than the reinvestment of wages and normal perquisites from the state sector. Whatever the case, it tends to involve 'straddling', the investment of funds deriving from the public sector in private businesses or agriculture, and the further use of state grants and subsidies to increase the latter's profitability.[20] It thus predisposes the members of the class to favour state intervention in the economy, since this expands the flow of funds and resources available for extraction through manipulation of political and bureaucratic processes.[21]

It would, however, be over-cynical and inaccurate to see the whole emergence of a statist ideology of development purely in these terms. Political ideologies are seldom simply naked and unashamed reflections of personal interest, and it would be incorrect to attribute motives of this sort to leaders like Julius Nyerere, whose modest style and personal honesty are well-known. For one thing, there is simply the fact that, in the absence of any other major focus of power and funds, the state *is* of necessity the focus for development policy. For another, while those who accumulate through the state include large numbers of politicians and businessmen without the necessary educational qualifications, recruitment to the upper reaches of the state bureaucracy is pre-eminently through education. Moreover, this is an education and socialization process which predisposes its members to think in terms of a clear dichotomy between the 'modern', representing development and personified by the educated elite of civil servants and technical 'experts', and the 'traditional' consisting of those lacking formal education and in need of transformation, through state-supported modernization.[22] It is worth noting that this ideology is at least as common among aid donor experts as among African civil servants, being among the more important conceptual bases of the 'aid project' as commonly defined and implemented.

Among the characteristics of this ideology is the fact that its uncritical acceptance of all that is 'modern' leads to the adoption of projects and programmes which have been subjected to 'optimistic' economic scrutiny, if any at all, and which increase the costs of state-provided services more than they increase output. While there may seem to be little connection between this 'service-ethic' ideology and the grubby realities of private accumulation through the state, such an ideology does underwrite and underlie expansion of the state sector and its control over the allocation of resources, thus providing the basis for expanded accumulation through straddling in its various forms. Moreover, the frequent failure of the policies involved and their impact on state budget deficits easily generate frustration and lowered bureaucratic incomes, which shift the balance from 'service to the nation' to manipulation of the state for personal ends. This would seem to underlie the very rapid transition of Tanzania between the late 1960s and mid-1970s, from one of the least corrupt African states to one in which corruption was widely accepted as a necessity for survival. But whatever the precise nature of the relationship, there can be little doubt that such an ideology predisposes nations to expansion of the state sector and budget, and to increased importation of the means for modernization.

Returning to the processes involved, with increasing budgetary problems the revenue net was cast wider and deeper. The industrial sector was largely exempt from taxation, through the various 'tax-holiday' measures needed to induce foreign investors to set up business. Where it was state- or parastatal-operated, this was further aggravated by a range of direct and indirect subsidies. Since the main source of personal income tax from wages was the state sector itself, any addition to the net tax revenue would have to come from the peasant sector and/or through indirect taxes. Sales taxes on non-food consumer goods and taxes on commercialized agricultural produce were among the means sought. It should perhaps have been obvious to the governments in question that this would have a negative effect on commercialized agricultural production. That it was not, seems to derive from two factors.

Firstly, the notion that the peasantry should be seen as economically rational had (and to a large extent still has) yet to penetrate the thinking of many government officials or that of some of their foreign advisers.[23] Secondly, for the state, its 'revenue imperative' took precedence over and overshadowed all other factors.

Revenue extraction from the peasantry has not been only a matter of central state taxation. Parastatal marketing monopolies have also been responsible for a significant proportion, and this is levied in a way which has a particularly direct negative effect on production incentives. That is, as an addition to the marketing margin and thus a deduction from the producer price.

In certain cases, this has been as a result of policies specifically intended to transfer state activities and revenues to parastatal corporations. It was in this light that consultants devised for Tanzania the transformation of marketing boards into 'crop authorities'. The latter would not only handle sale of crops, they were responsible for *developing* them. They provided inputs, extension and credit, and purchased, transported, processed and sold the produce. The idea was that they should subsidize development activities out of their 'surpluses' from crop-purchase. Thus the producers of a crop would pay for the developments which benefitted them.

That, at any rate, was the theory. Tanzania's crop authorities are generally believed to have played no small part in bringing about the country's stagnation in export-crop production. As Ellis (1983) has shown, the marketing margin rose steadily after the change, most especially for export crops, but also for food crops like maize. During that period, the physical services provided (collection of produce) deteriorated, processing facilities fell into disrepair, and the authorities still built up financial deficits so large as significantly to influence the rate of inflation. The crop authorities are usually considered to have arisen from Tanzania's socialism, it being conveniently forgotten that they were the 'brainchild' of a well-known consultancy firm and that many of their activities were actively encouraged and financed by aid donors.

Nor is Tanzania by any means alone in this respect. A recent work on marketing boards in tropical Africa mentions a widespread tendency to broaden the scope of marketing boards in the 1970s, turning them into 'crop development corporations' (Arhin *et al.*, 1985: 15–17). Malawi, Nigeria, Senegal and Zaire are cited, in addition to Tanzania. In most if not all cases, they have 'greatly disappointed the political leaders'[24], through a common tendency for non-marketing costs (those of 'development') to expand. Bates (1981) also gives examples of similar mechanisms for a wide range of states in east, central and west Africa.

THE 1970s

Already by 1970, the balanced or positive trade balances of early independence had given way to deficits. Of 26 tropical African countries for which data are available, 19 had trade balance deficits, while only 7 had surpluses. Moreover, the combined deficit amounted to $884 mn. (of which 42 per cent was due to Nigeria, where oil was not yet important) as against a combined surplus of $158 mn (fully two-thirds of which was for Zambia, then a prosperous copper exporter). But the size of the deficit was not yet sufficient to give rise to major international concern. For few countries was debt service over 10 per cent of exports (World Bank, 1982: Table 12).

With the exception of oil exporters, the countries of tropical Africa were thus not well-structured to withstand the impact of the oil price rise in 1973. To make matters worse, this coincided for many with large increases in food import requirements and enormously increased international food prices. The food deficits already emerging in a number of countries were accelerated by the drought which precipitated the Sahelian and Ethiopian famines of the early 1970s.

But 'help' was at hand, for the increase in oil prices had also led to the accumulation of huge quantities of 'petro-dollars', reinvested with Western banks, which then looked for investment opportunities at a time when Western recession prompted expansion to the Third World. The period of easy money began. Not only were loans easy to get, but interest rates were very low. For several years inflation-adjusted rates were actually negative, while the average for the decade was in the range 0.2–0.3 per cent, compared to 2.1–2.2 per cent during the 1960s (FAOb, 1984: 2). In addition to this, the official aid agencies of the West significantly increased their aid in both money and real terms (see Chapter 9), the large bulk of this going to pay for the foreign-exchange costs of one or other investment or development programme. Whether this took the form of factories requiring fuel, servicing and spare parts, roads needing maintenance, or rural development projects requiring equipment, inputs and increased presence of government officials in the rural areas, almost all of these programmes implied increases in recurrent expenditures, with fuel and spare parts as major foreign-exchange components.

This, however, was to emerge later. During the mid-1970s, credit was cheap and easy to get, donors were anxious to expand their activities, firms anxious to expand their markets. Correspondingly, African governments were most anxious to borrow, given the easy terms and that this allowed them to continue with their 'development programmes', that is, with the processes which had been developing over the previous years. The rapid inflation of the period was expected to eat up the majority of repayment, given the long-term nature of many loans.[25] The period after 1973 thus saw a continued increase in import spending, with a large proportion of this tied up in capital and intermediate goods and a significant proportion spent unwisely.

Export production was stagnant, however, and deficits were increasing more rapidly. Many governments took time to adjust crop producer prices upwards in line with the high inflation rates which followed oil and cereal price increases. Some of the burst of aid spending went to fuel the expansion of marketing boards to 'crop develop-ment corporations' with the negative effects noted above. Some of the countries worst hit by drought re-orientated agricultural policy towards food crops. Currently this is seen as a major factor in declining export production, but it was easily under-standable at the time. Between 1973 and 1974, Africa's cereal imports increased by 12 per cent in volume terms, but doubled in value. Tanzania's cereal imports increased in value by *23 times*. For these reasons, and those mentioned above, exports stagnated.

Then, with the second oil price shock in 1979, the situation turned around rapidly. Much tighter monetary and financial controls were instituted by creditors, interest rates were raised with a vengeance and credit became much more difficult to get from an international financial sector reacting against its previous over-extension. From the very low rates of the 1970s, real interest rates jumped to the highest level since 1945. From 1980 to 1984, inflation-adjusted rates of interest averaged well over 5 per cent, some *15 to 20 times* the comparable rates for the 1970s (FAOb, 1985: Table 1.3). The rise of the dollar and the decline in export commodity prices were further blows to export sectors already in severe disrepair.

By the late 1970s, several countries were already desperately short of foreign exchange. With 'increase inputs' largely exempt from controls and food supplies to the cities requiring increasing imports, the sector which suffered most was non-food consumer goods. In many cases supplies of agricultural implements also suffered. If these goods were freely marketed prices rose. If they were distributed under one or other form of rationing at controlled prices, large quantities leaked on to black mar-kets and prices still rose. In both cases this acted as a further disincentive to agricul-tural production for official sale. With the pressures of the 1980s, limitations on imports increased to new levels. That food imports ceased to increase from 1980 to 1982 has nothing to do with reduction in the needs of the populations concerned. It simply reflects lack of foreign exchange. The more recent increase in 1983–5 was largely composed of food aid.

From the above account it emerges that rapid growth of imports and government spending has been at least as important as stagnation in exports, and has been, to a large extent, responsible for the latter. The most rapid growth area for imports into Africa has been capital and intermediate goods for investment and development, and this poses questions not simply about levels and sectoral allocation of investment but about *choice of technology*. For both agricultural and industrial sectors in most trop-ical African countries, there is plenty of evidence of a focus on capital- (and thus import-) intensive techniques, which not only consume foreign exchange directly but increase vulnerability to shortages of foreign exchange. This certainly reflects choices made by African governments and by the state-classes controlling them, but it most certainly also reflects external and aid-donor pressures. Scarcely one of the capital-intensive projects could have been implemented without finance, approval and technical assistance from aid donors.

The impact on incomes and food supply

Even before considering agricultural and food supply policies, the above indicates that there is no shortage of reasons to expect a worsening of food supplies to the poorer sections of the African population, especially when taken in conjunction with the factors considered in Chapter 2.

The most important underlying factor has been concentration of incomes and resources, both geographically and by class, in a situation in which total income levels have risen slowly, stagnated or even declined. This can be related to the various inappropriate, wasteful and corrupt uses of resources considered above, but underlying that to the non-productive nature of accumulation by a state-class. Classical capitalist accumulation is a brutal process, imposing enormous costs upon those on whose labour it is built and those elbowed out of the way to let it pass. But it does lead to an overall accumulation of resources, through the reinvestment of profits deriving from production. Accumulation by the members of a state-class, however, does not necessarily lead to any social accumulation at all – and the evidence of huge indebtedness indicates that in tropical Africa, by and large, it has not. It is thus more akin to primitive accumulation – the concentration of resources by the powerful, through processes ranging from manipulation of state policies to outright robbery. This is classically seen as a prelude to a later stage of productive accumulation, though, in the African case, the conditions for the transition seem well over the horizon. To grasp this, it is worth summarizing some aspects of state-class accumulation.

In the first place, the primary source of accumulation is state revenue, or resources from external sources which (unless in grant form) have to be repaid through increased state revenue. In the crudest cases (kleptocracy) this occurs through direct misappropriation, but at least as important are manipulation and privileged access to a variety of subsidies, grants and quotas arising from state policy. Thus, even though the resources extracted from the state may be invested in 'productive' activities, it is only too likely that the profitability of these enterprises will depend on such mechanisms. This may involve cheap credit from state agencies (much of which may never be repaid), or quotas to import goods at controlled exchange rates (which can then be sold at considerable profit on the domestic market, whether or not fed through a local production process), or purchase of goods at controlled prices from state agencies, which can be sold at much higher black-market prices, and so on. Under such circumstances, there is no necessity for the formal production process to be well organized or profitable; the profits come from privileged access to free or cheap resources. Moreover, a significant aspect of the process is the 're-investment' of resources in maintaining or improving this privileged access, an essentially unproductive and wasteful process, classically involving buying drinks for, or paying bribes to, those who can provide favours.

Another aspect is the nature of and limits to the revenue base. While classic capitalist profit depends on the production of value surplus to wages and input costs, the extraction of revenue is not linked closely (in the short run) to levels of production. It depends rather on the power of the state and its ability to collect taxes in one form or another, this being limited by the political power of the population to resist. Evidently the latter is much enhanced by political democracy, which is largely lacking in tropical Africa. This refers not only to dictatorship or one-party states at national level, but to the characteristic political powerlessness of peasantries, who make up the vast bulk of the population. Their only available response to high taxation (usually levied on flows of produce, since income taxes are almost impossible to collect from peasants) is to route produce through channels where it is not available to the taxman (and the official purchasing agency). This they have widely done throughout Africa. Thus the revenue base is linked to production in the long run, but in ways which disrupt the flow of produce and only limit revenue *after* the damage has been done.

The other aspect of this is that, combined with the ideological factors considered above, it predisposes the state to wasteful development expenditure. In the first place, the ideology of modernization generates an uncritical attitude to state expenditures, so long as they are considered 'developmental'. Secondly, it expands the sources for accumulation by well-placed individuals. And thirdly, the revenue base only sets a limit to the process retrospectively.

It is here that foreign aid and lending add a further crucial dimension to the process, for these enhance the capacity of the state to overspend – and often still further enhance the possibilities for misappropriation.[26] One might suppose that, given the huge batteries of economic and other expertise which such agencies claim to dispose of, they would be more critical in their economic assessments. Cost-benefit and similar analyses are after all generally required. In reality, as discussed at further length in Chapter 9, this is far from being the case. For the moment it will suffice to point out that the World Bank, with literally hundreds of economists employed, routinely presents estimates of internal rates of return[27] which are grossly in excess of what turns out to be the case, and that their projects have been no minor source of African indebtedness.[28] In more general terms, it is impossible to overestimate the importance of easy donor funding in the 1970s as a cause of the crisis of the 1980s.

While there can be no doubt that such processes have increased income differentiation and tended to impoverish the poorest people and most marginal areas, it is worth considering briefly the possibilities for 'normal' capitalist development, since it is this at which the World Bank, IMF and other external agencies claim to be aiming in imposing conditionality. There can be little doubt that devaluation and many of the other measures worsen incomes and availability of food to many poor people in the short run (though for the reasons cited, I am less certain about food subsidies). But this would perhaps be acceptable if there was any sign that it would lead to long-run improvement. However, while at least some of the elements of conditionality seem reasonable in themselves – though applied with excessive rigidity – there is little sign that the objective is either achievable or a solution to the hunger problem.

In the first place, it is hard to see where the conditions for a transition to capitalist accumulation are likely to come from. While state controls are often justified in socialist terms, they are no monopoly of left-wing states and have proved highly resistant to eradication – predictably enough, since they are the means of accumulation of the ruling class. Even where it has proved possible to privatize state monopoly agencies, this has usually been done in ways which retain many of the monopoly elements, in effect selling them to politically well-placed individuals and groups. Indeed, the process of extracting revenue via the state is heavily dependent on the development of a complex variety of family, clan, tribal, professional and other networks, cemented with bribery and often lubricated with alchohol, which both tie it in position and increase its wastefulness through the plethora of unproductive expenditures involved. At the same time, the process is politically highly unstable in the larger sense, since it involves increasing competition for resources from a stagnant or declining base. This is often further aggravated by the presence (in the foreground or the background) of the military as the most concentrated focus of power and most profligate waster of resources.[29] In short, there is no reason to expect the legal and political stability which is requirement for productive capitalism.

Here it is worth commenting on a viewpoint which is sometimes heard, namely, that corruption and the black market are healthy signs of the emergence of a free market. Apart from the cynicism, this shows abysmal ignorance of the processes involved. The small peasant, selling a few baskets of maize to a private trader or in the local market, may be subject to market conditions (though invariably affected by non-market factors in the larger economy). The same cannot be said of the official who misappropriates and sells a few truckloads of maize, still less of the leader who takes a few warehouses full. In the first place, the bribes required to achieve this will vary according to his status and political contacts, being lower the better-placed he is. But the price at which the produce is sold will also vary with non-market factors. Other things being equal, kinsmen, business associates and those from whom future favours may be expected will receive the most favourable terms, and the poor and powerless, especially if outsiders of one sort or another, will receive the worst. Corruption is sometimes said to 'oil the wheels of development', though it would be more accurate to say that it oils the track. Not only does it distort the allocation of resources and lock the political system further into unproductive forms of accumulation, it also involves enormous direct losses to the countries concerned. This occurs not only in gross cases like Zaire, where the accepters of bribes become enormously wealthy. One of the more depressing consequences of state accounting and the lack of accountability of its personnel is that officials will often sign grossly unfavourable contracts without close scrutiny – and all for the śake of benefits as trifling as a three-month seminar in the 'donor' country, with possibilities for the purchase of consumer durables not available at home.[30]

Even if one could foresee a change in the local mode of accumulation, what are the chances that 'classical capitalist development' could either take place or solve the problems of increasing poverty and hunger? The small size of internal markets is one obvious problem, but that is partially related to the absence of political conditions for such a development path. One could thus (optimistically) assume that if these were provided, through the emergence of a substantial, politically represented, petty-bourgeoisie, middle-level commercial farmer and working class, this would itself provide a substantially larger internal market. But this would leave unsolved the foreign-exchange problem, since almost any form of development would require substantial inflows of foreign exchange for many years into the future. The limited possibilities for export-oriented industrialization have mostly been pre-empted already by far better placed NICs in Asia and Latin America, while their scope is constantly being limited by developed country protection. On the other hand, the prospects for a way out of the crisis through expanded primary product exports look even less hopeful (see Chapter 5). This pessimism is based not on the dependency argument about a 'necessary' decline in the terms of trade for primary products, but on the trends of the past quarter-century and analysis of the specific processes involved.

Moreover, there is still less evidence that this could solve the problem for poor and hungry people. Capitalist accumulation is based on the maximization of profit. Given the low value of most local currencies in present-day Africa, any secure strategy for profit maximization depends on access to and accumulation of foreign funds and resources – which in turn means operation in conjunction with, and subordination to, foreign interests, whose economic and political power is immensely greater and whose interests have little to do with the establishment of locally autonomous capitalisms, still less those of the poor.

Capitalism goes quite specifically for the areas of highest profit, and these are seldom likely to be the areas of greatest need. Profit maximization involves producing for markets where there is effective demand, not those where need is not backed by money. Capitalist farming has a clear tendency to establish itself in the areas of highest productivity and climatic reliability, with the exception of short-term high profit ventures which do more harm than good. Moreover, the concentration of land for capitalist farming worsens the shortage of land for peasants, while normally involving major reductions in labour-use per hectare. And even if the privileged access of the powerful as members of the state-class or its associates were miraculously to disappear, 'normal' market processes would still give the advantage to the rich and powerful.

Policies and programmes to increase food production

State policies towards agriculture in Africa have been overwhelmingly dominated since the colonial period by efforts to increase the amount of produce sold, and still remain so. This may seem obvious enough, especially as seen from Europe where close to 100 per cent of agricultural produce is for the market. Indeed, it is usually taken for granted in Africa, even though up to, or more than, half of food production is produced for own consumption by peasant producers who account for between 70 and 90 per cent of total populations. But it does indicate the nature of the state's prime concern – to increase the commercialized 'surplus'[31] for urban food needs or export and for the extraction of revenue.

Prior to the 1970s, the expansion of food-crop production tended very much to take second place after export crops in official thinking. This is partly a legacy of the colonial period, when the value of a colony to its metropolis lay in the production of export crops and the growth of markets for metropolitan produce. None the less, independence did not usually lead to very striking changes in the pattern, since independent governments also needed foreign exchange. It was rapid urbanization and its pressure upon food supplies which stimulated concern. Most notably, the series of famines and shortages which swept much of Africa in 1973/4, just as US policy was achieving a huge increase in the price of imported cereals, focussed the minds of many policy-makers on the need to maintain or increase food production. When the change came, it was again heavily dominated by efforts to increase marketed production of crops like maize, rice and (sometimes) wheat which were important staples of the urban population.

State agricultural policy can be roughly divided into those parts concerning prices and marketing and those which concern efforts to increase production or develop production methods directly. So the change took the form of both an increase in the producer prices of foodstuffs relative to those of export crops and a major expansion of the number of programmes and projects to increase food production.

PRICES AND MARKETING

Most governments of tropical Africa control, when they do not perform, the primary marketing of major agricultural products from peasants.[32] Large farmers and minor crops are more often, though not always, left to private markets. Most states also control the exportation of agricultural produce, though it may be actually performed by private companies, co-operatives or parastatal boards. The state thus normally has wide-ranging powers to set and control the prices of agricultural products, which allows it to set a number of prices and ratios, namely:

(i) *The terms of trade* between agriculture and the rest of the economy, that is, the general level of agricultural prices in relation to those of goods purchased by the agricultural sector (consumption goods, agricultural inputs, building materials etc). Since this will include whatever taxes or subsidies are imposed, it will affect the amount of revenue which can be extracted from agriculture on the one hand, and the general level of agricultural incomes, incentives and investment on the other.

(ii) *The relative prices* of particular products or groups of products, this being generally considered the more effective means of influencing the pattern of production.

(iii) In addition to setting the prices of agricultural products, the state can often control *prices of agricultural inputs or implements*, which may be done at a national level, or for specific projects, areas, or groups of farmers.

(iv) The state can also determine *the regional pattern of prices*. This is often done through transport cross-subsidies to equalize prices throughout a given country and offset the advantage which would otherwise accrue to producers situated close to major markets or ports ('pan-territorial pricing').

(v) Finally, the state has the power to set *the exchange rate of local for foreign currency*. Holding this above the level set by the market (overvaluing the exchange rate) reduces the local price of imported items but also the local prices received for exports.

While these various prices and ratios affect the agricultural sector, they are not always set with agricultural production as the main objective. Taxation of agricultural products or items purchased by the sector is among the major sources of state revenue. The major objective of pan-territorial pricing is usually equity rather than increased production. Overvaluation of the exchange rate is more often concerned to avoid the inflation which arises from increased import prices than with anything specifically related to agriculture. Controls over prices enhance the opportunities for state-class accumulation.

There are, moreover, other factors which affect producer prices, which are not under the direct (or perhaps explicit) control of the state. Corruption or inefficiency in statutory monopoly agencies concerned with agricultural marketing increases costs and (other things being equal) reduces prices to producers. Urban food subsidies will have a similar effect. Subsidized distribution of loans or agricultural inputs through such agencies will also impose costs, which are normally recouped through deductions from producer prices. The latter are indeed aspects of state agricultural policy, but also an example of how policies to affect production directly can have the reverse effect to that intended through their impact on prices.

In much of Africa, during the colonial period, it was widely thought or claimed that African peasant producers were not responsive to prices or wages for labour. In part, this was a justification for holding both at the lowest possible levels and making up for it by the use of force to increase production or involvement in wage-labour. Apart from direct coercion – like stipulations that all peasants in a given area should grow at least one acre of, say, cotton, on pain of fine or jail – one of the most common ways to achieve the required production or labour input was to set flat-rate 'poll' or 'hut' taxes, to be paid in cash, on pain of fine or a period of forced labour. This was most commonly used where crop prices were very low and labour requirements heavy. In cases where returns to labour were higher, there was seldom any need for such policies. Indeed, in the settler colonies of eastern Africa, the authorities spent more time trying to *prevent* Africans from growing high-valued crops like coffee, tea and pyrethrum, both to limit competition to settler production and to prevent this from drawing off labour which would otherwise be available for them.[33] But since all these policies involved trying to force peasants to do other than what they wanted to, and since this generated evasive action on the part of the peasants, it tended further to cement the view of officialdom that peasants were 'resistant to change' and needed to be pushed into it.

While there was some change in both outlook and practices during the 1950s and 1960s, these ideas often died hard. Ministries of agriculture were largely staffed with people who had spent their formative years trying to enforce unpopular policies upon peasants and experiencing their considerable powers of evasion. For many of them, the frustration engendered by this led to a feeling that peasants were simply 'against progress' rather than against the low returns of ill-thought-out policies, and predisposed them to think in terms of enforcement rather than monetary incentives.

For states faced with budgetary deficits, it was also tempting to assume that lowering producer prices to peasants would not affect production – and that if it did, the difference could be made up by enforcing rules and regulations. Apart from that, there has been a major tendency for African state policy-makers to take insufficient account of inflation, especially during the 1970s when it accelerated markedly, and to retain given crop prices in money terms, disregarding their declining real value.

Thus the 1970s saw a variety of different factors affecting agricultural producer prices. Foreign-exchange shortages led to falling market rates of exchange and the maintenance of official rates above them, reducing producer prices of export crops. Budgetary deficits led to increased revenue extraction from agricultural sales and purchases, this again being concentrated especially on the export sector (where large-scale exports made collection easier and the returns came in foreign exchange). At the same time, slow response to inflation made for further decline in real producer prices, while the burst of 'productive programmes' associated with concern over stagnating production and the easy donor finance of the period placed further downward pressure on producer prices.

Since prices of export crops were depressed further than those of food crops or other commodities for the internal market, this accounts for much of the stagnation in export production and sales. By the same token, one might have expected producers to shift into food crops, whereas, as shown in Chapter 2, where official markets were concerned, precisely the opposite occurred. It is concentration on official markets which has led authors like Hyden (1980) to talk of an 'exit option' from commercial production, assuming that peasants retreated to subsistence production. Available evidence (and common sense) suggests, however, that this was rather seldom the case – that peasants who had become used to a regular, if low, purchased consumption of basic consumer goods, clothes and agricultural inputs, would not simply give these up. What seems far more likely – and is supported by all the available evidence – is that peasants shifted out of export-crop production into food-crop production for non-official markets. To the extent that production did decline, this would more likely be the result of increasing prices and decreasing availability of agricultural inputs – and in certain cases, the frustrations caused by programmes for the compulsory production of either export or food crops.[34]

So agricultural price policies have played a significant role in the emergence of the general crisis, though they themselves have partly been determined by the crisis. Where food availability is concerned, there is no doubt of their negative impact on officially marketed quantities, and thus on food imports. Their impact on general food availability seems to have derived mainly from lowering general incomes and from the fact that the poor and powerless get the worst deal when corruption and black marketing determine the allocation of foodstuffs and access to resources.

Policies to increase production directly

Policies to increase agricultural production, whether of food or export crops, have focussed, at least since the 1950s, on the adoption of more valuable crops and higher-yielding varieties, together with increased use of purchased agricultural inputs, notably chemical fertilizers and insecticides. In many, but by no means all, cases, they have also aimed at increasing the area cultivated through the introduction of agricultural mechanization, either with tractors or draught animals. In money terms, this has focussed heavily on tractor mechanization. Since virtually every aspect of this increases requirements for purchased inputs, it is not surprising that attention has been concentrated on production for the market, as increased cash sales are required to pay for this increased expenditure.

While there have been some changes in recent years, towards a 'whole farm' or 'farm systems' approach, especially in research, it still remains the case that both research and extension policy are largely 'crop-based'. This means that extension programmes and the research upon which they are based consider only the particular crop for which they are trying to increase production and productivity. In many cases whole research stations are devoted to one (usually export) crop. This leads to a strong predilection towards both single-cropping and mono-cropping as opposed to inter-cropping and/or rotation, since both research and extension are concerned with the one crop and not the whole farming system. It also often leads to recommendations which are either unsuitable or unfeasible for peasant producers, since they ignore land or labour constraints imposed by requirements for other crops. For example, it can be shown that coffee yields are higher under pure stand than when interplanted with bananas. But in many parts of East Africa, bananas are both the main staple crop and yield a greater value per hectare than coffee, taking their value as subsistence into account. For small peasants with limited land, the choice is not whether or not to interplant coffee with bananas, but whether or not to interplant bananas with coffee. At the same time, the best pruning, fertilizer and insecticide regimes for interplanted coffee are significantly different from those for pure-stand. Yet in one such area in Tanzania in the early 1970s, the research and extension services not only gave no advice to the peasant who interplanted, but had not even done research on the combination.[35]

Programmes to increase food crop production are normally implemented by the extension service of the ministry of agriculture, but may also be under specific 'crop authorities', in which case the single-crop focus becomes even more extreme. They may focus on increasing the area cultivated through mechanization or the opening-up of new land, or upon increasing yields through the adoption of new inputs and varieties of crops. They can focus on large farms, state or private, or on peasants. They may involve extension and the provision of inputs and credit for increased yields, or they can take more highly organized forms such as 'settlement schemes' or 'inte-grated projects' under the direct control of a manager or managing agency. Before looking at examples of some of these, it is worth a brief excursion to consider 'emergency food production programmes'.

Emergency food production programmes are usually devised in the throes or after-math of a major food shortage. They tend to combine 'political mobilization' with more standard elements of agricultural policy. Thus, it may be decreed that all peasants in a given area should cultivate an acre of a specified 'famine crop', or that urban gardens should be cultivated with food crops. Efforts to maintain food availability within a given administrative region may also result in bans on 'exports' to other parts of the country. In addition to measures of this sort, the state may attempt to expand state-farm production or enagage in accelerated programmes for the resettlement of peas-ants in new areas, as in Ethiopia (see Chapter 4).[36] It is hard to tell whether such programmes have any effect on the level of production. It is quite normal for food production to increase in the aftermath of a major shortage, but arguably it is the experience itself which has concentrated people's minds rather than official exhorta-tions. My own opinion is that compulsory programmes for the production of specified crops almost invariably do more harm than good, wasting large amounts of official time and petrol to lower the morale of the population. The latter may spend so much time performing and evading the compulsory tasks as seriously to cut into their real agricultural production. In general, emergency food production programmes contain many of the same elements as ordinary policy, though they tend to exaggerate the faults through hasty planning and decision-making.

LARGE-SCALE FARMS

Colonial settler administrations focussed efforts to achieve a saleable surplus of basic foodstuffs on settler commercial farming, and to outward appearances with great success. Both colonial Kenya and Rhodesia (now Zimbabwe) exported maize and other foodstuffs, and in some years both still do so. For some years even after independence, most of the commercialized maize in Zambia came from a small number of large farms. But outward appearances can be misleading. It is true that settler farming has achieved some of the highest crop yields to be found in tropical African farming. This is hardly surprising since white settlers chose the most favourable areas to colonize and made sure that they were serviced with immeasurably better extension services, credit, roads and other infrastructure not to mention cheap labour. Regardless of yields, one large commercial farm will always produce a greater *saleable surplus* than, say, 2–300 peasant families on the same land, since the latter have also to produce their own food requirements. Moreover, that such systems could produce national surpluses had a great deal to do with the abysmally low incomes of the majority of the population and the fact that they mostly continued to produce their own food requirements. It is urban growth and rising incomes, not declining productivity, which have reduced the export supluses of these countries. Settler farming 'solves' the national food problem at the expense of the poorer sections of the national population and specifically those pushed out to become landless neighbours to settler farms.

Much the same can be said of a 'post-independence' variant based on large farming by Africans, but there are certain differences. Few Africans who are rich or influential enough to get access to a large-scale farm are full-time farmers. Many of them are fully occupied outside agriculture and operate as 'telephone farmers', ringing periodically to give orders and find out what is happening. Political influence is often a prerequisite for success in other ways – for example, in getting access to free or subsidized inputs, equipment and loans from state or donor programmes. As with settler farming, this diverts resources from peasants who certainly need them more and could arguably use them better.

State farms are perhaps the most problematic of all ways to 'solve' the food problem in Africa. State farming takes land from peasants just like private large-scale farming, but often without the productivity and almost always without the economic viability. Space does not permit a detailed examination here of all the problems with state farms, but they have not been successful anywhere in Africa, to my knowledge. In Mozambique, emphasis on state farms (which took over 90 per cent of all funds allocated to agriculture for some years) almost rivalled the country's many external problems in its contribution to food shortage (Raikes, 1984b). It was common for even recurrent production costs to exceed the value of output. On one state farm visited in 1980, they were from *three to five times* the value of output.

Zambia and Tanzania have placed less emphasis on state farms, but where they have done so, with similarly discouraging results, though they have not starved the peasant sector of funds to anything like the same extent. In Tanzania, donor-run wheat state farms produce most locally available wheat on a very large scale – after villagization had eliminated African medium-scale commercial production. State farms, with considerable donor assistance, also produce a significant proportion of officially marketed maize and rice – but not economically and only by dint of taking large stretches of land from peasant production for methods which are highly capital-intensive. Zimbabwe claims greater success for the farms run by its parastatal Agricultural and Rural Development Agency (ARDA), which runs large irrigated farms, with peasant outgrower schemes attached. These are claimed to be a means of developing peasant farming, but the methods used are far too costly and capital-intensive to be spread to more than a fraction of the population, and are probably not economic when capital costs are taken into account. The local population would certainly benefit more from having the state-farm land and assistance for a lower-level programme of improvement. These schemes, which were set up during the UDI period, are in structure much more like commercial farming, though at substantially higher capital costs.

'Project farms'. A sort of quasi-state farming has been taken up by a number of countries in recent years, with the assistance of aid donors – the project farm (or ranch), managed by an aid agency or a private firm on its behalf. These are commonly part of grandiose plans to increase food production in one fell swoop, through the application of the most modern methods. Even aid agencies which are virulently opposed to state farming as a policy of socialist governments, encourage such forms of production – perhaps because they are seen as a purely technocratic (and thus politically neutral) solution to the food problem. Alternatively, this may just be another instance of the disparity between rhetoric and reality in development aid.[37] Great faith is placed (or misplaced) in the fact that their managers are experienced foreign 'experts'. The tendency is towards use of the most advanced techniques available. This is what the experts or company have been hired to provide and how the firms make their living. Even if such a farm achieves viability during its 'project' period – and this is rather seldom – problems are likely to arise when it is transferred to a local parastatal corporation and can no longer benefit from direct (often subsidized) imports through the agency and the grant provision of experts.[38] Once this point is reached, the state has to decide whether to take on board the considerable costs involved, to look for another donor to take the farm over, or to let it collapse.

The greatest concentration of this type of project has probably been in Nigeria since the start of the oil-boom. Andrae and Beckman (1986) look at plans for producing wheat under irrigation in Nigeria for 'self-sufficiency', and at the succession of fraudulent, incompetent and just plain absurd reports and decisions by 'experts', justifying and pushing forward plans which never had the remotest chance of success. At one time it was proposed to produce 0.6 mn tons by 1990 (or about one-third of current imports); targets have been progressively reduced to about one-quarter of that level but appear grossly optimistic in relation to actual achievements. Not only has production been a fraction of that forecast, it is evidently grossly uneconomic. The Nigerian wheat irrigation schemes appear to be 'development classics' on a par with the Groundnut Scheme of post-war Tanganyika. They are relevant to wheat imports, not because they have any chance of achieving self-sufficiency. But the illusion that they may do so, assiduously spread by publicists, serves to still criticism over import-based consumption. In this respect, they play a similar role to Tanzania's dairy farms (see below).

Andrae and Beckmann also discuss some of the specific joint-venture and company-run farms which have been set up to produce wheat, and their disastrous results including pitched battles with the peasants expropriated to make way for them. Sano (1983) collects and discusses a number of other reports of large-scale food-production projects, showing not only their massive costs but also their minimal contribution to output. Their main effect seems to have been to drain the state budget and shift peasants off the large amounts of land expropriated. This has been particularly the case with a number of dam and irrigation projects. In view of the considerable criticism of World Bank policy in Nigeria, it is only fair to indicate that it opposed this large-farm focus, though its own focus on middle-large peasants suffered from many of the same problems.

Parastatal dairy farms financed by the World Bank in Tanzania follow the same model. Set up to produce a milk surplus which had been eliminated by dairy-building,[39] they concentrated on high technology and breeds of cattle in surroundings where this was totally unrealistic. Many of the farms failed to produce any surplus at all for official sale and none were even near to economic viability. But, as in the Nigerian case, the fact that plans had been made to increase local production obscured the real effect of expanded dairy processing, which was to increase imports of skimmed milk powder and butter oil.

While the evidence of waste and failure in such ventures is overwhelming, they remain popular with governments (if they can get credit for them) since they appear to represent a 'serious' effort to tackle the food problem. They are also popular with consultancy companies and equipment manufacturers in donor countries for obvious reasons. Since many aid agencies seem to be making increasing use of private companies for the operation of their aid programmes, it seems depressingly likely that this sort of 'project farm' will continue to waste state funds and to expropriate peasants from their land.

PEASANT PRODUCTION

A wide variety of different sorts of project and programme are aimed at increasing peasant production of food crops. They vary from national scale to small localities. Their focus ranges from one single crop to 'integrated projects' which attempt to change health provision, social services, education and the whole farming system in one go. Some projects aim solely at 'progressive' (i.e. richer than usual) peasant farmers, others (though in practice relatively few) at all in their sphere of operations. In terms of control, they range from projects comprising extension advice, with provision of inputs and perhaps credit, to 'schemes' under a hired manager who makes all important decisions and controls (or tries to) most of the production process. The most useful schemes are usually those run either by a local extension service or a non-governmental agency, which operate over a limited area, within which account is taken of not only the whole farming system but the culture within which it operates. But these tend not to show dramatic results and require considerable manpower in relation to capital inputs and levels of output increase. Not only are they less popular with national agencies and donors; one quite often does not hear about them except in their immediate area of operation. This sort of programme also makes mistakes – given the difficulties involved, they are impossible to avoid. But they tend to avoid the obvious and hugely wasteful mistakes which derive from stereotyped and oversimplified thinking, which marks many of the larger projects and programmes.

The discussion and examples which follow are thus not typical of all projects and policies to increase food production. They deliberately pick on issues and examples which show the more serious problems, since it is these which are most in need of criticism and change. Nor are all the issues to be considered necessarily 'mistakes', as, for example, the problem of new crops and varieties.

During the present century, there have been huge changes in the patterns of food crops grown by peasants in Africa. At the beginning of the period, millets and sorghum were the major cereals in many parts of the continent, together with a variety of root-crops and cooking bananas (technically plantains) in other areas. The period since then has seen a steady move towards cereals which have higher yields per acre, but which tend to be more vulnerable to drought. The first change was from millets and sorghum to maize, which had some effect in shifting the geographical focus of production. But more important has been the introduction of hybrid maize, since the mid-1960s. There is no disputing the value of this in places where the conditions are favourable, since, especially with fertilizer, it leads to enormous increases in yield, releasing both land and labour for other crops. But it does mean that areas of surplus production, and thus the attention of state authorities, become increasingly focused upon areas of high and reliable rainfall. There have also been advances in maize-breeding for drought-resistance, these varieties being more often 'composites' than hybrids. Two of the better-known are Katumani and Kalahari, developed in Kenya and South Africa respectively.[40] Valuable as the latter are, it seems, however, that attention and funds, in Africa as elsewhere in the world, are increasingly focused on hybrids. More 'advanced' and sophisticated methods of production are more attractive to breeders. But far more importantly, they have great advantages to seed companies and corresponding disadvantages to peasants. Hybrid seed must be bought every year, since if the product is sown it 'segregates' into the parent lines, with enormous loss of yield and homogeneity.[41] Furthermore, internationally the companies which control most seed production are oil majors and thus fertilizer producers, with a policy of breeding for maximum fertilizer use and response (see Mooney, 1979, 1983). This makes for serious problems where delivery systems for either seeds or fertilizers break down.

Controlled peasant production schemes are more common for export crops than food crops, since an important element of the control is exercised via marketing. None the less, a number have been started for food-crop production. Such schemes normally involve a specified area with contiguous peasant plots or block-farms, under the overall control of a manager. It is normally compulsory for scheme members to plant the crops specified by the management and to follow recommended practices with regard to weeding and application of fertilizer and insecticides. In some cases (block-farming) the plots are situated so that main cultivation can be done by tractor across the scheme as a whole. Two major problems are common on such schemes. One is that the production system has been worked out under more controlled conditions (say on a research station) than it is possible to maintain in the field, so that they turn out not to be economic. Characteristically input and machinery service costs are too high. The other is that it commonly turns out not to be possible to exercise 'discipline' regarding crops and production methods. Both these problems affected a series of 'pilot settlement schemes' in Tanzania shortly after independence. As a result, their contribution to production was negligible and their cost extremely high.[42]

One scheme which has achieved effective control is the Mwea Rice Scheme in Kenya. This is a large-scale scheme for production of irrigated rice on peasant plots through block-farming. That is, the plots are laid out as contiguous rectangular blocks, so that mechanized operations can be performed over large areas, which are then split up for the hand-labour operations. The scheme produces 70 per cent of the rice produced in Kenya for the official market (and duck-shooting for foreign embassy personnel) and is reputedly as near to economic as such projects are likely to come. What it achieves for its own members can be gauged by the fact that malnutrition is serious and persistent among the women who perform most of the weeding labour and among their children. The main reason for this seems to be that efficient operation of the scheme is deemed to make it impossible for them to have their own food-crop plots, while working hours are long and allow little time for subsistence cultivation when this also involves travelling off the scheme.[43] The more recent Bura irrigation scheme seems planned to run on lines similar to Mwea, though the enormous capital costs of the project make it unlikely ever to be fully economic (Mwea made use of the cheap labour of Mau Mau detainees during the 1950s). To date, it has been plagued with all manner of problems and seems likely to join the long and growing list of large irrigation projects in Africa whose main contribution is to national indebtedness.

EXTENSION-BASED PROGRAMMES

The development of peasant maize production in Zimbabwe has been in many respects a major success. In 1979/80, before Independence, peasants delivered less than 4 per cent of maize purchased by the Grain Marketing Board. Large-scale commercial farmers were overwhelmingly the dominant suppliers. Over the succeeding years, peasant deliveries have grown steadily and composed some 55 per cent of the total in 1986. This means twenty to thirty times as much as was ever delivered before Independence. This does not necessarily mean a similar increase in production. The major part of peasant maize production is for subsistence, so deliveries to the GMB may include produce which would previously have been eaten or sold unofficially. None the less, it is a very striking increase.

Part of the reason is simply that, with Independence, a number of previous impediments were eliminated. Another, that ten years of debilitating civil war had ceased. In addition, first steps were taken to alter the incredibly biased distribution of land, roads and marketing points, while the first small credit programmes for peasants were started. As part of this, the extension service recommends 'packages' of inputs for different crops, which are supplied through the credit agency. In terms of market deliveries, this is probably the outstanding 'programme' for food production to be mentioned in this book.

But there are still serious problems with this, seen as a means of solving Zimbabwe's hunger problem. Only some 15 per cent of peasants in the communal areas receive credit. This is not for lack of funds but because the criteria set by the Agricultural Finance Corporation are very strict. Only a small and select minority of peasants, most of them 'master farmers',[44] get loans. Not only are these larger than usual peasant farmers, but most come from the more fertile areas of the country. Between 75 and 96 per cent of total peasant maize deliveries to the GMB come from the two most favoured 'Natural Regions'.[45] Yet only 10 per cent of Zimbabwe's communal area peasants come from these favoured regions (and 20 per cent of those in resettlement schemes).[46] The vast majority of the country's peasants are untouched by this development. This is not to say that the remainder should be delivering maize to GMB, or even growing it for that matter. Production of sorghum for own consumption would often make more sense. But the problem is that credit and inputs are largely available only for marketed maize. This makes *economic* sense; peasants growing subsistence crops would not earn the cash to repay crop loans. But it does show some of the difficulties in trying to solve the problem of helping peasants to get enough to eat by attacking the very different problem of trying to increase the marketed surplus of food crops.[47]

Even in cases like that of Zimbabwe it is hard to discover what the effects of such programmes are on food-crop production. Where the figures of marketed production indicate success, some or all of this may come from the diversion of produce previously sold locally or consumed by the family. Where they show no increase in production, this may be because black-market prices divert a real production increase away from official markets. The following example indicates some of the problems.

TANZANIA AND ZAMBIA

After the major food shortage of 1973–4 (and the enormous increase in food imports), various measures were taken in Tanzania to increase food (especially maize) production. Emergency measures came first, followed by a decision to raise the prices of food crops, to extend the range of areas from which official purchases were made, and to include a number of new crops (sorghum, millet, cassava, peas and beans) among those purchased by the official National Milling Corporation. At the same time, the pricing system was altered so that producers in all parts of the country received the same price, regardless of distance from Dar es Salaam ('pan-territorial pricing'). Plans were made to expand state-farm production, including the setting-up of new 'project-type' farms (which fortunately came to nothing), compulsion was used in a number of cases to force peasants in the new villages to produce given areas of food crops, and a major peasant programme, the National Maize Programme (NMP), was started with World Bank help.

During the mid-1970s, when the NMP was in operation, maize deliveries to official markets rose by a factor of ten over a five-year period. Yet by common consent, the NMP was an almost total failure. One problem was its technical specification. Research on maize production in Tanzania had admittedly been limited, though at least fifty different sets of trials had been done over the previous two decades, giving widely varying results for different areas. Yet the combined resources of the Ministry of Agriculture and the World Bank came up with two levels of one fertilizer recommendation for the whole country (a country with probably as much ecological variation as the whole of Western Europe). Not surprisingly, it was uneconomic in many parts of the country. Moreover, the distribution of fertilizers and seeds was somewhat hit-and-miss, so that deliveries were made to areas where peasants were not interested, while, in others, desperately wanted inputs were unavailable. To make matters worse, with the nationalization of most supply channels – and their poor functioning – it was almost impossible for peasants to get hold of inputs apart from credit under the NMP. The conclusion seems inescapable that the programme had little to do with the rapid increase in maize deliveries, which came first and foremost from better prices and markets in areas previously not served by the NMC (and in some cases from switching from export crops). One effect of the NMP, in conjunction with pan-territorial pricing, was to shift the location of production to the peripheries of the country, thus greatly increasing the transport costs of moving produce to Dar es Salaam.

Because of this, the successor to the NMP (funded by the World Bank and USAID and known as the National Agricultural and Food Credit Project – NAFCREP) was concentrated geographically upon four regions in the Southern Highlands, bordering on Lake Nyasa, Malawi and Zambia. Two of the four regions and never produced maize for official sale before 1975, had started as a direct result of pan-territorial pricing, and would not be able to produce economically without it. It is curious that the World Bank, which has used much energy and pressure over the past decade to persuade Tanzania to drop pan-territorial pricing, should have focused its attention on these areas. The reason given was that these regions had the high and reliable rainfall which would allow the production of a steady surplus based on hybrid maize. Another reason, less often mentioned, was that the previous main surplus region, Arusha, was too near to the Kenya border for reliability.[48] NAFCREP was by all accounts better organized than the NMP. It was also a success in that the farmers were keen to accept hybrid maize seed and subsidized fertilizer. But while maize deliveries had grown rapidly during the period of the NMP, they stagnated during the NAFCREP period. The main reason for this seems to have been that maize and other food-crop prices declined in real terms after 1979;[49] the other that the disparity between official and black-market prices increased.

It was at about this time that Zambia started a maize production programme in the north of the country – and by all accounts a very successful one. Based on the same general mix of improved seeds and (subsidized) fertilizers, deliveries to the co-operatives increased by about 5 times between 1980 and 1985. This was achieved under a variety of donor schemes, each with its separate approach. Sano, who has been studying the development of maize production, finds prices, fertilizer subsidies, and simply the availability for the first time of regular markets, to have been the most important factors, together with reduced job opportunities in copper-mining for an area with a long tradition of migration to the mines. But how much was at all attributable to increased production in Zambia is open to doubt. This occurred during a time when commodities of all sorts were extremely hard to get in Tanzania – and when much of the black-market soap, matches, cigarettes and many other goods which could be purchased in markets in the Southern Highlands came from Zambia. It seems not impossible that some of the seed-fertilizer-induced increase in Tanzanian production was sold in Zambia. This contention is certainly borne out by the geographical pattern of buying in Zambia, which shows a clear concentration along the Tanzanian border and down the main roads from Tanzania.[50]

It would be interesting to know what has happened more recently on both sides of the border. Zambia has cut its fertilizer and transport subsidies, so that real official prices for maize from the north seem to have deteriorated badly. Tanzania has also cut its fertilizer subsidies, while a recent devaluation has further increased the price significantly. Whatever happens, it is a fair bet that it will be more affected by prices – both official and black-market – than by the plans of donors or government.

NIGERIA

The largest set of peasant-oriented food-crop programmes in one country has been undertaken in Nigeria, and is discussed, among others, by Williams (1981) and Sano (1983). Here the World Bank has been a major force and the main donor. Expenditure on these programmes has been modest by comparison with large-scale and irrigation projects (4 per cent and 8 per cent of capital appropriations for agriculture in 1980 and 1981, compared with 73 per cent and 76 per cent for large-scale projects – Sano, 1983: 39). But in other respects, there have been some serious problems. One highly dubious aspect has been the widespread introduction of maize into areas previously cultivated (and for good reason) with more drought-resistant sorghum. Another (as in Tanzania, Zambia and Zimbabwe) is that the programmes are defined on a single-crop basis and thus tend to promote monocropping. In Nigeria the maize variety favoured was developed specifically for single-cropping and interplanting was further discouraged because it 'made it difficult to determine the optimum package of inputs' (Sano, 1983: 53).

While the original outline stressed that the peasant-oriented projects were aimed to help 'small-holders' and 'the rural poor', when it came to implementation, project staff tended to concentrate on 'progressive farmers'. From data presented by Sano (1983: 57), of the 67,000 peasants affected by one of these projects, the top 24 per cent received 61 per cent of extension advice and around 80 per cent of all fertilizer. Since the latter was heavily subsidized, and regarded by most peasants as by far the most valuable project resource, the project clearly subsidized the rich. As Williams (1981: 23) concludes, 'the project explicitly aims to provide massively subsidized inputs on credit to two categories of farmers, the large and the larger'. Once again one must ask whether, even if these projects do add to aggregate production, they really improve the food situation of the poor.

One of the most serious technical problems of schemes to increase peasant food production is the tendency to assume that agricultural skills are basically a matter of using the right seed, applying the right dosage of fertilizer and insecticide, and following the recommendations of the extension service as to time of planting, crop-spacing and other aspects of the production process. Part of the problem lies in absence of data on which to base recommendations about these, but even if this knowledge was available, there are many other important aspects of farming, often requiring specific experience of the area in question. The notion that the peasants might know something about agriculture seems simply not to occur to most of those involved in peasant crop-development projects. Techniques like inter-cropping which are not currently used in Europe or the USA are not included in the curricula of agricultural universities (except as curiosities of 'traditional' agriculture) and are generally discouraged because they make it more difficult to specify crop spacing or fertilizer recommendations. That they are also a valuable practice, reducing risk, increasing the total production off a given area, and reducing erosion risk by maintaining soil cover seems not to count.

This is made the more harmful because projects so often concentrate solely on one crop and thus (whether intentionally or not) lead to monocropping. When all extension advice, inputs and credit are for the one crop, and so long as prices are high and inputs subsidized, peasant farmers will tend to go on growing it without rotation. African peasant farmers, cultivating in the manner learned over generations, may well be aware of resting the soil, as claimed by many who have studied 'traditional' systems.[51] All, the evidence I have seen indicates that when engaged in 'modern' farming for money, this awareness is much weakened. In Tanzania and Zimbabwe, I have asked (admittedly small and non-random groups of) maize-growing peasants about rotation of maize with legumes. In each case the answers were that they did not grow groundnuts, peas or beans because the return from maize was higher. Asked if they did not think it necessary to rest the soil, they answered that the extension service had said nothing about it and that was good enough for them.

Another problem with 'peasant' projects is their tendency, whatever the aims written in the original project documents, to focus on the larger and richer peasants and those from the more fertile areas. The really poor areas are too unreliable for economic use of fertilizers or hybrid maize, especially with credit. Small peasants are (thought to be) less reliable about repaying credit and it is certainly more expensive to administer a given amount to 200 small peasants than to 20 large ones.[52] This tends to be justified by reference to the 'trickle-down' theory – that advice given to large farmers will find its way from them to the smaller. But while this may be true of advice, it certainly is not true of inputs or credit, especially when subsidized. These obey a socio-economic law of gravity and tend to 'trickle-up'. In reality, it is often not true of advice either, since what is relevant to the rich peasant or large farmer is often of little use to the small peasant. This applies most obviously to tractor mechanization, but is not confined to it. Large farmers can more easily afford to take a risk to gain a higher expected return. They have the space to grow pure-stand coffee on one part of the farm and food crops elsewhere. They have enough land to keep grade cattle segregated from the herds of others, and the funds for veterinary treatment when required.

Also highly significant are differences in treatment on the part of extension officers. The rich non-peasant farmer will often be visited by a senior officer with university training, and the assistance he receives may include a thorough and serious discussion of problems raised by the farmer. The rich peasant can, to a lesser extent, avail himself of the better extension advice – and often avoid the worse. The poor peasant sees the extension officer, if at all, when there are orders to be given out, when the population is to be called out to listen to pep-talks on the need for more hard work, or at best for repetition of the same old standard advice which s/he has been ignoring for many years (and often with good reason). But even if the extension officer wishes to do a good job – and there are plenty who do try – lack of transport, office paperwork, the need to get round huge numbers of farmers, and the lack of relevant back-up research are enormous hindrances, together with the fact that training for the lower-level staff is often poor and based on rote-learning of standard items. Talk to lower-level extension staff, and they will almost always start by bemoaning the ignorance of their peasant clients and their unwillingness to listen. Extend the conversation and they will often admit that they do not follow the advice they give on their own farms, and that there are reasons why the peasants do not accept it. But they are responsible to their superiors, not to the peasants, and their superiors have told them what to push, just as they themselves have received the orders from higher up. And given the enormous respect which attends formal knowledge within such hierarchies, most are no more willing to accept challenge or argument from the peasants than they would dare to offer it to their superiors. The problem is not just that much of the advice, as presented, is too rigid to be relevant, or even plain wrong. Worse than that is the automaticity with which it is given and the reliance on higher authority, which positively discourages anyone from thinking, or using their minds creatively towards the solution of problems.

THE NEW SETTLEMENT OF MARGINAL AREAS

Among the most seriously famine-prone areas in tropical Africa are regions of low and variable rainfall on the margin between cultivable and grazing land. During the past two or three decades large areas of such vulnerable land has been cleared for cultivation, often with disastrous results. Ground-cover and trees are cleared, increasing erosion, while unsuitable systems of cropping lead to rapid soil degeneration.

It is commonly thought that population pressure is the basic cause of this movement, and to some extent it is. But there is also another extremely powerful force – new settlement by commercial farmers. O'Brien describes such a process in the Sudan:

> Forest and scrub rapidly began to disappear as new schemes were cleared, commercial charcoal-making expanded to meet growing urban demand for fuel, and people – especially pastoralists – displaced by the schemes sought new land on which to settle. Capitalist agriculture expanded by at least 5 million acres in the 1960s and 1970s, and most of the land taken had previously been prime seasonal pasture of nomadic herds. The pattern of capitalist farming in the rainlands itself deepened the predatory nature of its expansion. Tracts of hundreds of thousands of acres were clear-cut with World Bank assistance – reducing humidity and cloud formation and increasing soil salinity. The farms were mined for quick profit, before the soil gave out due to erosion and nutrient depletion. With the help of World Bank loans to the investors and technical facilities provided by the state, capitalists whose farms were exhausted in an average of five to seven years moved on to new fields. (O'Brien, 1985: 10–11).

This captures the essence of a widespread tendency. Capitalist farmers, or rather capitalists with tractors, move into a new area, concerned to make a profit in the shortest possible time. In some cases they are concerned to stay in farming only until they have accumulated the funds to invest elsewhere, in transport, trading or political influence, for example. In any case, they tend not to be concerned about the long-term future of the 'free' and 'empty' land which they colonize on the basis of mechanized shifting cultivation. As O'Brien remarks, it is the pastoralists who have been displaced who are 'often the direct agents of depletion of fallow and scrub, and who receive the lion's share of official blame'. Moreover, farming methods which take no account of soil regeneration are cheap in money terms, though hideously costly in life and land. They thus allow low-cost production, undercutting peasants and forcing them to cut costs and corners in order to compete – a sort of Gresham's Law of farming.

O'Brien does not simply present this as an example of irresponsible agricultural policy and land-use. He shows how it reflected the emergence of a particular sort of capitalist farming class, while government and World Bank policies encouraged this new settlement, for sorghum production, first for the local market and later (with Arab money) for export as livestock fodder. He also shows how this 'sowed the seeds of the famine' which hit the Sudan in 1984–5. The Sudan had largely avoided the Sahelian famine of 1965–73, although rainfall was lower than normal in Sudan too. It was the officially encouraged development of the intervening period which tore the covering vegetation off vulnerable lands. Ironically, it was precisely this development which kept current food production ahead of population growth and earned the Sudan the approval of FAO (see Chapter 2).

Similar examples can be found from other countries. In Tanzania, the opening of the Ismani Valley in the 1950s followed the same pattern. African capitalists, mostly with a previous basis in transport and trading, moved into an area which had not previously been permanently settled for lack of drinking water. The method of cultivation was the simplest possible form of tractor cultivation, with no attention at all to soil regeneration. Within periods variously estimated at 5 and 15 years, soils were degraded and yields too low for it to be worthwhile to continue cultivation. They are reportedly of little agricultural worth. Not dissimilar was a southward movement from Sukumaland in Tanzania for the cultivation of cotton in marginal areas, where the better than average rains of the late 1960s seemed to give good prospects. Poor rains in the early 1970s choked off this growth, possibly before it could damage the soils to the fullest extent. Thereafter lack of tractors and spare parts, combined with poor market prospects for cotton, prevented new cultivation in this area.

Diana Hunt records a movement of African capitalist farmers into Mbere in Eastern Province of Kenya. Here the point was not so much the poor practices of the new settlers, but that they pushed out the previous inhabitants who had little choice but to move into less hospitable areas, where they competed in turn with pastoralists. This seems to have been a relatively common trend around the margins of Kenya's densely populated high-potential areas. Either commercial farmers move into the marginal areas themselves, or their settlement of the better neighbouring land, together with demographic pressures, pushes the previous population into the marginal lands. Beyond them, pastoralists lose their dry-season grazing areas and are forced into less viable grazing grounds and greater overgrazing.

Sometimes the major impetus behind such processes is the local capitalist farmers themselves, Ismani being an example of this. In other cases, as with the Sudanese example, governments and donors stand behind. Clough and Williams (1983) describe a World Bank-initiated and financed project in Nigeria, where maize was to be introduced into an area with rainfall low and variable enough to indicate sorghum production to most observers, including the local peasants and capitalists. In a number of other examples, government and donor projects and programmes aim at introducting monocropping (with improved seeds, fertilizers and insecticides) as a means of increasing production as rapidly as possible.

This indicates one of the great dangers with the 'conventional wisdom' about food crisis and the conclusion that donors must move in as rapidly as possible with a technological 'fix' to increase production rapidly. Firstly, there simply is not the knowledge available to provide such a fix. There is, however, enough knowledge to show that standard government and donor practices are far from ideal and often definitely harmful. When one combines this with an increasing trust that 'the free market' and 'private enterprise' can solve all problems, one has the makings of a disastrous cocktail in which states and donors provide the finance (and often subsidies) for local and foreign capitalists to go out and mine some of the most vulnerable soils in areas crucial to the continent's long-term food supply.

Conclusions

This chapter started out by looking at the extent of the current general economic and social crisis in Africa and indicating some of the reasons for its emergence. Significant in this process have been policies which increased import requirements without corresponding growth of exports, these themselves leading to policies with negative effects on exports. Both the initial policies and their continuation have been aggravated by foreign aid and lending, especially during the mid-1970s with petrodollar-induced 'easy money'. This has also been related to a process of class-formation which has increased differentiation without corresponding social accumulation, resulting in increased poverty at the bottom end of the scale. Food subsidies during the 1970s and early 1980s shielded some of the urban population from this, though not the very bottom strata, but at the cost of severe disruption of agricultural markets, diversion of produce on to unofficial markets (leading to official food shortages and imports), and increased rural poverty. The effect of more recent IMF-induced policy changes has been to shift the balance against urban populations, though any advantages to rural populations have been largely lost to the increased severity of the debt crisis, falling international prices of export commodities, and devaluation-induced inflation in the prices of purchased items. The general effect of the crisis on food availability has been through declining real incomes at the bottom end of the scale and the disruption of markets and infrastructures, leading to wider variations in prices.

Turning to agricultural policies, these have tended to favour richer peasants in the climatically more favoured areas, this being increased by the emergence of a 'state-class' based on the manipulation of state controls and distribution mechanisms. The policies defined for, and the advice given to, peasants have often been irrelevant or unfeasible, since they have too seldom been based on study of whole farming systems, let alone the societies within which they are embedded. The final section illustrated a particularly disastrous process occurring in some of the most drought-prone areas of the continent. The general conclusion is that agricultural policies to increase production directly have often aggravated the effects of the more general crisis, not only through further increasing differentiation but also through generating policies which take little account of the long-term requirements for soil and fertility maintenance. The following chapter continues the story, looking more specifically at hunger and who is likely to suffer it most seriously, how, when and where.

NOTES

1. The fact that debt service is lower in relation to exports in Africa than elsewhere reflects firstly that Africa has received a larger proportion of its loans on concessional terms, and secondly that its debt is more recent, some of it still being covered by grace periods.
2. Cited in *Telex Africa*, 9 July 1985: 18–19.
3. OECD, 1984. The top five debtors were Nigeria, Sudan, Ivory Coast, Zaire and Zambia. Because of inflation, this measure exaggerates the real growth. On the other hand, there is considerable disagreement between different sources as to the true size of the debt, the one sure thing being that it is larger than as given in the different sources, because of the incompleteness of figures for private debt.
4. The figure is from a different source, a UN report cited in the *Kenya Daily Nation* of 16 October 1987, and is thus not directly comparable. In their informative book on the Third World debt crisis, Körner *et al.* (1984) show the very considerable disparity between estimates of international indebtedness by the World Bank, IMF and OECD (Table 1.5). They also make clear that this arises from incompleteness in the figures, especially of private debts.
5. In reality, there is significant smuggling of export-crops across national borders, in cases where this pays. Examples include cocoa from Ghana and Benin to Ivory Coast, and coffee from Tanzania and Uganda and Kenya.
6. This is Nyerere's phrase, but would be echoed at least in the policy statements of a wide variety of governments.
7. This implies lower average standards of living because at least some of the intervening increase in real incomes arose from the monetization of subsistence production. This involves including in the assessed national income items which were previously ignored or underestimated, thus exaggerating the real increase in welfare. This affects the poor more than most, both because their incomes tend to have a higher subsistence component and because the intervening period of growth has usually been accompanied by an increase in the degree of differentiation – which has not been reversed during the more recent income decline.
8. Other factors include increased imports of military hardware and the poor levels of maintenance which swell the import bill for spares and service.
9. The others were Mauretania 26%, Burundi 22%, Uganda and Mali 21% (all figures rounded).
10. To avoid confusion, while Körner *et al.* use the 'English billion' (one million million), I have translated into the internationally more common American billion (one thousand million) used in this book (which they refer to as a 'thousand million').
11. They may well have positive effects on production, but these will never be attributed to them, so they continue to be considered as drains on the foreign-exchange balance by IMF and World Bank economists.
12. See Raikes, 1985a, 1986a, Coulson, 1985 and many others for further details.
13. Reuters report in the Kenya *Daily Nation*, 21 May 1987.
14. Because it allows them to keep the assets in question on the books at a value (however imaginary) related to those promises, rather than writing them off and reducing their apparent financial soundness.
15. Nyerere's phrase, but representative of the formal aims of a wide range of countries.
16. The relevance of citing Bates in this context is that he is among the major critics of this process in an African context. For a discussion of the above from a different viewpoint, see Kitching (1982).
17. A classic example of this is a fertilizer factory built in Tanzania under a German loan, and billed as assisting agricultural development. In reality, it raised fertilizer prices well above import prices ('the most expensive fertilizer in the world', according to one World Bank report), reduced the range of products available and consumed large amounts of state funds in a variety of subsidies. Local training and employment above unskilled level have been minimal and only fifteen years after it was built the plant is in such poor shape as to be little

more than scrap. One could cite dozens of similar examples from almost every country in sub-Saharan Africa.

18. One especially pernicious habit of donors is to insert a 'wedge' in the local budget by starting projects which require an increase in the staffing of local state agencies and paying for this on a declining scale. Thus in the first year the agency will pay, say, 100% of these extra costs, reducing this to, say, 60%, 30% and 0 over the succeeding years. I must admit to having been involved in recommending such procedures, before the implications occurred to me.

19. Among the problems with this term is that of where the limits should be drawn. It is obvious enough that top political and bureaucratic leaders should be included, together with their close business associates, but how far down should one extend the definition of 'ruling class'? Should it include middle-level bureaucrats who take subsidized loans to start their own businesses, or those who misappropriate one (or a few) trucks of maize from the marketing agency? And if it should include them, why not the policeman who takes a cash bribe to forget a speeding ticket or overlook the overloading or faulty condition of a vehicle? It should be noted that the problem lies not in the term but in the phenomenon, since the methods of surplus extraction are similar from top to bottom. Another problem, which some see as of great importance, is that one should not refer to such a class as 'ruling' when it is subordinate to international capital. But while accepting this, I do not see the local state as purely 'governing on behalf of' international capital. It is clearly subordinate (as the debt crisis and conditionality among other things demonstrate). But it has its own aims, which are only partially congruent with those of international capital, itself not a fully homogeneous entity.

20. The useful term 'straddling' was coined by Michael Cowen to refer primarily to the reinvestment of wages in government service in advanced peasant agriculture (in Kenya). The term has been expanded by Iliffe (1983), and following him Orvis (1987), to cover all situations in which non-agricultural income is used for investment in agriculture. Though this latter use is useful for certain purposes, my concern here is specifically with the use of state resources, though Cowen's original use is expanded to cover non-agricultural investments.

21. On the processes by which this is achieved, see Bates (1981).

22. One of the clearest statements of this ideology can be found at the beginning of Hyden (1980), though it underlies huge numbers of books on 'political development' and even more works on agricultural extension in the 'diffusion of innovations' tradition. For a criticism of the latter see Raikes and Meynen (1972).

23. I have described such phenomena for Tanzania (1983), as has Coulson (1978). From Shaba in Zaire, Schoepf (1985: 36) found that 'government officials [continued to] repeat the colonial view that peasants are target earners and only produce for the market if coerced'. While such views usually have their origins in the colonial period, this is not just a colonial hangover. It tends to reflect a structural similarity in relations between policy-makers and peasants, in which the former are concerned to 'persuade' the latter to increase market production, while failing to offer prices which make that an attractive option. To assert that the peasantry is economically rational, in the sense implied here, is not to assert that they are small capitalist firms (as some economists imply), merely that they respond in the expected manner to changes in crop prices.

24. I find this sort of 'official reportese' peculiarly misleading when, in many cases, the political leaders have been disappointed all the way to the (Swiss) bank.

25. In early 1975, I can recollect a friend citing a 'liberal' World Bank adviser, a friend of his, as having advised the Tanzania Government to borrow all it could, for this reason (and no doubt quite sincerely).

26. That is, they often supply resources in forms which lay them more than usually open to misappropriation through political control.

27. The internal rate of return is a common measure of the expected viability of a project. It is defined as the rate of discount (roughly compound interest) which reduces the net present value of expenditures and expected returns to zero. For the project to be considered viable, it should be above the going rate of interest. This can be achieved by fiddling with the figures in a number of ways. One of the more common, for agricultural projects, is to set totally unrealistic estimates of the expected yield of crops. Another is to assume the life of a tractor as, say, ten years, when three years is good going under the conditions on most aid projects in

Africa. If all this fails, estimates of 'indirect' benefits can be added, in which case the measure is referred to as the 'economic rate of return'.

28. For example, the World Bank's Integrated Agricultural Development Programme in Kenya, which included the worst small-farmer credit programme ever implemented in that country, and was largely responsible for the effective bankruptcy of several co-operative unions. Another example was a two-phase project to build cashewnut processing factories in Tanzania. The second phase went ahead, even though total cashewnut production had already fallen below the throughput of phase I factories before its completion. As of 1985, virtually none of the factories was in operation – partly because of low cashewnut production and partly because IMF/World Bank conditionality denied Tanzania the funds needed to run them. For further examples see below (and references).

29. Examples like Rawlings and the late Sankara may be exceptions to this rule (but perhaps not in the long run). Even if they are, they do not invalidate the general rule.

30. I have been involved, with colleagues, in turning down such an agreement (much to the disgust of the firm concerned) when local officials were quite willing to accept it, even after the implications had been discussed at some length. The country in question was at that time (and is still) in a period of desperate foreign-exchange shortage.

31. The marketed production of peasants is commonly referred to as a 'surplus', on the assumption that subsistence production takes care of all necessary consumption. This is seldom true under current conditions when most peasants include purchased items, from tea, kerosene, clothes and soap to agricultural implements and inputs, as standard items of normal consumption.

32. These are generally considerably more subject to control than their European or US counterparts, in some cases being little more than parastatal organizations. In return for this, they are commonly granted statutory monopolies for a given crop or crops, within a designated area. While the history of co-operatives in Europe is one of attempts to evade monopolies by private middle-men in order to increase returns to farmers, that of African co-operatives is characterized more by attempts to evade co-operative monopolies, for the same purpose.

33. In Kenya, most of these crops were forbidden to peasants outright for most of the colonial period, except for a few deliberately limited schemes. In Tanganyika the policy was more varied, though discouragement was often practised. In few cases were more than a few highly controlled and limited 'schemes' allowed throughout eastern Africa, the excuses usually given being either that pests and diseases from African produce would infect or infest that of the settlers, or that it would provide a basis for theft of settler produce. In Kisii District, Kenya, where an early, but small, coffee programme was permitted, this drew great scorn from local colonial officials, who pointed to the *much higher* quality of African produce (Barnes, 1976).

34. This seems certainly to have been the case in Tanzania (see Raikes, 1986a), and very probably in Mozambique. I have no evidence from elsewhere, but it seems generally the case that programmes of compulsory production seldom achieve their aims for more than one or two seasons, if that.

35. See Raikes (1976). In Kenya it is illegal to interplant other crops with coffee, and peasants in Central Province were actually fined for growing maize and beans among their coffee during the severe food-shortage year of 1984. In some other parts of the country (Kisii District) the same formal rules apply, but are so universally ignored that enforcement would be virtually impossible. Here there is no doubt at all that interplanted maize reduces coffee yields. One can often not even tell that there is coffee in the field until the maize is harvested. But this again reflects a switch to both maize and other crops in a situation where returns to coffee are declining, delays in payment are long, and the need for subsistence food or ready cash to buy it with is steadily increasing over time – while it is also illegal to uproot coffee. Even here, extension advice continues to focus on pure-stand coffee.

36. Such a programme was Tanzania's 'Kilimo cha Kufa na Kupona' (roughly 'agriculture as a matter of life and death'), in the aftermath of the 1973/4 food shortage. This combined forceful mobilization (made easier, though the food situation was made worse, by the concurrent villagization of the population), compulsory production of food crops in urban gardens and a series of state-farm projects, many of which came to nothing. It *did not include* ensuring

adequate supplies of the most basic agricultural hand-tool, the digging hoe, which was almost impossible to obtain in many parts of the country. I attended a Regional Development Committee in the period when this programme was under formation. After a brief pep-talk from the Regional Commissioner (a soldier) on the seriousness of the situation, the discussion among the elderly men present soon focused upon the idleness and depravity of modern youth, especially young women who wore 'mini-skirts' (i.e. above the bottom of the knee) and wigs, and upon the desirability of expelling unemployed members of this group from town and forcing them to grow food crops under supervision. After some hours on this congenial topic – and the disappointing conclusion that the legal means for its implementation did not exist – a poorly worked-out and expensive programme for the production of cassava (which the local population would not eat) with tractors, was passed virtually without comment. In discussion of the problems of the state-class and kleptocracy, one should not entirely over-look the problems of gerontocracy.

37. Such programmes are often funded by the 'soft loan' or 'export credit' departments of donor agencies or governments, which are highly responsive to suggestions from (their own) private firms (See Chapter 9).

38. Regardless of the competence or otherwise of a foreign manager, he (seldom she) will nor-mally have access to a number of means of trouble-shooting (via parent firm or aid-agency headquarters) not available to his local successor. The same back-up will make it easier for the foreign expert to resist pressures for corruption.

39. Costs of modern pasteurization and especially packaging were so high that, combined with governmental unwillingness to increase the consumer price of milk, this meant halving the producer price and drove pre-existing commercial producers out of business, or into local markets.

40. The yield increase in a hybrid results from the 'hybrid vigour' or 'heterosis' deriving from the crossing of two or more separate 'pure lines' under controlled conditions. A composite results from a less sophisticated process of mixing and selecting from the mixtures of a large number of different open-pollinated (not pure) varieties and strains. The resulting broad genetic base leads to high adaptability, but also much greater variation in yields – which are significantly lower than those of hybrids.

41. In theory, and in one field seen in Tanzania, this produces about one quarter recognizable hybrid plants, something like half from the parent lines (small spindly plants with finger-sized cobs) and one quarter sterile plants, with neither male nor female flowers.

42. Discussed in Raikes, 1986, from Cliffe and Cunningham, referenced there.

43. King-Meyers, 1979, cited in Wisner, 1983, along with several other reports making the same point (but without effect on the management of the scheme). Also Alanagh Raikes, personal communication, 1985, after a consultancy visit to a centre dealing with malnourished women and children, the situation at Mwea being among the worst seen. That things have not improved much is indicated by an article of 7 August 1987 in the Kenya *Daily Nation* indicating that 'the farmers (at Mwea) are among the poorest people in the country', and mentioning also problems of cholera, typhoid and bilharzia in relation to irrigation. The article does indicate, however, that the prohibition on maize cultivation for subsistence has been *de facto* relaxed.

44. Master-farmers must be full-time farmers (not part-time labour migrants) and must produce a certain specified proportion for sale, both implying larger than normal farm size. In addi-tion, they must have demonstrated modern farming ability and passed an examination to qualify for the certificate.

45. Zimbabwe's 'Natural Regions' are largely defined in terms of rainfall, though with some allowance for soil quality. Natural Regions I and II, with 750 mm per annum and upwards, are designated for intensive farming, cover 20% of all agricultural land and are two-thirds to three-quarters controlled by large-scale (mostly white) farmers. Weiner *et al.* (1985), Table 2.

46. 'Communal areas' are the previous 'reserves', where the vast majority of the African rural population live. Resettlement schemes are the much smaller areas which have to date been transferred from large farms to peasants.

47. For further details, see Weiner *et al.* (1985).

48. That is, given disparities in exchange rates and the superior availability of many consumer

goods in Kenya, Tanzanian agricultural produce tends to stream over the long border when-ever there is a shortage in Kenya.

49. One reason for this reduction was the very 'success' of the high prices of the previous years. The NMC was having great difficulty in storing and off-loading large quantities of produce and notably the 'new products' which it has started to buy. In particular, almost 50,000 tons of inedible Serena Sorghum (its bitter tannin layer is highly unpalatable and cannot easily be removed in milling) had to be sold at below the producer price to the EEC for cattle-feed. Serena, which had been presented some years previously as a major solution to the food-shortage problem (it is high-yielding and tolerant of drought) also comes in a more palatable white version. This, however, is extremely vulnerable to bird damage, no small problem in an area of Africa where flocks of grain-eating quelea can be tens or even hundreds of millions in size.

50. Maps of deliveries by buying post for districts near to the Tanzania border in the possession of H.O.Sano (personal communication).

51. See Richards (1983), for a number of examples, and Richards (1986a) for a more detailed study in one area of Sierra Leone.

52. Such evidence as there is on relative rates of repayment tends to the opposite conclusion, that the richer and more locally influential a borrower, the more easy it is for him (again seldom her) to avoid repayment. At a rather higher level, evidence from Kenya provides further support for this conclusion. In an article entitled 'AFC's untouchables – leaders among major debtors', the *Financial Review* 3 August 1987 shows that a significant propor-tion of the Agricultural Finance Corporation's Sh 1.9 billion of debt in arrears is owed by cabinet ministers and other influential figures, and that efforts to improve the situation have achieved a higher flow of general managers than of repayments.

4

Food Shortages & Famine

How do they occur, where & to whom?

Introduction

This chapter tries to get nearer to the complex issues of what generates food short-ages, under what conditions and who suffers most from them. To do this, it starts by distinguishing between different sorts of food shortage, followed by a discussion of the implications of seasonality. In considering factors generating food shortages, another distinction is made, between long-term factors which predispose to shortage and the more immediate factors which precipitate a particular shortage or famine. Having set out some of the patterns and processes and discussed one well-known analysis of famine, the chapter finishes with some examples of widely varying cases.

Types of food shortage

There is no one 'correct' way of categorizing different types of food shortage. How one discriminates and gives labels depends on the purpose of the exercise.

One exercise which I shall not enter into is the debate on how many people are 'hungry' in Africa, since one can spend enormous amounts of space arguing the validity of different definitions, assumptions and estimation models. In a recent sum-mary of work in this field, Svedberg (1987) concludes that 'the prevalence and severity of chronic undernutrition in Sub-Saharan Africa cannot be answered on the basis of the data available today' (p. 74). For a variety of reasons, he finds FAO and World Bank estimates (the latter showing 50 per cent undernourished and 25 per cent severely so) to be upwardly biased. He points out (p. 77).

> a . . . model . . . so lacking in robustness that slight (and plausible) changes in . . . the values of the main exogenous parameters . . . produces estimates of the prevalence of undernutrition ranging from below 10 to over 50 per cent of the population of Africa, is of little practical use.

For similar reasons, such models are of little use for estimating changes in the level of undernutrition. Whatever the figure, all are agreed that the problem is serious and most agree that it is growing.

If one is concerned to distinguish food shortage on the basis of severity, the simplest distinction is between *malnutrition* (poorly balanced diet, lacking certain essential nutrients)[1], *undernutrition or hunger* (not enough even of basic staple foods) and *starvation*. One can delve into details of where the dividing lines are to be drawn and how one measures different types and degrees of malnutrition or hunger, but that is not relevant to the present discussion. An alternative classification, in terms of people's own perceptions, might start with dull or monotonous diet, proceeding via hunger (not enough to eat) to hunger crisis and famine. In the more serious cases, the problem will not 'just' be lack of food, but more general misery, uncertainty and social break-down. Wijkman and Timberlake (1985: 12) cite such names as 'forget your wife' and 'sale of children' by which the Tuareg commemorate past famines, giving some idea of the gruesome reality of such social breakdown.

The most relevant distinctions for our present purposes relate to incidence over time and who is affected. Bryceson (1984) categorizes food shortages in terms of their incidence over time, distinguishing between *constant* or long-term food shortage, *seasonal* shortage, and *exceptional* shortfall or famine. Of course, combinations are possible. Constant food shortage will characterize the absolutely poorest strata, both urban and rural, and is clearly related to lack of income, since it occurs even when there is plenty of food around for those who can afford it. All the same, its severity will usually vary seasonally, while further groups of the population may experience food shortage at specific stress periods. Rural seasonality is generally more obvious and predictable than urban. Exceptional food shortages will usually also be seasonal in their incidence (though the most serious usually involve more than one season) and will tend to hit the poorest income strata, intensifying tendencies which occur in most years.

While a large number of specific factors determine who suffers from food shortage and who has secure access to food, three important factors stand out:

- access to land or other productive resources;
- availability and security of employment;
- the effectiveness and stability of social networks.

Those who suffer worst from food shortage are primarily those who have no (or insufficient) land for own production of food, those who must compete for the worst, most poorly paid and least secure jobs, and those who by custom or through lack of jobs are forced into dependent relationships to kin or non-related households, and specifically those whose rights within such relationships are weakest. The level of savings is also an important factor, since those especially vulnerable to famine are often those whose savings are least or are held in forms whose value falls drastically (in terms of food) precisely when most needed.

Seasonal cycles

Any agricultural production process depends on a climatic sequence which activates the biological-chemical processes of plant growth and generates seasonal patterns of growth, output and labour requirement. Characteristically, agricultural production processes intersperse periods of heavy labour with others spent waiting for growth processes to work themselves out. Generally speaking, the lower the rainfall and the shorter the rainy period, the more marked this distinction is likely to be. Similarly, the more likely it will be that the harvest, and thus availability of food and income, are concentrated at one period of the year. In almost all cases, however, there will be some time-concentration of both labour and output.

The period prior to the main (or only) harvest is often one of major food shortage, since by then stocks from the previous harvest have fallen to their lowest level. This period can also be one where labour requirements are at a maximum (the main growing season), the problem sometimes being aggravated by increased incidence of disease.[2] It is not always the case that these different cycles within the production pattern coincide; the most serious food shortage may occur after the peak labour period. But it is often enough the case, for one to be able to speak of periods of 'seasonal stress'.

It is precisely the poorest households and individuals which suffer most from seasonal stress. As food prices rise towards the period of greatest shortage, it is they who have too little to eat, and who must buy – from those who have the surplus to be able to hold stocks. The poor produce less and are often forced to sell some of the year's food-needs at harvest time, when prices are lowest, to meet pressing needs or obligations. The same groups will be forced to take wage-labour at peak periods for own-farming (thus reducing food production later), or borrow at the least favourable rates, from sheer necessity. Periods of seasonal stress will thus often be the occasion for secular changes in status (loss of land, indebtedness etc), which render their victims less capable of withstanding subsequent crises.

SEASONAL VARIABILITY AND RISK

In distinguishing seasonal variations in food supply from major shortfalls giving rise to famines, it is common to consider these as cycles of different amplitude. In some cases, the terms intra- and inter-seasonal cycles are used to distinguish seasonal variations within a year from the differences between years. This is both incorrect and misleading. For precisely what distinguishes major shortages and famines is that they *do not* occur as part of any regular or predictable cycle. Inter-annual climatic variation is random (stochastic) in the short run. There may or may not be long-term cycles and trends in climate; opinion seems to be divided and I am in no way qualified to judge. But there seems to be no relation between one year's climate and the next.[3] One can predict the frequency of given variations in rainfall, with sufficient data. One cannot predict the sequence. One is therefore not concerned with two cycles of different periodicity interacting, but with the non-predictable variation *around* a regular seasonal cycle. This unpredictability is of the greatest importance as regards both its effects and the strategies adopted to deal with it.

Risk has major effects on the way peasants plan and carry out their daily lives, production and social relations. Economists stress the effects of risk minimization on production, which normally involve a reduction in average expected production in return for an increase in the minimum to be expected. Thus low-yielding but drought-resistant crops or varieties may be preferred to those which yield more on average but are vulnerable to water stress. Inter-cropping and staggered planting are means to reduce risk, though the former may also increase overall output per hectare.[4] Dividing activities between cropping and livestock herding is another strategy aimed at minimizing risk, though in some cases it brings its own risks (see below). Another strategy is to have one or more members of a household employed or engaged in business outside agriculture. Here again, one can note the disadvantage of the poor. Rich households, which can afford to accept losses in one year, can risk higher-yielding varieties, increasing average year-by-year yields, incomes and reinvestment. This is one reason why richer peasants often are more 'progressive', i.e. more willing to adopt new crops and methods.

But risk will not only affect production. Kinship and other social networks are also affected.[5] Religion seems often to play a greater part in social and personal life, where life and livelihood are significantly affected by unpredictable climatic variation. Greenough (1982 b: 792) cites a number of Indian studies indicating that 'an ideology of reciprocity and redistribution' and 'local arrangements to adjust scarce resources to communal needs . . . turn out to be strongest in just those environments threatened by chronic crop failures'. But the religious or other beliefs which underlie redistribution may also be bound up with practices which reduce the labour or other resources available for production.[6]

There is some reason to believe that seasonal stress has more serious effects upon the poor under market relations than in societies characterized by pre-capitalist relations. Reduced to its simplest, the argument is that many pre-capitalist societies operated some form of redistribution, whereby assistance was provided to the poor and indigent and to victims of misfortune. Among the reasons for this were the necessity to ensure the survival of the social group in question and legitimization of the authority structure and tribute, upon which such redistribution normally depended. Another (particularly relevant to pre-colonial African societies) was that, with abundant land and low levels of technology, availability of labour and manpower for other activities like defence (or attack)[7], rather than of resources, defined the wealth and power of the society. These factors in turn generated a social ideology which further underpinned the necessity or desirability of redistribution.

By contrast, in commodity-based societies, the transfer of political functions to national state level reduces or even eliminates the obligation of leaders and the wealthy to assist the poor. With the development of commodity markets, food takes on a price which is dependent upon supply and demand and thus increases in periods of shortage. Prior to the penetration of commodity production, food shortage provides the opportunity for leaders to legitimize their control and accumulate socio-political obligations by providing food to dependants and clients. Under market conditions, it provides the opportunity to make a profit by selling food at high prices and to accumulate economic obligations when people lack the cash to pay. Hoarding and speculation replace redistribution as characteristic responses of the wealthy to periods of shortage.

Thus periods of seasonal stress become the *occasion* for major and irreversible changes in the fortunes of a family. It is at these times of year that new loans are taken on, that family labour is sent out to work for food or wages with which to buy it, and that assets are sold if neither of the above suffices. Moreover, land, cattle or personal possessions are then sold at the worst possible terms of exchange. The 'survival strategy' for getting through one bad patch may involve reduced capacity to survive the next. From another viewpoint, this can be seen as one of the mechanisms for differentiation and class-formation, since seasonal variation strengthens the position of those who control resources and risk still further enhances the distinction.

Redistribution has certainly not always saved the poor from destitution or starvation, however. When crises became really severe and there were questions of who would survive, it was normally the core membership of the group which did. Leaders and patrons then reneged on part or all of their obligations, starting with those of lowest social standing and least capacity to react. The old, the very young, the sick, widows and orphans and those outside the main clan or family would be the most vulnerable. The Tuareg names for famines cited above ('Forget your wife', 'Sale of children') give some indication of this.

Of course, one cannot write off redistribution mechanisms because they fail to work in famine situations in which, virtually by definition, all or most social relations break down. At the same time, one must be cautious of over-romanticizing them, as some have done with the redistribution mechanisms of pre-colonial Africa. But here another factor enters. Romanticism tends to see these mechanisms as something inherent in 'African culture' rather than as being related to particular economic, social and environmental conditons, which are in any case never going to return in their pristine form (if indeed they ever took it).

It would also be an oversimplification to assume that, with the penetration of market relations, all others have disappeared. For most of tropical Africa the generalization of market relations is far from reaching this point. Not only are extended family relations significant, specifically with regard to food shortage. Patron-client relations of one sort or another are also very prevalent. Here a dominant figure, the patron, provides support and assistance of various sorts, to one or more dependants, the client(s), who in return have certain obligations to him. These will normally be neither purely economic nor purely political-social, but a combination in varying proportions of the two. The client who receives assistance in time of need (food, cash, a cow for bride-price, etc.) may be required to make some form of repayment in cash or kind. S/he may be required to work for the patron and will usually be expected to support him in elections or factional struggles. While the relation between patron and client can vary widely, there is usually some element of reciprocity, if only that the patron needs the client's political support and/or that his social standing depends on his capacity to maintain his clients. Quite often patron and client are of similar family, clan, or language-group, which again implies some degree of obligation on the part of the patron. But patronage relations can still be extremely exploitative, and are central to the process of state-class formation discussed in the previous chapter. This to some extent blurs the distinction between 'pre-capitalist' relations based on reciprocity and redistribution, and market relations with their tendency to sharpen seasonal stress. It also emphasizes that security of food supply is not simply a matter of access to income or resources from land or wages, but also of social networks. This implies not just the existence of richer and more powerful others to whom the vulnerable can turn in time of need, but the social relations which generate wealth and influence at one end, poverty and vulnerability at the other.[8]

To summarize so far, while a certain proportion of the urban and rural poor are subject to almost permanent hunger, even they will experience seasonal variations, while a much larger number, especially in the rural areas, will experience seasonal food shortage. The seriousness will vary with climatic and other factors, to be considered below in relation to the precipitating causes of famine. Before looking at that, it is worth considering some of the long-term factors generating hunger and food shortage. This discussion will start with colonialism, not because it is my purpose to attribute all the current ills of Africa to that phenomenon, but because colonialism set in motion a number of highly significant social processes which have certainly altered the nature of societies and thus of food shortage. There can be no doubt that there were famines in Africa before the colonial period, and before the centuries of slave-trading which preceded it. But the nature of both food shortage and the responses to it has changed drastically, and many of the processes initiated then still continue.

One can distinguish different sorts of effects of colonialism:

(i) Processes related to the political fact and processes involved, like the imposition of national and local boundaries, the imposition of administration and taxes, and in some cases, the alienation of land for settlers.

(ii) Processes related to the introduction and expansion of commodity production, both for export and for the local market. These include changes in the response to food shortage, as discussed above, the beginnings of accumulation, and urbanization.

(iii) Processes arising out of the social services first introduced (albeit at very low levels) by colonial governments, one of the most significant being reductions in mortality and accelerated population growth.

The colonial incursion itself and the accompanying violence and disruption had disastrous effects on food production and the general quality of life (as did the slave trade before it). In many parts of Africa, this was compounded by the first introduction of rinderpest, a livestock disease which killed up to 90 per cent of all cattle in the worst affected areas. This led to widespread loss of human life and significantly affected the colonization process. It was, for example, possible for settlers to claim parts of the Kenya highlands as 'unpopulated', because of the weakening of pastoral societies by rinderpest. Wars of colonial conquest and the 'requisitioning' of food led to food shortage, epidemics and significant population decline in East Africa (Kjekshus, 1977). In most parts of the continent, people lost land, had their crops and villages burned, suffered forced labour and were generally impoverished. Local social and political structures were either smashed or forced into submission.

The following phase, that of the imposition of 'order' and early incorporation into commodity production, had less unambiguously negative effects, but local societies were weakened in a number of ways. Political structures were replaced by, or subordinated to, those of the colonial power and re-oriented towards its aims. Taxes in cash were imposed, having the specific aim of forcing peasants to migrate for labour or produce export crops to pay them. Unlike the previous tributes, they were seldom related to harvest levels, thus weighing more heavily in years of shortage. The capacity to redistribute food in time of need was thus diminished, while the responsibility to do so was transferred to the colonial power. The imposition of administrative boundaries severely limited movement, previously a common response of pastoralists to drought. The latter were further weakened by expropriation of dry season grazing lands for settlers or cultivating peasants.[9] In the major settler colonies of eastern and southern Africa, this was accompanied by alienation of huge tracts of the best and most fertile land. The extreme in this sense was South Africa, where whites control 85 per cent of the cultivable land and virtually all the good land, but enormous proportions were also taken in Zimbabwe, Mozambique, Angola, Kenya and other countries. Even where the overall proportion alienated was smaller, as in Zambia and Tanganyika, this was invariably the best and best situated land, the limitation often being the lack of more land considered 'suitable for European domicile' (i.e. with good soils, high and reliable rainfall and the altitude to provide a cool climate).

Migrant labour took men away from farming without regard to the effect on production or women's labour burden. Transfer to export crops diverted land and labour from food-crop production. This often affected the quality of the diet more than the quantity. Nutritious but labour-intensive crops (often legumes) were dropped out of cropping patterns. In other cases cereals gave way to cassava, reducing cultivation labour at the expense of lower protein content and more processing work for women.

The degree to which this increased the frequency of food shortages is hard to tell, though it is highly likely that it did so and reduced the general quality of diets. Colonial powers and colonies varied enormously in the seriousness with which they took the task of providing food in the event of grave shortage, as did particular colonies over time. Even where they took the responsibility seriously, it had different effects from previous redistribution systems. During the period 1948–56, the colonial government of Tanganyika provided famine relief to Dodoma Region, in the dry centre of the country, in almost every year.[10] There is evidence that this led to dependence on food relief and deterioration in local mechanisms for coping. Provision of food by an impersonal state agency has different effects from loans or gifts from a known patron in the local society. Both will ultimately have collected the cereals (or funds to buy them) from peasants' labour. But in the case of the state, its funds are seen to be limitless and its capacity to provide in future in no way linked to peasant activities. Moreover, the state is seen largely as an enemy whose exactions are to be avoided and whose limited benefits are there to be milked. The problem is that doing so can weaken peasants' own survival strategies.

Commoditization of peasant production was one major effect of colonialism, and has accelerated in the period since independence, especially in former settler colonies, where the requirements of settlers for cheap labour acted as a brake on peasant production and accumulation during much of the colonial period. Production of cash crops initiated or hastened the accumulation of land and other resources by rich peasants, businessmen and the politically-favoured. Initial efforts to generate export-crop production through increased labour input were later supplemented by 'modern-ization', increasing productivity through use of improved seeds, fertilizers, chemicals and machinery, and leading to increased emphasis on the more fertile and better-watered areas.

Alienation of land, accumulation of holdings by commercialized peasants and popu-lation increase put enormous pressure on available land with varying effects. In highland and other fertile areas, the effect was enormously increased population density, leading either to fragmentation of holdings or to the emergence of a clear division between the landed and landless.[11] In other cases, pressure stimulated move-ment into marginal areas, or the previous dry-season grazing reserves of pastoralists, thus further reducing the grazing lands available to the latter. Another response was urbanization, further stimulated by the factors mentioned in Chapter 3. While colonial alienations, changes in land-tenure systems and commodity production initiated the process, it has, if anything, accelerated since independence.

This generated three main social groups of poor and vulnerable people: those whose holdings had declined in size to the point where they could barely (if at all) sustain minimum subsistence (whether directly or through sale of cash crops); those who had been pushed into increasingly marginal areas; and those who had been pushed right off the land into urban poverty. In the first two cases, a common response is (male) migration for labour, whether full- or part-time. Evidently the degree to which these processes operated has varied widely, with the pre-existing availability and quality of land (and level and reliability of rainfall), with the degree of concentration of land and other resources, with the level and impact of investment in both agriculture and industry, and with population growth. But apart from this, there is one large section of society whose diet and general conditons probably suffered more than cost.

Women and children

It seems likely that women ate less well than men in most pre-colonial societies in Africa. With few exceptions, these were heavily male-dominated. Male household heads decided (as they still generally do) on the overall disposition of resources, labour and products, and generally to their own advantage. Their ideologies quite specif-ically assigned an inferior position to women and undervalued their labour. It is true that, in many cases, man and wife or wives operated in different spheres, with women having partially independent control over the production of food for themselves and their children. True also that women were generally those who controlled (and had the skills in) gathering – an important source of food, especially in times of shortage, and still important today. All the same, it seems likely that overall male dominance would have been reflected in better male diets.

Several aspects of colonialism and commodity production have had especially detri-mental effects on women, notably through increasing their labour obligations. Labour migrants were mostly male and this left the migrant's farm work for the women remaining behind.[12] Both absence of male labour and increasing shortage of land have led to longer periods of cultivation and shortened (or completely absent) natural fallowing, increasing the burden of weeding, generally an overwhelmingly female activity. Cash-crop production shifted the emphasis of farming towards earning cash, most of which accrued to the male household head. In many cases, this was accompanied by a shift in labour patterns in which the man took care of cash crops (and the income from them), while the woman took on an increasing burden of work producing food for the family. At the same time the shift of land to cash crops left less on which to feed the family. Shifting from production of grains to cassava increases the calorie yield per acre, and lowers the labour requirement for *cultivation* and for this reason has been a common response to land and labour shortage where climate allows.[13] But two-thirds to three-quarters of the total labour input consists of processing (crushing, steeping and washing to get rid of the cyanide in the 'bitter' cassava which is most common since wild pigs eat the 'sweet' varieties). This work is done almost entirely by women (Guyer, 1984).

In addition to this, household tasks require increasing labour. In many, if not most, parts of tropical Africa, forest and bush cover has been drastically reduced by clearing for cultivation. This means longer trips and more work collecting firewood. Another effect of deforestation is to increase run-off during rainy seasons, accelerating soil erosion and increasing the seasonality of stream-flow, so often implying longer trips to collect water in dry seasons. Where corrugated iron replaces thatch for roofing, insulation deteriorates, requiring more firewood in cold areas (though reducing the labour burden of water collection). Clothes and kitchen implements mean more washing and this means collecting more water, often taking up to several hours per day. Schooling means the loss of child labour and throws the burden on to women.

Most of these factors imply extra work rather than any direct decline in nutrition for women, though reduced land and increased household labour requirements probably reduce food production. But there is clear evidence of a negative impact on child nutrition. Jonsson (1985) finds that infant nutrition is affected by the frequency of feeding as well as by the amount ingested and shows that increased labour burdens for women mean decreased frequency of feeding and poorer nutrition (from a survey in Iringa, Tanzania). Similar findings are reported from Nyanza, Kenya, in an area where women spend much time away from home processing and selling fish.[14] Longer working hours and days generally reduce the frequency of infant feeding (and its quality if children are left with siblings to be fed on unhygienically prepared and often over-watered baby food), negatively affecting nutrition.

There are also plenty of reasons for supposing women to be less well fed than men in general. Men still control the disposition of household resources – the more so, the greater the proportion of cash in total income. Women's rights in land are generally weaker – in some cases almost non-existent. In many parts of the continent, meat is eaten (if at all) largely by men, often as a snack to accompany drinking in bars, a heavily male province.[15] Wheat products like bread and chapatti are especially attractive to single men who lack the skills, facilities and inclination to cook. Within households men tend to eat first and best. Food taboos more commonly limit women's consumption – as, for example, that against pregnant women eating eggs in parts of Tanzania.

Perhaps the most vulnerable sections of tropical African society are widows and the grass-widows of long-term migration. True, there are husbands who work long hours at more than one job, live frugally and send home what they can. But there are others who do not, who send a small proportion of their income irregularly, and there are those who disappear for years on end and send nothing at all. But the wives of migrants are often better-off than widows, because they retain their husband's rights in land. Widows quite often lose these and are expected to live as 'dependants' within the houses of kin. Women heads of household generally experience greater difficulty in getting access to credit than men. In cases where a migrant husband retains connection with the household, it is often he who takes major decisions about what should be grown, when and where. Extension agents usually concentrate on male farmers and co-operative committees are normally mainly composed of men. In situations of famine, it is usually men who leave the house to look for work, or for a loan to tide the family over. Women and children wait behind. Widows with children have more limited mobility, unless they can get others to take charge of the children – probably not easy when others have no food either.

Whether more women starve than men during famines, I do not know. I suspect that it is not possible to generalize, since it would depend on the means of survival. One would expect women to have superior skills in gathering and gleaning, in preparing food to avoid waste and in eking out what there is to best effect. There seems to be evidence that women, especially with children, fight harder to survive – perhaps because they have more practice in it from 'everyday life' and because their social-ization places more stress on it.[16] Women are possibly less hampered by pride and consciousness of status from seeking means to survive. These factors would seem to offset some part of their general disadvantage. But this is largely speculation. It apppears also that famine-relief feeding programmes give preference to women and children, presumably seeing them as more essential to the long-term survival of the society.

Much of what refers to women can also be said of 'dependants', a large proportion of whom are women, which serves to indicate that the term does not mean quite what it seems. It is true that these are people who lack a secure 'entitlement' (see below) to income or food, being dependent upon provision by others. But this does *not* mean that they are dependent on the labour of others, except in the most formal sense. Wives, widows living with relations, poor relations and orphans more commonly work very hard, at the most heavy, menial and unpleasant tasks. What distinguishes them is that their work is not regarded as 'productive' and is thus not rewarded as such; they are said to depend on the largesse of their patrons. This fiction is maintained by the acceptance of a distinction between 'productive' and 'unproductive' labour, held by a curious coalition of traditional patriarchs, African middle-class employers of their poor relations, and economistic Marxists.

Summary

This has been a brief overview of some of the major long-term processes in the develop-ment of current patterns of hunger. Apart from the generally worse situation of women and dependants, one can distinguish four major groups of the hungry.

(i) The very poorest urban strata are probably exposed to the most constant hunger. It is not constant, but varies irregularly and with factors like the state of casual employment and the arrival of food import shipments rather than agricultural seasons.

(ii) In the higher rainfall agricultural areas, one has the emergence, at the bottom of the pile, of poor households with insufficient land for full subsistence, some of them being pushed into full landlessness, as seasonal shortages, especially in bad seasons, force the sale of resources for survival.

(iii) At the drier end of the scale, one has people being pushed not on to smaller plots of land (though sometimes that too) but on to steadily poorer and more climatically unreliable land, and the pastoralists from whom they have taken this land. It is here, and in higher rainfall areas where excessive population and deforestation have led to serious deterioration in soil fertility, that one finds the most serious famines – except that war and similar social breakdown can spread the latter over much wider areas.

Though famine and exceptional food shortage are more dramatic manifestations of hunger, they are not necessarily the more significant in terms of the overall effects or generation of illness and suffering. Saunders (1985) presents evidence that general levels of malnutrition are worse in areas of higher rainfall than in the marginal famine-belts. Comparison of nutritional studies within Tanzania seems to yield a similar conclusion.

The next section turns to the more immediate causes of famine and exceptional food shortage. But before this, a brief note on one often forgotten but quite important factor.

ROOT-CROPS AND 'MINOR CROPS', GATHERING AND GLEANING

Most of this book has concentrated on 'major' food crops and especially cereals. As Guyer (1984) points out, this is clearly incorrect, when root-crops contribute an estimated three-quarters of the calories supplied by cereals. This does not affect the argument of Chapter 2 – if anything it strengthens it, since she shows that yields of root-crops are probably underestimated in official sources and in potential yields still more so. A significant proportion of cassava, the most important single root-crop in Africa, is left unharvested because of the time-consuming labour of processing, because it can be stored for a considerable time in the ground, and because it is often grown as a 'fall-back' crop, to be harvested and eaten if others fail. One would expect production (the amount harvested) to vary inversely with yield and with the state of the overall harvest. Cassava and other root-crops thus provide significant insurance against food shortage, in the more humid areas where they grow.

Also widely ignored are 'minor crops' – a variety of roots, legumes, pulses, leaf-crops and fruits which seldom reach official markets. Here, the insurance and dietary variation which these crops provide has probably been reduced with the spread of commodity production as land and labour are transferred to the 'important' crops.

Still more is this the case with gleaning (picking up grains and other products dropped during harvesting) and gathering (the collection of wild fruits, nuts, leaves and roots). The latter in particular have been a very important source of food – generally more important than hunting among 'hunter-gatherer' societies.[17] Gathering is still an important source of food in many cultivating societies. A number of studies of child nutrition have found its status to vary as much or more with wild fruit seasons as with main harvests, and vitamins often come, for large parts of the year, from wild greens. The process of commoditization, with privatization of land, clearing of bush and in some cases application of agro-chemicals, reduces this unacknowledged source of food and insurance against hunger. Reference was made in Chapter 3 to over-estimation of income increases from monetization of subsistence activities by ignoring the associated losses. This is an important case in point.

Immediate factors in the generation of major food shortages

Just as increased food imports do not always imply declining local food production, it is not always the largest harvest shortfalls which generate the greatest food shortages. A topical example of this is the Sudan, which is thought to have had a record harvest in 1985/6 (FAO, 1986e-No. 2, March). Despite this, emergency assistance was still needed for the huge numbers devastated by the previous year's famine. During this famine, their incomes and livelihoods were largely wiped out, leaving the survivors without means to pay for food, even at lower prices.

In 1986–7 the EEC has recently granted 5.5.million ECU for a 'market intervention', the purpose of which was:

> to counteract the effects of the domination of the market by private traders, who, finding a local market flooded, are able to pay lower and lower prices for cereals, which can then be sold at much higher prices elsewhere (and not necessarily in the Sudan or even the Sahel) . . . By buying sorghum itself, the EEC hopes not only to ensure the movement of food to areas of hunger, but also to prevent a price collapse in the area of surplus, thus ensuring that farmers earn sufficiently high incomes to invest in next year's crop. (*Telex Africa*, 1986, 274: 2).

Apart from showing an apparently useful and relevant use of EEC aid money, this example shows how private markets push goods towards money not need, and indicates some good reasons for government and/or aid agency 'interference' in their operation. Because it demonstrates so clearly the non-coincidence of harvest failure and food shortage, the example makes a good bridge to one major analysis of famine and food shortage.

ENTITLEMENT TO FOOD

The concept of 'entitlement to food' was introduced by A.K. Sen in a book about the causation of famine (1981). Sen's major purpose is to show that the standard (FAO) analysis of famine in terms of (national) 'food availability decline' (FAD) is not only insufficient but often just plain wrong. This he shows theoretically and by looking at four major famines, from which he concludes that in only one out of the four was the aggregate decline in food production sufficient to be considered a major cause of famine. In the other three cases, the decline in national food production was significantly *less* than in previous or subsequent years in which famines had *not* occurred. What did happen in every case was that specific sections of the population had their entitlement to food wiped out, and many starved as a result.

This is in line with a number of other analyses, like Garcia (1984), Tarrant (1980), Timmer, Falcon and Pearson (1983), and is partly implicit in that of previous chapters in this book. So one can ask how the notion of 'entitlement' adds depth to the conclusions already drawn, and whether it can help one come to more usable policy conclusions. Firstly, what does it mean?

The notion of 'entitlement' is similar to property title, a legally accepted right. The purpose of using it in this context is to stress the importance of factors determining access to food and to find a term to cover this, which is broader than ownership and which can also cover non-market forms of access to food, like obligations to kin and clients. As I shall show, the notion is not very satisfactory for dealing with the latter. Sen himself concentrates heavily on 'market entitlement', that is, access to food deriving from having produced it or from having money or goods to exchange for it.

One of the advantages of the notion is that the entitlement to food can be seen as a quantity which varies over time. Among the most useful aspects of Sen's analysis is to demonstrate some of the patterns of variation in entitlement to food in relation to real or expected harvest shortfalls, and to demonstrate how famine relates to this rather than to the aggregate amount of food in the system. The method is basically to trace the processes leading up to the famine, indicating changes in relative prices, wages, employment levels and peasant production, to show how it comes about that those who suffer do so.

This works well enough with market entitlement since changes in relative prices alter the *amount* of food which, say, a given amount of money will command, without changing the nature of the entitlement. But it works less well for non-market forms of access to food which also fit less easily into the notion of 'entitlement' itself, without stretching the word's meaning. In such cases, it is not only the amount, but the very entitlement itself which tends to disappear or to be weakened.

Three main processes seem to emerge from Sen's survey of recent famines:

(i) Workers (urban or rural) suffer from food shortage when the price of food increases in terms of money wages; or the level of money wages (and probability of getting work) falls; or both.

(ii) Peasants face food shortage with harvest failure because this not only wipes out food but their direct entitlement to it as producers. But how serious the food shortage will be is still dependent upon market entitlement. Peasants also produce for sale, so that their access to food will be affected by the production and exchange-value of non-subsistence products.

(iii) Pastoralists and people partially dependent upon livestock, living in low-rainfall areas, are especially vulnerable since the livestock which form their insurance against drought lose their exchange-value against cereals in times of drought.

In relation to *the Great Bengal Famine* of 1942–4, in which from 2 to 4 million people are thought to have starved to death, Sen shows how, although the harvest failure of late 1942 was not exceptional, it was followed by the worst famine of the century. He also shows how British colonial policy made the situation far worse than it need have been, actually moving food *out of* Bengal well into the period of serious famine.[18] Those who starved in the greatest numbers were specific groups of landless rural workers. War-boom-induced inflation in 1942 and failure of the main harvest at the end of 1942, followed by pre-emptive hoarding, drove rice prices rapidly upwards, while wages and job opportunities of agricultural labourers, fishermen and transport workers sank. Peasants, including share-croppers, were less affected, since falling harvests were partially compensated by increasing prices. The resident population of Calcutta was even less affected, though large numbers of starving rural people migrated there. The hardest hit of all were rice-hullers, an extremely poorly paid (female) occupation, which can be considered as the last stop before total destitution.[19]

A number of points stand out in this account:

- It was expectation of shortfall, as much as the reality, which triggered off speculative hoarding and an explosive increase in rice prices.
- The poor harvest led to reductions in employment and real wages for exactly those rural groups most vulnerable to shortage.
- Wholesale failure by patrons, kinsmen, fathers and landlords to fulfill their obligations to dependants led to social breakdown and desperate migration to Calcutta, where state failure to import food in time led to massive death and suffering.

- Although a large number of people starved in the city, most of them were not of the city, but famine-induced migrants.
- Most of those who starved were landless and dependent on low-wage, insecure casual employment.

The Ethiopian famine of the early 1970s is Sen's second example, from accounts from Holt and Seaman (1976) and others. Following them, he divides this into a major famine, affecting Wollo and Tigrai Provinces in the north in 1973 and a smaller one, affecting Harerghe in the south-east in 1974. Relief supplies aimed at assisting the Wollo famine started to arrive after that had subsided, but in time for the peak of the Harerghe shortage. Had they been diverted (which they were not), they could significantly have reduced the suffering involved.

Although the famine in Wollo and Trigrai was precipitated by drought and harvest failure, Sen finds that aggregate (Ethiopian) food availability decline was not the cause. Countrywide production fell, but hardly enough to precipitate a major famine. Not only was the famine confined to the two provinces, but *within* Wollo, 80 per cent of those in relief camps came from three sub-regions containing about 23 per cent of the Province's population. These were sub-regions in the eastern lowlands and marginal between cropping and pastoralism.

Apart from this, food continued to move out of Wollo well into the famine period, indicating that loss of entitlement was the significant factor, not physical shortage. Peasants had lost their direct entitlement along with their crops. Other people lost wage incomes as employment fell, or dependent status as patrons pushed them out. Most likely to become destitute were small cultivators, both owners and tenants, together with evicted farm servants, dependants and labourers. Less in number, though proportionately even harder hit, were pastoralists whose herds had been decimated, the exchange-value of what remained having fallen drastically. Others hard hit were urban casual male labourers, women in service occupations, craftsmen, beggars and dependants. This loss of income helps to explains why cereal prices did not rise much during the drought period.

The second drought primarily affected pastoralists and while the exchange-value of livestock for grains fell in the northern case, it totally collapsed in Harerghe. Thus in both cases, they were doubly hit, by livestock deaths and deteriorating terms of exchange against grain. In Wollo another factor was that large amounts of grazing land had been taken from Afar pastoralists for irrigated cultivation. As so often, this was crucial dry-season grazing land, the key to their system of grazing far larger areas, and its loss thus crippling.

This is a very characteristic factor in African food shortages. Historically, pastoralists have occupied many of the drier plain lands of tropical Africa, in part because this is one means of using land with too little and too risky rainfall for cultivation alone. Livestock are favoured as a form of savings. Not only are they more convenient, movable and self-regenerating than sacks of grain (over the above seed requirements); they are used in a wide variety of non-market relationships like bride-wealth, while livestock ownership is an important component and indicator of social standing. But they are poor insurance in the worst food shortages, since on every such occasion the amount of grain which can purchased for one animal falls drastically as a flood of livestock hits markets. This always happens since, apart from purchase of food, shortage of water and grazing means that, if they are not sold, they may die in any case.

While this relation between livestock and grain prices has been in existence as long as there have been markets, there are other ways in which the situation of herders has deteriorated. Almost all systems of dry-land livestock herding depend on movement between drier wet-season grazing areas and better-watered areas in which there is sufficient grazing and water for the dry season. Almost every group of herders in tropical Africa has seen its dry-season grazing reserves enormously reduced by settlement and clearing for cultivation.

The Sahelian famine is seen by Sen as the one case where an argument could be made that food availability decline was the cause of the famine. That is, there was a series of poor harvests from the late 1960s, culminating in a disastrous one in 1973. Production did decline in aggregate, according to the FAO figures cited, though, after preparing Chapter 2, I am less inclined than Sen to take these at face value. Indeed, there is independent evidence that overall food availability decline was less than assessed by FAO.[20] On the other hand, it would seem that in this, and possibly the Ethiopian, case, insufficient account is taken of the impact on savings of a multi-year drought. But Sen still finds it incorrect to propose that FAD caused the famine. There was little relation between the degree of FAD and the severity of famine (all the countries produced enough for survival if it had been better distributed). The problems were made significantly worse by the enormous increase in world cereal prices, resulting from US policy. And most of the victims of the famine were pastoralists and members of the small sedentary population in the marginal areas. Pastoralists suffered the double loss of entitlement as noted above. They lost many stock to lack of grazing and water and suffered from exchange-rate losses on what they sold.

Sen also discusses overgrazing in the Sahel, which he agrees has been a major contributory factor. His explanation is mainly the 'private herds, common grazing' argument,[21] which is valid but not sufficient. He mentions the development of cash-cropping as a factor but mainly in the sense of breaking down pre-capitalist patterns of redistribution and in making the grazing of stubbles more difficult. Others (Klaasse Bos, 1981, Oxfam, 1984) have pointed to the enormous expansion of land under cash crops, which continued in most of the Sahelian countries, right through the drought. Swift (1977) notes the increasing penetration of pastoralist areas by merchants, notably through purchasing control of water-rights. While the growth of export-crop production took place to the south of the Sahel in areas of more secure rainfall, it displaced peasant cultivators northwards and thus contributed to both soil deterioration and overgrazing in the Sahel.

While Sen's analysis is very useful, it is not without its weaknesses. It is strongest where it treats the processes occurring immediately before and during a famine, and weaker on the longer-term factors lying behind it. There is, however, nothing to stop one combining the two. A more difficult problem is the use of the term 'entitlement' for less secure and accepted claims on food (or other resources) which derive from social obligation rather than wages, produce sales and ownership of property. The former can be analysed in terms of supply and demand to yield quantitative changes in prices and other ratios. The latter cannot, since it is the 'entitlement' itself, rather than its value, which tends to weaken or disappear in conditions of famine. Moreover, it does this less in relation to market processes than to the breakdown of non-market processes in the face of overwhelming strains on society.[22] On the one hand, one might assert that, almost by definition, famine involves the breakdown of social structures and obligations. But it may well also be that severe food shortage can sometimes be prevented from turning into famine, if the protective mechanisms do not break down. In this respect, two very dissimilar examples, both from Asia and by the same author, are instructive.

Greenough (1982) shows how, historically, the prevailing ideology in Bengal has been strongly premissed on the notion that the wealthy and powerful had the obligation to assist the poor in time of need. This indeed was among the major justifications for their privilege. None the less, he shows from historical evidence that the wealthy reneged on this obligation in every case of serious famine. One might imagine that this was the breaking-down of a pre-capitalist mechanism in the face of commodity production, but Greenough shows that it happened during a major famine in the mid-nineteenth century. Still more significantly, he shows that the wealthy and powerful, having maintained themselves by throwing their dependants to the dogs, were still able to re-establish the ideology of assistance to the poor within a relatively short time. This demonstrates the need for caution in referring to the supposed safety-net provided by redistribution. But one must be careful not to over-generalize from such an example, thought-provoking as it is. Greenough himself, in another context, refers to a wide variety of studies of village communities in South Asia which have tended to demonstrate:-

> a surprisingly common ideology of reciprocity and redistribution, which served, and to some extent still serves, to adjust scarce resources to communal needs. These local arrangements and the quasi-welfarist ethos behind them turn out to be strongest in just those environments most threatened by chronic crop failures. (1982b: 792).

In this case, it would appear that the ideology had a real component, but then so it seems to have done in the case of 'normal' crop failures in Bengal. On the other hand, there would also appear to be a significant class difference between the two cases. In the Bengal case, a wealthy class maintains an ideology of provision of plenty to those who never experience plenty, but dumps them in the event of crisis and then rapidly rebuilds the ideology. Social distance spares psychological concern. One would expect the ideology and organization of a small village community to be somewhat different.

When all is said and done, this is an area in which the scope for 'hard analysis' is limited. Each society and community within it will differ in its response to food shortage, the way in which it takes care of the hungry (or does not). One relevant factor will be the degree to which commodity production has penetrated. Others will relate to the importance and type of kinship, patronage and other non-market relationships. The degree of social and linguistic homogeneity will also be significant. Finally the specific history of the area will be important.

But that does not mean that it is unimportant for policy. Almost without exception, close observers of, and relief-workers involved in, famines stress that the greater the degree to which normal social structures can be kept in operation, and people assisted at their homes – rather than in camps, or on the outskirts of towns to which they have trekked – the better the chances of avoiding catastrophe and the quicker people and areas will recover. This puts a premium on the timely reporting of famine conditions, but, as some of the examples below indicate, an even greater premium on timely response to that reporting. As Gill (1986) makes crystal clear, hundreds of thousands of lives could have been saved in Ethiopia, if donors had responded to a well-founded prediction and appeal from the Ethiopian Relief and Rehabilitation Committee (see below). One of the reasons why they did not was that an FAO mission was reaching rather different conclusions at the same time, these being based largely on the FAD model which Sen criticizes.

Before proceeding to some examples, one can briefly summarize some of the contributions of Sen's entitlement model and some of the conclusions which can be drawn from it.

(i) While there is no doubt that poor harvests and low production are very important factors, they may be restricted to particular areas of a country, yet still have a devastating effect.[23] Moreover there is no simple relation between the extent of harvest failure and the severity of the famine – that depends on a variety of other factors, including the degree of speculation and the government response.

(ii) One thing which emerges clearly is the effectiveness of the private market – in aggravating the situation. To some extent it seems that the more integrated the national cereals market, the more effective it is in sucking food from the worst-hit areas and generating price spirals. One of the major contributions of Sen's analysis – and of its application to the details of famines – is to give a picture of the operations of such processes.

(iii) Another clear conclusion is that inappropriate and delayed responses by governments can contribute, and have done so, to the severity and devastation caused by famines. This will emerge still more clearly from the Ethiopian example below.

(iv) While the analysis cannot handle non-market entitlements very effectively, it does at least point to their importance. There is, moreover, no way in which microeconomic analysis can deal satisfactorily with processes whose major characteristic is that they do not conform to its basic assumption of atomistic, self-interested market behaviour.

(v) Sen does not have very much to say about long-term processes, or about those which generate the patterns of entitlement or access to resources which characterize famine-hit societies prior to its onset. This would argue, not for rejecting his analysis, but for combining the two.

The preliminary conclusions which can be drawn seem to be as follows:

(i) Reporting systems for the prediction of food shortfalls need to be based on far more disaggregated data than is normally the case. Moreover, monitoring of prices should be an essential component. (Both are in fact quite often included).

(ii) Perhaps even more important is the timeliness and appropriateness of government response. Arrangements for food imports and distribution are only one respect of this, important as they are. The Bengal Famine was much aggravated by policies bearing a more than passing resemblance to IMF structural adjustment.

(iii) Given the dangerous power of markets to make situations worse, there is a clear need for some form of state agency to allocate food on other than market principles. This is a difficult problem since, as shown in previous chapters, such agencies have contributed more than their share to the long-term problems in Africa. Nor have they generally performed especially well in situations of severe food shortage (though this is not always the case). But in spite of these problems, given the overriding importance of preventing massive deaths from starvation, it does seem necessary that some form of state agency should be given powers to gain access to enough of the nation's grain supply to allow it to perform this function. It must also be accepted that this is inevitably a loss-making activity, to be paid for with an explicit subsidy rather than, as so often at present, simply allowing the deficits to pile up.

(iv) Whatever the failings of social networks as forms of safety net, there is a very clear case for action by states to assist people where they live, before major breakdowns and desperate migrations occur. Since the problem is lack of entitlement, not simply lack of food, it would make sense (if the funds and organizational structures are available) to maintain 'shelves' of counter (harvest) cyclical public works projects in specific drought-prone areas, to be brought into play as and when the need arises. Once again, huge problems would attend any effort to operationalize such schemes under current crisis conditions in Africa.

Some African food shortages

This section presents some brief sketches of food shortages in different African countries, having as its purpose both to illustrate some of the points made above and to pick up some others not covered. No claim is made to comprehensiveness. The purpose is to raise issues for discussion. The examples are divided into two categories: the best-known famine areas, and the rest.

MAJOR FAMINE AREAS

Ethiopia suffered major famines in both the early 1970s and 1984/5. The former led to the overthrow of Emperor Haile Selassie's corrupt and exploitative feudal regime. The second, as is well known, occurred under the 'Marxist-military' regime of Colonel Mengistu Haile Mariam. It is generally accepted that the areas most affected by famine, in both cases, have been subject to serious environmental degradation and overpopulation in relation to resources. Redda (1984) provides some historical perspective on this, pointing out that famines and observations of serious environmental degradation had been made by external observers as early as the sixteenth century and at intervals since then. The basic causes appear to have been disruption by war and invasion and the heavy exactions of feudal landlords, who not only kept the population too poor to invest in improvement, but underpinned an ideology which inhibited development and apparently still does so. Redda cites local religious tradition, with 180–250 days per annum of strict fasting and a plethora of religious holidays which reduce the working days per month to 8–12.[24]

Turning to the post-1974 period, Redda indicates that the peasantry of the northern areas did not benefit to the same degree from land reform after the revolution, as did those from the south, and stresses the importance of increasing levels of military attack against liberation movements from the Ethiopian regime – in precisely the areas worst hit by drought. He describes how six major military offensives in the north of the country, between 1978 and 1984, not only killed and wounded large numbers, but sent refugees streaming over the Sudanese border. Crops and houses were burned by the troops, roads and infrastructure destroyed, standards of living pressed sharply downwards and the capacity to withstand drought much weakened. (Poluha and others stress that the liberation movements also destroyed grain supplies).

Neither Redda nor other sources I have seen have very much to say on specific agricultural policies in the years before 1984,[25] though Gill (1986) refers to the Agricultural Marketing Corporation (AMC) as a 'monument to socialist planning . . . buy(ing) from farmers at prices that provide little incentive to producers . . . to sell to those favoured by the administration' (p. 48) – in short a standard African marketing board. But certainly the famine of 1984/5 did not start in that year. There had been a series of poor harvests from 1982.

Peter Gill's highly informative book (1986) outlines what happened in 1984 in considerable detail. Contrary to widespread opinion, he shows clearly that the Ethiopian Relief and Rehabilitation Committee made a quite accurate estimate of the overall shortage and the amount of grain needed to cope with it by March 1984, though they appealed for only half the amount needed, assuming that it would not be possible to transport more. None the less, had that appeal received an adequate response, a very considerable proportion of the suffering could have been averted. In the event, the response was virtually non-existent. This had to do with the strong aversion of Western governments, especially the USA, to the Mengistu regime, reflected in doubts cast upon the reliability of RRC estimates. Gill, though clearly not an enthusiast for Mengistu's grim-faced Stalinism, shows that the losses and corruption in Ethiopia's famine relief were a fraction of those in the Sudan and Sahelian countries, while even Western agencies had grudgingly to admit the competence of the RRC (after the event). In early 1984, a further obstacle to acceptance of Ethiopian estimates was the alternative, and very much lower, estimate of shortage put out by an FAO/WFP mission, 'which seemed to endorse all the RRC's workings but shrink from its conclusions' (p. 44).[26]

While this was used as a stick to beat the Ethiopian estimates, it was itself held up for some months by bureaucratic obstruction at the top of FAO, so that even this inadequate amount was not appealed for until June or July. And even then the response from the international 'donor community' was further foot-dragging for more months until television and international public pressure led to a sudden change in October – by which time it was too late to prevent massive numbers of deaths from starvation. The above contains only a fraction of the political ill-will, bureaucratic infighting and foot-dragging, and bland self-satisfaction, which Gill describes and which condemned so many. With reports of another Ethiopian famine looming in late 1987, it is to be hoped that Gill's book will have been widely read and its message absorbed, so that public opinion, if nothing else, can push the authorities into more prompt and adequate action.

Apart from this, the civil war seriously inhibited efforts to distribute food once it arrived, not to mention the continuation of bombing raids on the drought-afflicted areas throughout the crisis (Wallace, 1985). She and Gill both note the reduction in effectiveness of relief efforts through the unwillingness of official (state and international) agencies to channel food to representatives of the Tigrayan and Eritrean liberation fronts. Those who have looked at the relief programmes of the liberation fronts have compared them very favourably with those of the Ethiopian state. This is scarcely surprising, since there is a clear contradiction between bombing people into submission and providing them with effective famine relief.[27]

Limited evidence regarding loss of entitlement comes from an eyewitness account by a relief worker. In the area in question, people stayed in their villages so long as they had any money to purchase food. When all had been spent and all chattels sold, houses were pulled down and the wooden frames sold at the roadside for the pitiful sums they would fetch as firewood. Finally destitute families set out on the roads with only inadequate clothing and remaining silver pieces, both family heirlooms and their absolutely last security. Those who succumbed were those with least chattels, smallest houses and least silver and, as always in such situations, items had to be sold for a fraction of their normal value. But I have not seen more detailed evidence on who was hardest hit and the more precise processes involved.

One highly controversial response by the Ethiopian Government to famine in the north, has been large-scale resettlement in the south and other lowland areas. Most criticism has concerned its compulsory nature and the fact that it is obviously in part a measure to weaken liberation movements in the north. Gill (p. 143) refers to this as a 'perfectly valid option' in view of the overcrowding and massive suffering in the north, and the non-availablity of donor funds for any alternative. But his argument is weakened by the figure he quotes of $5000 to settle one family, since this by far exceeds the cost of rehabilitation *in situ*. Furthermore an Ethiopian professor, who spent one summer on such a scheme, refers to clear-felling of vegetation in a fragile environment and to other disasters of short-sighted military-style planning, indicating more evidence of destroying the environment than of rehabilitation.[28]

Sudan. Until the major famine of 1984/5, the Sudan was among the few countries thought by FAO to have kept the growth of food production in line with that of population. Although in the same general climatic zone as some of the Sahelian countries, Sudan avoided the worst of the famine of 1972-4, though there were severe localized shortages, especially in the west of the country.

This was the period when dry-land farming of sorghum was expanding rapidly with 'low levels of fixed investment and ecologically damaging cultivation practices', as cited in Chapter 3 (O'Brien, 1985: 24). Since this was primarily aimed at the domestic market, it kept cereal prices down and contributed to limiting wage increases, thus assisting a healthy rate of expansion of urban investment and employment. But this had been achieved through the diversion of resources from export-crop (cotton) production which, together with increasing imports of consumer and capital goods, produced a large balance-of-payments deficit. The Sudan was allowed to build up the second largest, and later the largest, debt in tropical Africa ($6.6bn in 1985, OECD, 1986: 208),[29] but this generated strong pressures for 'structural adjustment' towards exports. This included devaluation and other aspects of IMF conditionality, while the World Bank and EEC funded projects for the resuscitation of cotton on the Gezira irrigation scheme. At the same time, Sudan was being promoted as the potential 'bread-basket' for Saudi Arabia, which generated both lending and huge purchases of land in the Sudan. One Saudi prince owns 1.2 million acres (nearly 2,000 square miles) and several others have very large holdings.

Devaluation and the Saudi link meant diversion of sorghum from the domestic food market to export to Saudi Arabia as livestock fodder, increasing the local price of food significantly. This brought a previously hidden crisis into the open.

Another important factor has been that the large number of refugees from Ethiopia and Chad already in the country strained available transport facilities. Another was the mismanagement and corruption of the Numeiri presidency and the rapid concentration of resources, both urban and rural, in the hands of local and foreign merchant (or pirate) capitalists. This contributed both to rural poverty and to the chaotic transport situation which inhibited famine relief movements of food, together with the refusal of merchants and transporters to move grain into famine areas for less than outrageous payments (Gill, 1986, Chapter 13).

FAO figures appear to show a massive decline in food availability. From an average coarse grain production in the period 1974-83 of 2.35 million tons, production fell in 1984 to a mere 1.29 million tons. This 45 per cent reduction was followed still more startlingly by a record crop estimated at 4.4 million tons for 1985 (nearly 90 per cent above the ten-year average and three and a half times the 1984 level) (FAO, 1985e, 1986). I suspect that detailed research would show less dramatic swings in production, but there seems little doubt that production did fall very significantly in 1984. None the less, as quoted above, the record harvest of 1985 did not eliminate the need for emergency food aid in 1986. The affected populations were penniless and could not afford to purchase the abundant grain. The capitalist producers and traders were not interested in their problems and were concerned to export grain if possible.

The Sudanese example demonstrates a number of the points made above. Firstly, there is the pattern of rapid and ecologically damaging penetration of vulnerable areas previously grazed by pastoralists (among the major victims of the famine). Secondly, there is the demonstration that, even where grain is abundant, it does not help those without entitlements. D'Souza and Shoham (1985: 521) cite grain prices as having increased in money terms by a factor of four, while livestock prices fell to a tenth of their previous level – a catastrophic loss of value to herders. These authors mention the negative effects of overgrazing and over-cultivation on yields. But, unlike O'Brien, they make no mention of the dynamic behind it. And behind all this lies the enormous magnitude of Sudan's debt crisis, a large part of it a monument to the favour with which the US and international agencies looked upon Numeiri's corrupt and incompetent, but fiercely anti-communist, regime.

The Sahelian region is normally taken to comprise five former French colonies, Mauritania, Mali, Niger, Chad and Burkina Faso (previously Upper Volta). Senegal and the Cape Verde islands are sometimes included, though Senegal lies mostly in areas of better rainfall. Much of Sudan and Northern Ethiopia also lie in the Sahelian climatic zone. The Sahel is perhaps the classic case for many people of 'natural' or 'population-induced' famine. Certainly all the ingredients are there, population pressure pushing cultivation into marginal and vulnerable areas, livestock increases and overgrazing, erosion and advancing desertification.

But there are other factors which have contributed to the process of degradation and increased vulnerability to drought. Civil war has been one of these (in Chad). The relative lack of concern by military regimes (Mali and Upper Volta until 1984) another. A longer-term factor is the encouragement of export-crop production, often with monocropping, both during and since the colonial period. The development of commercial farming, concerned with short-term profit rather than long-term viability, is another factor. Yet another has been the continued focus of aid funds outside agriculture, and within it, on large projects of dubious worth. Thus, in a report to an EEC/Club du Sahel Conference in October 1983, it was stated that 'less than 4.5% of the total (aid) is allocated to the development of rainfed cereal farming, which accounts for 95% of cereal production, and only 1.5% has gone to ecology and forestry' (Courier, No. 83: 44). This is the case in spite of a torrent of reports from various agencies, indicating that these should be given top priority. One reason, though not the only one, has been the diametrically opposed viewpoint expressed (and pressed) by the IMF and World Bank, that export crops should be given priority. Another is the fact that what aid donors say and what they give funds for are two very different things. A third is that much the same can often be said of Africa governments.

Sen's analysis of the famine of the early 1970s stressed that, while there had been a decline in food availability, it was not the major factor turning this into a famine. This was rather the disappearance of entitlements for vulnerable groups. The same thing clearly happened in 1983–4. Many other analysts stress that there has been a long-term decline in per capita food production, attributing this to increasing production of export crops. Klaasse Bos (1981) shows that the development of export-crop (especially cotton) production continued in the 1970s, and this is confirmed by FAO (1984a). Taking 1974–6 as 100, the volume indices for 1984 were 222, 190 and 140 for Burkina Faso, Mali and Senegal, though for Chad the index had fallen to 75, largely as a result of civil war.

Production of groundnuts, the other major export crop of the area, has fallen but not as a result of policy changes. One reason is falling prices, with competition from soya beans (see Chapter 5). Another is soil exhaustion, disease and falling yields from monocropping. Yet another, in all probability, is the fact that, with increasing urban populations and incomes, domestic demand for human consumption has bid production away from exports. But in general the Sahelian countries continued to devote major energies and funds to export crops. An exception to this is Niger, where the emphasis has been placed on food crops, and where export-crop production has declined drastically. But Niger is a major exporter of uranium and thus has considerably less severe balance-of-payments and budgetary problems than its neighbours. Wijkmann and Timberlake (1984: 42) indicate that Niger's policies have significantly improved the country's capacity to withstand drought and to provide relief for the pastoralist population and their herds. The other exception is Mauritania, again a mineral exporter but one in which local food production from a small strip of cultivable land in the south of the country covers far less than half of total consumption of cereals and appears to be declining (Mitchell and Stevens, 1983, FAO, 1986g).

Within the Sahel as a whole, all too much of the available investment in agriculture has been spent on the development of large, very expensive and dubiously productive dams, many of them with full or part funding from the EEC.[30] Meanwhile, failure to find any solution to the problems of dry-land farming has led to increased pressure on resources, deforestation, accelerated erosion and in consequence reduction in river-flow upon which the large dams depend. Aliou Ba and Crousse (1985: 389) describe recent developments in the Middle Valley of the Senegal River:

> Over the past ten years or so this region has been subjected to the repeated assaults of drought and desertification. The rainfall ... which formerly ranged from 600mm ... to 300mm ..., has now declined to less than 100mm in many areas. This has led to a corresponding decrease in the river floods. In 1983 and 1984 it was necessary to resort to the ultimate solution of damming the river with an earth dam below Rosso in order to retain the water needed by the crops of the Middle Valley. Much damage has been done to the ecosystem as a result. Dunes have now reached the river itself and sandstorms are increasing in frequency. Designated forests which have previously been preserved have had their cover reduced by over half. And survival for the people of the region entails much more effort. The food production of the Senegal Valley, formerly the granary ... of Mauritania and Senegal, is now decreasing year by year.

This is the river along which many of the large dams are being built. Engineers are beginning to worry about salinization in relation to the almost 600 milllion ECU Manantali-Diama dam complex, before irrigation has even started. River-flow has declined to the extent that the planned barrage to prevent sea-water inflow is thought to be insufficient. Again Tinker, 1984 (*Courier* No. 87: 59) refers to the 'hundreds of millions of dollars which have been spent introducing irrigated ricefields along the Niger river (with) almost complete failure'. This refers to Office du Niger (Mali) plans for rice-growing in the internal delta of the river. As early as 1962, plans were made for 3 million ha of rice. To date 45,000 ha are cultivated and yields are well below target. As of 1983 they were well under one ton per hectare![31]

Much of the famine problem also relates to the ineffectiveness of efforts at relief. Part of this derives from the long time which it takes for food aid to arrive after decisions have been made. Another part stems from corruption and inefficiency and from the difficult road and transport situation. Another can be illustrated from the case of Mali.

Since the mid-1970s, the EEC and other donors have been pressing Mali to decontrol and privatize its state monopoly of grain marketing, citing the inefficiency of the agency, OPAM. Another prong of the strategy has been to use EEC food aid counterpart funds to subsidize producer prices. Both of these have had as aim to improve incentives to producers and increase grain deliveries. There is little evidence that this has happened, though it is hard to tell when the drought of 1983–4 intervened soon after the policy began to be implemented. But the experience of these years shows that privatization has its problems, where famine relief is concerned. A report of late 1984 looks at the drought in Mali and indicates the normal features of cattle prices falling drastically in terms of grain. It continues on the programme for distribution of emergency relief:

> Government stocks of food aid are distributed by OPAM, the state marketing board. But although OPAM transports food to the regions, final distribution is in the hands of private traders, who now control 90% of the market in foodgrains (in 1980 OPAM gave up its

monopoly in grain). An estimated 20% of millet stocks disappear into the hands of black marketeers who sell them across the border for higher prices than they can get in Mali. Traders refuse to risk their trucks on the almost impassable roads to reach the areas where famine is at its worst. They can do good enough business by speculating on local shortages. (*The Economist Development Report*, November 1984: 5).

One should perhaps point out that from the description of the Malian state machinery given by Francois (1982), there is no reason to expect OPAM or other state bodies to have been much better.

A comparison of Mali and Burkina Faso on this score demonstrates that practical policies cannot easily be read from official ideology. Mali's state corporations were largely set up by the socialist Keita regime in the 1960s. This was removed through a military coup to install the current repressive military dictatorship, which was much preferred by Western countries and donors, because of its anti-socialism. None the less, and in spite of frequent nudging from donors, the Malian state has maintained its control over the economy. 25–30 state corporations still account for about 70 per cent of the national product. Indeed, current plans for 'privatization' seem largely to involve the purchase of shares by private individuals, and it is hard to see what positive effects this could be expected to have (except on the incomes of the share-holders). Even that is reputedly moving slowly.

On the other hand, the military-socialist Sankara regime, which took over Upper Volta by coup from a previous military regime in 1984 and renamed it Burkina Faso, has initiated a campaign to reduce the scope of, and employment in, the state sector, without reference to the IMF or World Bank. The reason given is that state agencies weigh too heavily upon the resources of a small poor country. Whether this will survive the general *penchant* of soldiers for top-heavy and rigid institutional structures remains to be seen. The prospects look dimmer since the removal and murder of Sankara by a 'brother' officer in 1987.

Whether the necessity of serious and sustained action to combat erosion and desertification in the Sahel is at last impressing itself upon governments and donors is hard to say. The rhetoric has been there for years, so it provides little guide. Burkina Faso has announced self-sufficiency in food as a major priority, is trying to cut foreign-exchange spending as a means thereto, and has instituted large-scale campaigns for tree-planting in the arid northern areas – at least some of which are reportedly successful.[32] Elsewhere it is harder to see signs of change. As of 1984, it was reported that most of the funds spent on anti-desertification in the Third World were used 'on roads and buildings, in water-supplies and on research, training courses and meetings' (*Courier* No. 87: 72). It is perhaps easy to criticize when the problems are so huge. The technical requirements for stopping the advance of the desert are often seen as being so severe as to require drastic depopulation – which is clearly unthinkable in human and social terms, given the numbers involved and the fact that this is their only source of livelihood.

Mozambique – In 1975, a popular and idealistic government came to power, determined to improve the hitherto miserable lot of the people under a more than usually backward, unproductive and brutal colonial regime. At least for the first several years, levels of commitment, hard work and public morality were exceptionally high by any standards.[33] Now the economy is in ruins and tens of thousands of people have died of starvation in recent years, while the vast majority of the population have experienced serious food shortage for years on end.

Mozambique has been almost uniquely unlucky with the weather for most of the past decade, those years without drought having usually been blasted by torrential rains or high winds. Again, the government has had a particularly difficult task. Most of the Portuguese left rapidly after independence (in spite of efforts to persuade them to stay), taking with them most of the skilled manpower and almost all the private farming, supply, marketing and service infrastructure, having neither trained nor employed more than a handful of Africans above the level of unskilled labour (much of it forced). As if this was not enough, Mozambique's support to Zimbabwean freedom fighters led to attacks from the Smith regime, and then from the South African-backed MNR. The latter have wrought havoc in large parts of the country in a savage campaign apparently not concerned to mobilize political support but simply to disrupt.[34] As recorded by Tickner (1985) and a number of more recent newspaper reports, the MNR have not only destroyed villages and food crops, killing large numbers in the process, but made famine relief immeasurably more difficult by mining roads and railways – often going quite deliberately for food relief supplies. Under such conditions, even the best of state policies would have been unable to achieve much progress. Moreover, famine did not come first to Mozambique with independence. Portuguese colonial policy simply ensured that the outside world did not hear about it.

And yet, at the same time, government policies have been responsible for some of the problems. An understandable but ill-advised effort to supply wheat bread and other 'middle-class' foodstuffs to much of the city population at heavily subsidized prices, made for an almost immediate food deficit, since commercial farm production had fallen with the departure of the Portuguese. An attempt to increase production through state farms proved costly and unsuccessful. This, together with an ambitious investment programme, used funds which might otherwise have been used to keep light industries working or to import consumer goods and tools for the neglected peasant sector. Much of the policy towards peasants involved moving them, often compulsorily, into 'communal villages', with generally negative results on production and livelihood. Problems of policy direction were compounded by those of a cumbersome bureaucracy, inherited from the Portuguese but made still worse by the imposition of soviet-style top-down planning.

Urban food requirements in Mozambique have been met (or not met) largely through imports for most of the period since independence, though food imports were also substantial during the colonial period. A variety of different rationing systems have been tried, some of them reasonably effective except that there was just not enough food to go around. On the other hand, years of shortage have taken their toll on morale and organization, so that black marketing and misappropriation have increased in recent years. There have been a number of localized famines in rural areas, in which loss of crops and incomes is compounded by the inability of the state to respond for lack of produce to ship and because of MNR activity.

One of the worst cases occurred in Tete Province in 1984 when an estimated 10–20,000 people died of starvation. The Province, as such, probably did not even have a food deficit – a factor which led to delays in identifying and reacting to the crisis. The north part of Tete Province (Angonia) is a well-watered highland area, which produces a significant surplus of food. Virtually none of this moved to the southern part of the Province, which is much drier and poorer, and was affected by drought. By far the major proportion was sold over the border to Malawi, because of the absence of most consumer goods in Mozambique. Even if the entitlement of the people of south Tete had not been wiped out with their crops by drought, the exchange entitlement of their Mozambican money had been wiped out by the long-standing lack of goods to buy with it.[35] This has been a problem in most parts of the country, and has played no small part in disrupting production, through the lack of incentives and means of production and through the necessity to spend hours, days and even weeks getting hold of basic necessities, queuing for official supplies and/or trading them.

OTHER AFRICAN COUNTRIES

This section will look at disparate aspects of food shortage in a number of different countries, usually making no attempt to encapsulate the overall food situation. Indeed, the conditions of different areas and population groups are so different that it would be meaningless to speak of an overall food situation. Moreover, the purpose is to demonstrate examples of particular processes, rather than to present a summary of the 'average'.

Nigeria is a huge country, with climate ranging from humid forest to the semi-arid northern boundary with Niger. The studies I have read refer primarily to the north. Shenton and Watts (1979) look at the impact of colonialism on the incidence of famine. The area they consider is one of limited and variable rainfall, so that food shortages have recurred at relatively frequent intervals. They show, however, that the pre-colonial societies had well-institutionalized means for coping with food shortage. Drought-resistant crops and intercropping reduced the risk of harvest failure, while agricultural production was supplemented by gathering during the dry season. Seed for the subsequent harvest was safeguarded by religious tradition and a ceremony in which the divinity of agriculture forbade its use before the first rains of the next season. Apart from this, there were a variety of systems of redistribution, some operating between equals and/or kin, others controlled by the ruling families and involving collection of taxes in grain and their storage in royal granaries for distribution to the needy in times of shortage.

Colonial incursion worsened the situation in the process of conquest, through the enormous social dislocation which it brought and the series of shortages and epidemics which accompanied it. But so did the imposition of colonial order. Taxes were transformed from grain to cash and made rigid instead of varying with the harvest. With the granaries gone and ruling houses subjected to superior colonial authority, the capacity of leaders to provide for the needy declined, as did the obligation to do so. Taxation in cash led, as intended, to labour migration and export-crop production to earn the cash, taking land and labour from food production, with little effort to assist an adjustment of productivity in food production. This significantly changed the impact of drought, and serious famines became more frequent, as Shenton and Watts document, showing how changes in the naming of famines demonstrate the relevance of factors other than climate or harvest shortfall. In 1942, there was a famine without serious drought or shortfall, which they compare to the almost contemporaneous Bengal Famine. In this case, large-scale conscription for war-time porterage and military service were the main factors, together with campaigns for enforced groundnut cultivation.

Although Shenton and Watts refer to a serious weakening of the peasant capacity to produce food enough for themselves and local surplus needs, it may be that production patterns adjusted with time to maintain the growth of food production. Beckmann (1986) refers to the capacity of the peasantry to perform this function until the early 1970s. It seems likely that an explosive increase in urban consumption was the main reason for the national food deficit which emerged in the 1970s. Rising incomes from the oil-boom intensified the sorts of factor mentioned in Chapter 3, as did state subsidies on foodstuffs for the urban areas. At the same time, foreign companies and the government combined to hasten the spread of wheat consumption, with construction of wheat mills and subsidies on wheat flour. This was followed by further combinations of companies, aid agencies and governments (national and state), seeking respectively to fill the growing food import gap and to profit from selling modern technology and management to achieve this aim (Oculi, 1979, Andrae and Beckmann, 1986). The vast sums spent on this have done little to increase aggregate food production, while hastening the concentration of control over land and other resources, impoverishing the majority of poor peasants. Apart from this, World Bank projects for the development of (rich) peasant food production focused on maize production in areas vulnerable to drought where this increased the risk of harvest failure. Fortunately the peasants preferred sorghum (Clough and Williams, 1985). But the extension service *penchant* for monocropping seems to have been accepted and to have increased long-term vulnerability to variations in climate.[36]

Kenya has a series of different food problems. Geographically, the country consists of a consolidated 'island' of fertile and high-rainfall land in the centre and west, covering some 20 per cent of the land-surface, surrounded by 80 per cent which ranges from marginal for cropping to arid semi-desert. Population growth, but also the accumulation of large-scale holdings, has led to increasing over-crowding of many high-potential areas and an increasing push into the surrounding marginal lands. Per capita national cereal production appears to be declining, though not to the degree proposed by FAO, and to be increasingly variable. Where the country exported food in most years a decade or two ago, the balance has now been reversed, with imports more common. The world's highest population growth rate is part of the story, but urban growth rates have been very high and the growth of food consumption even more rapid. Food subsidies and the growth of a middle class have led to an enormous increase in the demand for staples, fresh fruit and vegetables, meat and dairy products, eliminating most of the food exports of the colonial period. As late as the 1970s, there were plans to export meat to the EEC from 'disease-free zones', which generated costly and counterproductive donor-aided efforts to control foot-and-mouth disease.[37] But there has been no surplus of fresh meat to export for some years now. The same is true of dairy products, though policies aimed at expanding the market (free distribution of milk to schools) are also significant in this case.

Colonialism brought the usual problems to Kenya, disruption of societies, rinderpest, famine etc. – aggravated by the alienation of huge tracts of the most fertile land for white settlers. In spite of severe restrictions on peasant production of high-valued crops, to limit competition and labour loss to the settler sector, the dynamic of capitalist growth did lead to development and accumulation, accompanied by significant impoverishment, in the highland areas.[38] The marginal and pastoral areas experienced significant loss of land as well, and an exclusion from markets, which much aggravated stock increases and overgrazing (Spencer, 1983, Raikes, 1981). At the same time, low incomes and a small urban population allowed the settler sector to export foodstuffs (notably dairy products) in addition to export crops.

Independence transferred some settler land to Africans, both peasants and large-scale farmers, while removal of legal barriers led to an explosive increase in peasant production of export crops and dairy products. Food-crop production has also expanded quite rapidly, but sales have not kept pace with the growth of urban demand. At the same time, four sets of problems have also increased. Pastoralists in the huge part of the country with inadequate rainfall for cropping have lost land and water rights, while their herds have increased. Peasants on the edge of the cropping areas have expanded or been pushed into the pastoral margins and are increasingly subject to harvest shortfalls and hunger. In the more heavily populated highland areas increasing population leads to fragmentation of already small farms, or to landlessness. The adoption of cash-crop production (export or other) leads to cash income at the expense of nutritional standards in cases where farms are insufficient to provide both.

One area of localized shortage is Kitui District, a dry area where cropping has much increased in recent years, pushing livestock grazing back. Erosion and soil deterioration have been serious problems. The climate of the area is such that harvest failure can be expected quite frequently, and despite the drought-resistant qualities of Katumani maize, this has led to several serious food shortages recently. Cattle are a major form of savings and security, but in every case of food shortage, their value in terms of grain declines seriously. In 1981, and almost certainly other shortage years, this was happening in Kitui, while the official grain monopoly was having problems with storage capacity in the west of the country, after an excellent maize harvest. So one has the characteristic combination. Peasants are pushed into vulnerable areas by population increase and by the expansion of commercial farming, while the need to earn cash pushes them into farming practices which intensify the problem. This generates vulnerability to climatic variation, while the marketing system fails to move food in, where there is no effective demand (until aid donors assume the cost by sending food aid).

North-Eastern Province, where the conditions are still more arid and the population are Somali pastoralists, has also been a focus of food shortage. This is often put down to climate, grazing practices and loss of stock (to drought). Other factors have also been important. Since independence, the Kenya army has been in the Province, mopping up separatists or bandits (according to viewpoint). Especially during the mid-1960s, they killed large numbers of livestock, leaving the population with far less than usual as they went into the major drought of the early 1970s, and thus leaving a larger proportion destitute in its aftermath. Drought relief focussed upon moving people into camps near to the Tana River and on attempts to settle at least some of them as irrigation farmers. That few of these were successful has as much to do with their organization as with the relative inexperience of pastoralists with irrigation. It is true that most were concerned to rebuild their herds and move out into the range as quickly as possible, many leaving wives and children behind until they could establish themselves. But the pastoralists had in fact been practising flood irrigation in good years for many decades, something which seemed to have passed the local authorities entirely by.[39] The development of irrigation has been bedevilled by all sorts of problems, and has so far proceeded very slowly. This is in some ways just as well, for if it takes a large proportion of the river bank, its implications for the Somali pastoralists could be very serious. The river and land near to it are crucial dry-season water and grazing, allowing them to use a much larger (and totally uncultivable) area on a seasonal basis. With large and small-scale irrigation schemes, there is a serious danger that their access may be restricted in the future.[40] If and when that happens, famine is likely to become a regular feature.

In the highlands, the problems are rather of overcrowding, emerging landlessness and poverty. There is almost always grain for sale if one has the money for it, which conceals the fact that there is a food problem in these areas. There is a substantial and increasing number of households which are landless, or whose farms are too small to give an adequate livelihood, and among whom malnutrition is a serious problem.

Kisii is a highland district in western Kenya. Drought is no problem, since it has perhaps the highest and most stable rainfall in the country (1,600–2,400 mm per annum). In times past, drought elsewhere in the country meant increased incomes for Kisii farmers, since it increased the prices of their products. But Kisii also has the highest rural population density in the country, almost 400 per sq. km. or 4 per hectare (parts of other districts are even more crowded). It also has the highest population growth rate in Kenya, over 4 per cent per annum. At the beginning of the colonial period, the population was less than one tenth of the present 1.2 million, there was land in plenty, and the people combined cattle-herding on communal pasture with agriculture based on shifting cultivation, most of this work being done by women. The common land has long gone and the land-surface is covered with small farms, divided by hedges. Since the normal pattern of division on inheritance is up and down the hill, most are long and thin. As little as ten metres across is not uncommon, so that hedge-rows take up to 20 per cent of the total space. For many of the smaller farmers, it is already impossible to make subsistence stretch until the following harvest (even with two maize harvests in parts of the district), so for much of the year they must rely on wage-labour. Since agricultural labour is mostly casual, unreliable and poorly paid, this puts a premium on education, and almost all children start in primary school, though many drop out along the way for lack of funds. But where ten or twenty years ago primary school education would yield a reasonable job, today there are large numbers of unemployed secondary school leavers.

Kisii was one the first districts in Kenya where Africans were allowed to grow coffee during the colonial period. It is also a major producer of tea, and a number of cash crops for local sale, such as sugar-cane and sweet bananas, fruits and vegetables. But at least on the smaller farms, there is beginning to be evidence of reduction in cash-crop areas, partly because the land is needed for maize, the main staple (though also partly to do with marketing problems for coffee). There has been a clear process of differentiation in the district, but it does not show up in land-holdings. Steep slopes and very high prices make land in Kisii an unattractive investment for accumulation. As elsewhere in Kenya, the characteristic mode of accumulation has been straddling, but where the proceeds are invested in agriculture for production, this has usually been elsewhere in the country.[41] When the rich now buy land in the district, this is more commonly to fulfill obligations to family or to build houses and establish a base for political power. More moderately endowed persons in employment purchase land, when they can afford or find it, to provide for family subsistence and the future. Those who sell land are almost invariably poor, those who have got into debt or, in the most irresponsible cases, men who sell off land to maintain a level of consumption above that of production. This is a not uncommon cause of violent intra-family quarrels and murders in Kenya, if the frequency of reports in the press is any indication.

The food situation of poor families is made more serious by the fact that, while much else has changed out of recognition, the sexual division of labour has lagged behind. Men no longer do much herding labour, since cattle numbers have declined and most are kept in paddocks. Some of this has been taken up by wage-labour and some with cash-crop production, but food-crop labour remains to a very large extent women's work which, together with household labour and wood and water collection, gives them a substantially longer working day. Intensification (especially of food-crop production) is hampered by this, and because men control both the overwhelming proportion of cash income and decision-making. Many men look upon the provision of food as women's sole responsibility, being unwilling to spend cash income on inputs for food production. Yet women's possibilities for earning cash are greatly inferior to men's. Hybrid maize and use of fertilizer are widespread in the district, but it is precisely the poorest families, which need them most, which have not adopted them. Nor is this simply a matter of absolute lack of income. The minimum package of seed and fertilizer which can be purchased in local shops costs about 20 shillings, which is two or three day's casual labour wage – and for many men, not more than a week's drinking money. But the (considerable) extent of male drinking may itself be partly a symptom of the crisis. For even with intensification, it is hard to see how really serious sustained food shortage will be avoided in the next generation, without a very significant expansion in the availability of off-farm jobs. Though full-scale drought is highly unlikely, years of reduced production are not (1984 was such a year). Those who go hungry are those heavily dependent on casual labour, since wages and availability of jobs fall, while food prices rise.[42]

Other examples come from out-grower schemes, in which peasants produce tea and sugar under contract to state (tea) and private (sugar) firms which control the processing. Buch-Hansen and Secher-Marcussen (1982) have claimed that these schemes significantly increase the incomes of participating peasants, but a series of nutritional studies conclude that the children of tea and sugar contract growers are worse nourished than those of neighbouring non-contracting peasants (Government of Kenya, 1979). For other contract growers, they conclude that increased cash incomes do not improve child nutrition, though they do not worsen it. [43] Leaving aside the contracting issue, this does provide evidence that increased cash income does not necessarily improve family nutrition. The case of the Mwea rice scheme and the malnutrition of children of women working on it has already been mentioned above.

Uganda lies at the conjunction of two separate rainfall systems and the southern part of the country has among the most reliable rainfalls in Africa. That there has been widespread food shortage (and shortage of other goods) in recent years has more to do with the ravages of the Amin regime and civil war than with climate. One study in south-eastern Uganda in the period just after Amin found improved levels of nutrition compared with the period prior to the Amin regime. With the breakdown of cotton marketing, peasants had switched to production of foodcrops, and had increased own consumption.[44] One should be cautious of generalizing such examples, however. Evidence from a rural area of Tanzania indicated that a similar market breakdown had led to increased drinking, with negative effects on both incomes and child nutrition.[45]

Karamoja in the north-east of Uganda is an arid area, peopled mainly by pastoralists, with drought a common occurrence and famine increasingly so in recent years. Mamdani (1982) and Gartrell (1985) agree that colonial policies bear much of the blame for this, but disagree in other respects. Mamdani argues that colonialism 'destroyed not only the basis of a pastoral way of life but also hampered the transition to an agricultural mode of existence' (p. 66). He sees cultivation as a more 'developed' form of production than herding, and blames the colonial government for preventing its development. Gartrell takes issue with this rigid conceptualization, pointing out that similar assumptions underlay much of colonial government policy. There is no reason to suppose pastoralism to be 'prior to' or more backward than cultivation. It is a means of coping with environments too arid to sustain life by means of settled agriculture. Karamoja is such an area (with a one in three chance of a reasonable crop and a serious drought once every four years) and oral historical records show that the Karamojong took up transhumant herding *after* migrating there, evolving a system which allowed survival under these harsh conditions.[46] Far from keeping the Karamojong from cultivating, the colonial government made a number of attempts to settle them, a common policy of governments in Africa (colonial and post-colonial) – and usually an unsuccessful one.

As elsewhere in East Africa Karamoja was devastated in the 1890s by rinderpest which decimated herds, followed by epidemics and other disasters which reduced population, cattle numbers and capacity to withstand droughts. This was followed by loss of land, imposition of administrative boundaries and controls on movement, the forbidding of grass burning and compulsory destocking campaigns. As in Kenya (Hedlund, 1982) and Tanzania (Raikes, 1981), official policies, aimed at 'improving' herding by drilling wells, in fact led to concentrations of stock around them and accelerated overgrazing. But as Gartrell shows (and as a series of government programmes have demonstrated by negative example), forcing pastoralists to settle and grow crops is not the answer. This is usually intensely unpopular and also usually unrealistic. Even if parts of the rangeland can be found where agriculture can be carried on, these are invariably dry-season grazing and their excision from the system reduces the viability of vast areas, to provide better incomes for a small minority.

More recently, Karamoja has again been a focus of famine in several recent years, this having largely to do with the almost total breakdown of civil authority in northern Uganda. Not only has constant fighting and raiding contributed to food shortage. It has very severely hampered efforts at relief.

Tanzania has been in a situation of almost constant urban food shortage since the mid-1970s. But food shortage to the extent of famine seems less common than in most other countries. There was a serious food shortage in many parts of the country in 1974–5, but the National Milling Corporation, the otherwise unimpressive state grain marketing monopoly, is generally thought to have distributed food to the needy areas quite efficiently.

The area which became Tanzania was ravaged by slave-trading, internal wars, colonial incursion and rinderpest in the last quarter of the nineteenth century, these being followed by a series of secondary disasters, including famines, which led to significant population decline (Kjekshus, 1977). Tanganyika regularly imported basic foodstuffs during the colonial period, and especially in the years following the Second World War, one reason for this being the expansion of plantation production and labour force without corresponding efforts to increase surplus food production.[47] All the same, at independence, the country was more or less self-sufficient in basic foodstuffs.

Tanzania is a country of great ecological diversity and the widely differing histories and social structures of its 120 or so different language groups make generalization hard. Loss of land to settlers was a problem in some areas, but never to the same extent as in Kenya. The poorest and most hunger-prone areas were probably those chosen by the colonial regimes as labour reserves for the plantations, often previously foci of the slave trade. Bryceson (1985) has collected data on officially recognized food shortages, which tend to show a concentration in the dry centre of the country and the south, but no very clear pattern emerges. One reason for this is that the definition of a food shortage depended on the subjective judgement of the officer reporting, and these varied widely.[48] The sheer number of shortages reported increased from the 1920s to the 1940s and then declined, but this seems more likely to reflect differences in definition than a real trend. Bryceson (1978) stresses the concern of the colonial government to stabilize peasant subsistence, if necessary through famine relief, as a means to increase cash-crop production. One could also note that Tanganyika, a League of Nations Trust Territory and area of relatively minor interest to the British, was not subjected to the full rigours of settler colonialism. A 'poor relation' colony, the mass of its people fared better than in the richer relations, like Kenya and Rhodesia.

Both during and since the colonial period, however, another process of concentration has being going on. Handeni District in the north-east of the country is dry with a variable climate, and is currently one of the most food shortage-prone areas in the country. In the last years of the previous century, it was a food surplus and export area for the slave plantations of Zanzibar and relatively wealthy. But with a change from cultivation of sorghum to less drought-resistant maize, harvest variability increased and the focus of surplus production moved to areas of more stable rainfall. Since independence, there has been a further concentration process with the adoption of hybrid maize, which requires a higher and more reliable rainfall than local varieties. At the same time, sorghum production has been made less viable in many parts of the country by huge flocks of grain-eating birds.

As in most other tropical African countries, malnutrition is common especially among children, and has been for many years. Maletnlema (1985) estimates that 50–60 per cent of pre-school children suffer from protein-energy malnutrition, while more specific forms of deficiency are also common. Beyond this, it is hard to generalize.[49] Several series of figures exist which purport to indicate regional infant mortality, morbidity and other factors relating to nutrition, but the results are too divergent (not to say contradictory) to give any clear picture. It can be said, however, that there seems to be little relation between the officially recognized level of food production and nutrition. Nor are *average* nutritional levels apparently closely related to assessed levels of regional income. One reason for this may be that many of Tanzania's richer areas are dependent upon bananas for basic nutrition, these being much lower in protein than grains, especially millets and sorghum, which form the staple food in some of the poorer and drier areas. But average figures tell one relatively little in any case. None of this material shows trends in nutritional levels, since most relates to single years. There seems, however, little evidence that Tanzania has as yet reached a point of secularly declining food production. Levels of marketed production, both in the period 1974–9 and in 1984–5, indicate considerable powers of rapid increase in output after a serious shortage has led to price increases, though the arguments of Chapter 2 caution against reading too much about overall production from this.

Trends in cereal marketing since independence have been discussed in Chapter 2, and the point made that shortfalls relate more to a breakdown in the official system – and to rapid urban growth – than to declining per capita production. What may not have emerged is that it is misleading to generalize about the country as a whole, since this hides local variations. Different parts of the country rely on different staple foodstuffs, maize, sorghum, cassava and bananas being among those which predominate. Climatic patterns vary and markets are segmented. Price data for local markets show complex patterns of movement and steep price gradients over relatively short distances.

It is claimed that a major food shortage can be expected approximately once every five years, though it is far from clear what this means.[50] From Bryceson's (1985) tabulation of food shortages, this is the rough *frequency* of years in which more than half the country's administrative regions experience food shortage. But again one is back with the question of what this means. Certainly it bears little relation to figures for marketed production, or to general impressions. 1981 is often referred to as a year of severe shortages, yet both 1980/81 and 1981/2 are recorded as having fewer regional food shortages than any year since 1951.

1984 was a year of serious shortage in much of the north of the country. Production was estimated to be 95 per cent of average, but in Kenya it was estimated to be only 70 per cent of normal. With many goods only available or much cheaper in Kenya than in Tanzania, Kenya's shortage sucked large quantities of cereals from north-central Tanzania across the border, worsening the situation where it was already worst. The 'epicentre' of food shortage in Tanzania was Shinyanga Region, where there was serious drought, and where maize prices rose as high as 2,500/- per bag, compared with an official price of 360–450/-, this in turn drawing maize northwards from as far as Ruvuma on the southern border. Production certainly fell, but the 'Kenya effect' probably had a more important impact on the availability of food and its price. Available evidence indicates, however, that the much-reviled NMC did a reasonably efficient job of moving produce (at official prices) into the area, while the 'free market' was moving it out to Kenya.[51]

While the above examples only scratch the surface of the situation, they may give some idea of the complexities involved. Further examples could be taken from Zaire, Zimbabwe and Botswana. For Zaire, Schoepf (1985) and Newbury (1984) show how forced cultivation and heavy levels of exploitation before and since independence have worsened income and food levels, in a country with the potential to provide plenty for all. Zimbabwe is another case where settler colonialism led to export-surpluses combined with hunger, and where the current rapid expansion of peasant maize production leaves out the dry southern areas where food shortage is a serious problem. None the less, its large maize surplus did allow Zimbabwe to distribute famine relief and prevent starvation in 1983–4, even though many lost life savings in cash and cattle (Leys, 1986).[52] A recent OECD study congratulates Botswana for its effectiveness in 'dealing' with famine during the same period, referring primarily to the distribution of famine relief. Cliffe and Moore (1979) tell a very different story, showing how colonial and post-colonial policies dominated by the interests of large cattle-owners, including the ruling family, have led to increasing marginalization of small peasants and increasing vulnerability of the environment, with overgrazing and replacement of grass by inedible bush. This has increased the frequency of food shortages, though increases in mineral exports have allowed imports to prevent starvation.[53]

In these examples I have tried to pick out some of the more relevant issues relating to food policy and food availability in some tropical African countries. One thing which emerges is the impact of colonial policies in reducing capacity to withstand drought and harvest failure.[54] The disjunction between aggregate production and access to food comes out clearly, and would do so with still greater clarity if South Africa was included. Here is one of the world's major maize-exporting countries, but one in which malnutrition is rife among the African population, most especially in the 'homelands' in which those pushed aside to make way for white commercial farming have been dumped. As might be expected, apart from the general ravages of the apartheid state, South Africa's approaches to development for the homelands demonstrate a peculiar insensitivity to the real problems.

For instance, in the Ciskei, one of the most overcrowded and run-down of all the homelands, with malnutrition at extremely high levels, the 'Keiskammahoek model' has been the 'development' response. This is an irrigation scheme where, at extremely high capital costs, a few scheme members produce milk from pedigree cows, with the most advanced equipment, under the close supervision of a management company. Well over 80 per cent of gross revenue is taken for central services and company remuneration. The scheme members (in effect little more than cow-hands) receive rather less than the minimum urban wage for unskilled labour. They in turn are assisted by wage labour at yet lower rates. In 1980, only 36 settler families were covered by this scheme, out of a population of over 650,000. Moreover, all were man-and-wife families, regardless of the fact that a high proportion of the total population consists of female-headed households (families of urban migrants who are not allowed to join their men by apartheid regulations). In spite of the fact that milk is part of the traditional diet of the Xhosa population, incomes are so low that most, if not all, of the milk is 'exported' to South Africa. Not only are 85 per cent of the South African population cooped up in the worst 14 per cent of the land, but money spent on increasing productivity is used on an expensive showpiece which provides little for the few settlers and actually removes resources from the rest of the population, producing food, which they cannot afford, for export (Southall, 1981).

Conclusions

The issue of development policy and development projects will be taken up again in the two final chapters of this book. For the present, it is worth drawing some conclusions from the different sections of this chapter.

i). Food shortage takes a number of forms, the most basic distinction being between sustained under- and malnutrition, and seasonal or occasional food shortages. The distinction is underlined by the fact that, as shown by Sanders (1985: Fig. 1) sustained malnutrition (as represented by protein-calorie deficiencies) is more serious in the better-watered areas of the continent than in the drier areas which are the main foci of famine.

ii). Agriculture is subject to cycles in which food shortage and a number of other aspects of stress are concentrated at particular times of year. Seasonal stress is likely to hit the poor with more than normal severity, compounding the disadvantage of low income level and being the occasion for deteriorations in status.

iii). Social systems differ in their ways of coping with or responding to such problems. There is some reason to suppose that the penetration of markets weakens these and sharpens the processes whereby seasonal stress leads to secular deterioration.

iv). Seasonal cycles are subject to unpredictable variations and this generates risk as an important condition facing peasant households, affecting both cultivation techniques and other response strategies. Social mechanisms deemed 'non-economic' by external observers may be crucial to survival under conditions of high risk – even if they do reduce production.

v). Poor households are more vulnerable and thus follow strategies more concerned with risk avoidance than is characteristic for larger and richer households. Since this often reduces average income over the years, it will have the effect of widening income and social divisions.

vi). The complexity of the sequence of climatic, biological-chemical, labour and waiting processes involved in even the simplest peasant agricultural production is such that detailed study of the *processes* involved is an absolute pre-requisite for the formation of relevant policies.

vii). Despite the importance of climate and production, food shortages do not correlate at all well with harvest shortages, most especially at the national level. Study of the way in which entitlements to food move is a better predictor of food shortage or famine and a better guide to policy than 'food availability decline'. Even within the marginal areas where famines are most common, one can delimit specific social groups which are most vulnerable and socio-economic processes which make them so.

viii). Where there are landless rural labourers they are often the most vulnerable group. Elsewhere, the most vulnerable rural groups in terms of numbers are cultivators who have been pushed into marginal areas of low and variable rainfall by demographic and/or commercial farming pressures.

ix). Pastoralists or cultivators who depend on livestock as 'famine insurance' are among the people most seriously hit by famines in Africa. An immediate factor is the tendency for cattle prices to collapse in periods of food shortage. While overgrazing of common pasture land under demographic pressure is part of the long-term problem, it often derives in part from expansion by cultivators into their best (dry-season) grazing. Limits on movement, limits on sale of cattle, the breakdown of social mechanisms controlling grazing and inappropriate development policies are other factors.

x). In almost every case, problems of famine are exacerbated by late and/or inappropriate state action. War and civil war are probably the most important causes of famine, assisted by the huge amounts spent by some African governments on arms imports and the military (typically far more than on agriculture).

xi). While famine is a major problem in certain specific parts of the continent, far more people are affected by long-term malnutrition, interspersed with outright hunger at 'crisis' seasons and in bad years. This tends to be ignored in discussions of the 'hunger problem' since it is so obviously the result of low income and maldistribution of access to resources rather than stagnating aggregate production.

xii). Both where famine is concerned and in terms of long-term malnutrition, women are generally more vulnerable than men, this having been aggravated in many cases by the development of commercial farming. Two reasons for this have been income shifts within households from women to men and labour shifts in the opposite direction. In addition, male domination at all levels leaves women least protected in periods of shortage.

xiii). While colonial policies can be shown to have initiated many of the processes increasing the incidence of harvest failure and vulnerability thereto, there is no evidence that the post-independence period has seen any change of direction. If anything, the reverse is the case, as expensive and large-scale projects to increase agricultural production hasten the concentration of control over resources, while accepting or even encouraging harmful and short-sighted forms of production.

xiv). The answer, however, cannot be a return to previous patterns. Populations have increased enormously and political-economic forces have emerged and are consolidating which make this totally unrealistic. Nor can one rely on 'peasant wisdom'. While peasants know more about their agriculture than external experts, they have also been affected by the penetration of commodity production and the 'Gresham's Law of farming' which comes from competition from high-profit short-term 'mechanized shifting cultivation' (or analogous processes in different climatic areas).

xv). The more one knows about a concrete food situation, the less easy it is to generalize.[55] This points with still greater force to the need for careful local study before rushing into projects or policies to improve the food situation.

xvi). There are major gaps in the above, one being integration of the discussion of emerging class forces within Africa and the sorts of pressure exerted from outside. The following chapters look at external forces prior to considering the internal situation again in the final chapters.

NOTES

1. To paraphrase Price Gittinger et al. (1985: xi), the four major nutrient deficiencies in the developing world are protein-energy malnutrition (PEM) which, as the name implies, results from inadequate intake of food; iron-deficiency anaemia; vitamin A deficiency (which causes blindness); and iodine deficiency (which causes goitre and cretinism). While the first is a general problem arising wherever there is hunger, and can thus be considered as either malnutrition or undernutrition, the other three are specific micronutrient deficiencies often

relating to particular characteristics of soils, crops or diets. Unlike PEM, they are thus
sometimes susceptible to treatment through a 'technical fix', such as food fortification with
iron, vitamin A or iodine.

2. See Chambers, Longhurst and Pacey (1981), introduction and especially Figure 1. p. 7.
3. There may be links between one year's *harvest* and the next, but that is a different matter. An
obvious example is where famine forces people to eat their seed-corn. Bantje (1985) has
shown, for a part of Tanzania, how one year's coffee harvest affects the next year's maize
output, through its effect on the capacity to buy hybrid seeds and fertilizers. This is probably
also common. In addition, some tree crops, like coffee, are subject, under certain conditions,
to cyclical yield variations (the so-called 'overbearing and die-back' cycle).
4. Staggered planting, planting different sections of the crop at intervals, can have serious
negative effects, since it provides optimal growing conditions for certain insect pests. In
southern Mozambique, where an enormously variable climate forces peasants to adopt this
form of planting, this is said to be partly responsible for very heavy infestations of maize with
stalk-borer.
5. Watts (1983) and Richards (1986a) have provided detailed accounts of the whole range of
risk-avoidance strategies for areas in Northern Nigeria and Sierra Leone respectively.
Richards (1986b) summarizes, compares and contrasts the two studies.
6. This seems the case with Ethiopian Coptic Christian practices (see below), though whether
they have much redistributive impact I do not know. But in many places where the ideology
prescribes redistribution, it will also require attendance at ceremonies (funerals, weddings,
circumcision etc) which are time-consuming. The time conflict is often aggravated by new
agricultural practices, which impose new labour requirements during periods of ceremony.
7. Much of the labour was rather a matter of womanpower.
8. Richards (1986 a and b) and Watts (1983) are again illuminating on this issue, though they
take opposing viewpoints, Watts seeing evidence of major breakdown in coping mechanisms,
while Richards sees a system which is still working. While this clearly has much to do with
differences in the areas studied, as Richards (1986b) points out, it also clearly reflects the
different outlook of the two authors. In general, I am more in agreement with Watts, seeing
Richards' account as a fascinating description of a somewhat exceptional case. Another
point of some importance is the relationship between reciprocity in this sense and the
centrality of patronage and mutual back-scratching in both the process of state-class formation and its unproductive nature. On the latter, see Berry (1984).
9. Settlers were usually the direct beneficiaries, but the imposition of order weakened
or smashed the military power of pastoralist societies, allowing a process of encroachment
and settlement by cultivators on the margins of their land, which has accelerated since
independence.
10. See Bryceson (1978, 1982). For various reasons, the colonial government of Tanganyika saw
itself as being, and was in fact, more concerned with 'native' welfare than most. At the other
end of the scale, one could place the British in Zimbabwe and Malawi, or the Portuguese in
Mozambique.
11. Kenya's Swynnerton Plan was meant to achieve the latter and avoid fragmentation through
consolidation and registration of land rights. In fact (see below), it seems to have had varying
effects, the emergence of a landless class being slowed but not eliminated by continuing
fragmentation with inheritance.
12. It seems true, as Kitching (1980) points out for Kenya, that a considerable male labour
surplus was created by limitations on such activities as cattle-raiding (and defence from it),
and that prior male involvement in crop production had been limited to certain specific tasks,
notably clearing new land. But he does not use this as an argument against increases in
women's labour burden, specifically referring to its increase with changes in the pattern of
cultivation.
13. This, plus the fact that it can be stored in the ground for substantial periods, also accounts for
its adoption by many governments as a 'famine crop', compulsorily planted in case of failure
of other crops.
14. Hanne Thorup, personal communication.
15. In Kisii District, Kenya, where many poorer families run out of the staple maize long before

the next harvest, a major health preoccupation among middle-class men is gout, attributed to excessive consumption of roast meat (and no doubt of the beer which washes it down).

16. Regarding socialization, this emerges from (non-systematic) discussions in Kisii, where both sexes seem agreed that it is the function of women to cope, and to hold the family together (even if that means getting beaten for withholding savings from a husband's demands for drinking money). I would hesitate to generalize from a small and non-random sample from a small and in many respects non-typical society, but this seems consistent with what I have come across and read about elsewhere. See also note 19.

17. The reason that such societies are referred to as 'hunter-gatherer' rather than 'gatherer-hunter' is only partly because it rolls more easily off the tongue. Those who define the ideologies of such societies are usually male, and it is usually also males who have presented them to anthropologists. Since men did the hunting, it is hardly surprising that they would have tended to exaggerate its importance within the diet.

18. Sen's analysis of the Bengal Famine has given rise to a vitriolic debate in the columns of *Food Policy*. So far as I can judge, the points cited emerge unscathed.

19. This raises some interesting points not mentioned by Sen. Greenough (1982) refers to wide-spread reneging by patrons on obligations to clients, specifically mentioning women and children. This fits well enough with Sen's finding. But Ali (1984) cites a number of sources to the effect that both in this and in other famines, while more women were pushed into destitution than men, a higher proportion of them survived. The two can be reconciled by reference to the superior 'survival skills' of woman. Lest one be tempted to over-romanticize this, these would include gleaning and scavenging, together with a reported increase in adult and child prostitution in Calcutta in 1943. But also relevant seem to have been the refusal of mothers with children to give up, and the prominence of women's groups in providing such relief as there was.

20. Garcia (1984) finds the total deficit of the five worst-hit Sahelian countries in 1973 (the worst year) to have been no more than 0.8 million tons (which he compares to 400 million tons of grain fed to livestock in the industrialized countries in that year). Others, like Klaasse Bos (1981), find clear evidence of a *long-term* decline in per capita food production.

21. Also referred to as the 'tragedy of the commons'. That is, if grazing land is communal, while herds are private, no individual herder has the incentive to limit stocks, since if he does, he will simply lose out to others. This argument is generally true for present-day pastoral systems, but assumes the existence of an atomistic 'market' system without social controls. In historical fact, this has usually resulted from the breakdown of pre-existing systems of control under pressures from land-loss and commercialization.

22. One speaks of reneging on obligations, and this may give the impression that it is an easy thing to do. In cases like the rich Bengali elite cited by Greenough, this may (but may not) have been the case. In others it will clearly not be. Roger Leys (personal communication) cites the case of a local chief in mid-south Zimbabwe who was in a state of deep depression after the 1984 near-famine, over his inability to perform what he conceived to be his duty – to provide for his people. In this case, because of relatively effective famine relief, starvation was kept to a minimum, but loss of cattle and other assets was still very serious.

23. There are some who would pick on certain formulations by Sen to claim that he almost entirely dismisses the importance of harvest failure. I do not think this is the case. But more importantly, the purpose here is to see what one can make from the most sensible version of the model rather than to look for weaknesses to pick holes in.

24. When I read this it seemed so high as to be exaggerated. Since then I have met Eva Poluha who has done anthropological fieldwork in the past few years, in an area just south of the Blue Nile dividing line between famine and non-famine areas. She confirms 200 days of fasting per year (no meat, rather than no food however) and a huge number of other religious limitations on what may be done on what days, severely limiting labour input and aggravating labour peaks.

25. Gill cites two studies of drought in Ethiopia. Mesfin Wolde Mariam (1984) looks at the period 1955–77, while Shepherd (1975) looks at the 1973/4 famine. I have unfortunately seen neither, though they are included in the references.

26. Like Gill, I met a member of this mission, who stressed that the AMC was actually trying to

reduce stocks at the time for lack of storage space, and trying to sell them to the north, with limited success. Both he and Swedish colleagues who heard reports from a visitor in early 1984, stressed that no-one was predicting a famine. The latter seems now to have been clearly untrue in light of the RRC report of March, and some voluntary agency reports going back to 1982.

27. On the other hand, there may be some partisanship in these accounts, something less likely on the other side because of the unattractiveness of the regime to most observers.

28. My apologies for mislaying his name. But perhaps, since he still works in Addis Ababa, he will not regret this too much.

29. By 1987, Sudan had climbed to first place in the league, debt was put at $11 bn, debt-service at 105% of exports and debt arrears at $2.6 bn. (*Daily Nation*, 4 October 1987, *Sunday Standard*, 15 November 1987).

30. See Chapter 9 for further discussion.

31. So far as can be deduced from figures of acreage and production presented on different pages of an article in *The Courier* (No. 84, March–April 1984: 16, 17).

32. Mike Speirs, personal communication, after a visit to Burkina and discussions with involved persons in October 1986.

33. This was certainly the case in 1980/81, when I spent some 8 months in the country – this in spite of already serious shortages.

34. At least in the south and centre of the country, the MNR seems to concentrate solely on terror and destruction, acting primarily as an arm of the South African state. There seems never to have been any question, even from its right-wing backers, of doing anything to relieve famine in its areas of operation.

35. Ironically, part of the Angonia surplus may have been among the maize purchased in Malawi (the only country in the region with a surplus that year) and flown back to southern Tete as famine relief. Malawi's surplus is attributed by some to its 'right-price' marketing policies. The dismal poverty of most of the population and strict limits on urban migration are probably just as important.

36. For further discussion of Nigeria (especially the north) see Shenton (1986), Watts (1985), Andrae and Beckman (1986). Watts (1983) is particularly highly recommended by people with experience of Nigeria but, to my great regret, I have not read it.

37. Because they hampered efforts to control tick-borne diseases, which are a much more serious problem, except where exports to industrialized countries are concerned (Raikes, 1981).

38. Considered by a number of authors, notably Kitching (1980).

39. Seen during a consultancy visit of 1983. The extension officer with whom I was travelling was most surprised to see flood irrigation and assumed that it was a new phenomenon. Discussion with people working in the fields showed that it was not. It is perhaps worth explaining that most extension and other government officers in the area are non-Somalis.

40. Further down the Tana River is the enormously expensive Bura Irrigation Scheme which, despite its lack of production hitherto, already denies pastoralists access to a large riverine area.

41. One of the richest men in the district is reputed to have a 25,000 acre farm in one of the ex-settler areas (though I have not been able to check this), and many other investments. This farm alone would suffice to provide subsistence and a small income for, say, 5–10,000 of the poorest families in the District.

42. See Orvis, 1985a and b, 1986, 1987 for more detail and analysis.

43. Heald and Hay (1985) take issue with Buch-Hansen and Secher-Marcussen on the incomes from contracting, citing the invalid comparison of figures from non-comparable surveys and (for tea) failure to take account of non-contract cash crops. For sugar, where plots are very small and there is no room for cane and food crops, let alone other cash crops, this is not an issue, though declining cane prices have led to reduction in production and incomes since the study was made.

44. Michael Whyte, personal communication.

45. Finn Kjaerby, personal communication regarding areas in Hanang District.

46. A note of caution here. Gartrell includes raiding the cattle and fields of others as part of this survival strategy. While this may be true in the narrow sense, in others it seems a rather

extreme case of 'lifeboat ethics', except that it involves stealing someone else's lifeboat.
47. Bryceson (1982) discusses the evolution of food policy in colonial Tanganyika.
48. From my own experience of reading archival records, these can vary very widely indeed. In Mbulu District, the reporting of food shortages coincided precisely with the posting of one agricultural officer, and ceased immediately on his departure. The reverse was the case with cash-crop reporting (Raikes, 1975).
49. Perhaps even that far, given the variations in definition and the strictures by Svedberg cited at the beginning of this chapter.
50. It does not mean, as claimed by Maletnlema (1985), that one can expect a five-year cycle. Not, that is, unless one ignores weather entirely and bases the argument on a cycle in, say, price policy.
51. R. Gommes, personal communication. For further discussion see Bryceson (1982, 1985 and forthcoming Ph.D. Thesis) and Raikes (1985b, 1986a and b).
52. Here again political factors come in. Famine relief to areas supporting the opposition ZAPU, and where guerrillas/bandits were being put down militarily, was very significantly less adequate.
53. Further studies of food systems and food shortages can be found in issues of the *Review of African Political Economy*, Rotberg (1983), Pottier (1986), while others are referenced in Leftwhich and Harvie (1986).
54. On this issue see also Rotberg (1983), with interesting accounts from Kenya, Malawi, Zambia and Tanzania.
55. That is to say, consistently throughout this book, I have found it easier to summarize and encapsulate situations about which I know nothing except from the writings of others. This may, of course be because other authors are better at summarizing than I am. Even so, it points to the dangers of oversimplification involved.

PART TWO

International Dimensions

5
Trends in World Food Production & Trade

During the past century, the structure of world agriculture has changed out of all recognition. Since the Second World War, the pace of change has accelerated. With the coming of bio-technology, this acceleration seems likely to continue in the future. The purpose of the present chapter is to provide an overview of some of the more important trends and structural changes. Firstly, this helps to advance the explanation of processes occurring in Africa. Secondly, it allows some assessment of the possibilities for alternative policies. And thirdly, it provides a framework for the consideration of EEC policies which follows.

While the details of agricultural development during the present century (or even the past decade) form a mosaic of bewildering complexity, many of the basic underlying trends have been consistent over time, though, in combination, they have had highly contradictory, in some cases downright absurd, effects. Significant among these have been a continuous increase in production and productivity, a steady concentration in the control of resources (both within agriculture and from agriculture to agribusiness) and a continued release of labour. International trade in agricultural products and inputs has grown enormously, as has the international scope of agro-based investment. Concentration of resources has led to increasing disparities in production, productivity and incomes both within and between countries. This is reflected in a relative weakening of the situation of Third World countries and especially the poorer, Africa being prominent in this latter category. Furthermore, most of these processes have been developing exponentially.

Since 1945, these processes have been accelerated by 'policy', that is, protection of agriculture and efforts to increase its productivity.[1] The developed countries (notably the EEC and the USA) subsidize agricultural sectors containing a small minority of the population. This primarily benefits the richer farmers, accelerates technological development and generates surpluses of produce which require further subsidies to export. In most Third World countries, and especially the poorest, peasant agriculture, accounting for upwards of 70 per cent of the total population, is the major base for extraction of surplus (including tax revenue) for the remainder of the economy. As discussed in previous chapters, this reduces prices, incomes and deliveries of produce on to official markets. Small wonder then that agricultural production has grown more rapidly in the developed countries. The effects of internal policies have been reinforced by trends in and policies for international trade. Previous tropical export staples have been hit by synthetic and bio-technological substitutes, while developed country protection has hastened the process and eaten into Third World domestic markets.

While the main focus of this chapter is Western 'developed' agriculture, its purpose is also to consider some of the implications for the Third World. Moreover, the dividing line between 'developed' and 'underdeveloped' agriculture does not coincide precisely with the (ill-defined) boundary between developed and underdeveloped countries. Agriculture is more advanced in parts of the Punjab than in, say, southern Germany. Some peasants in highland Kenya produce more, on larger acreages and with more advanced techniques than many in the Abruzzi Mountains of Italy. Argentina is classed as underdeveloped, Portugal as developed, but the agriculture of the former is generally larger-scale and more technologically advanced than that of the latter.

The chapter starts by looking at long-term trends in production and noting some of the related structural changes. It proceeds by looking at international trade in agricultural products, starting with long-term trends and proceeding to the period since 1970, with a brief section on the different international context of the two recent 'famine periods' in Africa. This is followed by brief notes on some of the major food commodities, finishing up with a discussion of the implications and prospects for Africa.

Long-term trends

Change in the structure of Western agriculture has been intimately bound up with the development of capitalism, and increasingly dominated by the dynamic and development of industrial capitalism. At the risk of gross oversimplification, it may help to explain some of the processes which have occurred, to set out a three-stage model of the development of Western agriculture. It is not claimed that this corresponds even roughly to the actual development of agriculture in any one country or period (which indeed would be hard since the 'agricultural revolution' or revolutions of Europe alone have been spread over a period of some 200 years and have taken very different forms). The point is not so much to show the actual stages through which agriculture developed as to indicate some significant processes involved.[2]

FARM-LEVEL INTEGRATION

The first major acceleration of growth and productivity in European agriculture corresponds roughly to what is sometimes referred to as the 'agricultural revolution' and is related to the break-up of feudalism and the enclosure of common lands. In terms of farm organization, it can be seen as primarily a matter of the integration of crop and livestock production, together with new related technologies. This included the introduction of crop rotations including fodder-crops and increased storage of fodder, which increased crop yields both directly and by increasing livestock numbers and the availability of manure. It also allowed an increase in the horse population, stimulating advances in agricultural mechanization, since horses could pull greater weights faster than oxen. Another important factor, closely related to a shift from feudal to capitalist forms of land-tenure, was soil improvement through improved drainage. While this advance was related to increased production for commercial sale, it remained largely self-sufficient for inputs, while agricultural equipment was produced by local blacksmiths rather than capitalist firms. At this stage, the commercial farm could be considered an independent economic unit, albeit integrated into the market through the sale of its products.

This is the technical side of the process. But enclosure started the major exodus of population from agriculture, by impoverishing or directly expelling large numbers of peasants, while improved machinery and animal draught substituted for farm labour, accelerating the process. That is, the process related to the beginnings of the indus-trial revolution and urbanization both by expanding the food surplus and by accelerating the formation of a landless working class. But this occurred at a stage when landed interests were still dominant within the economy and society as a whole, and when international trade in agricultural products had hardly begun, being severely limited by protection. This maintained prices, profits and rent for the farmers involved and their landlords. It was these aspects among others of this stage of agri-cultural development which prompted Marx and other observers to see the future of agriculture under capitalism as the development of independent agricultural units similar to capitalist firms. Major features of this were seen to be the extinction of the peasantry and the establishment of farm units, working with wage-labour and follow-ing the investment criteria of capitalist firms. It was recognized that the actual pat-tern of nineteenth-century agricultural development was more varied. Small peasant production continued to grow and to be drawn into market production. Export-oriented agriculture, as in Poland's 'second serfdom', was sometimes based on con-tinued, if not sharpened, feudal servitude. But the end-point was expected to be farms not unlike capitalist farms.

SUBORDINATION TO INDUSTRIAL CAPITAL

A major structural shift occurred from about the mid-nineteenth century. The main factor precipitating this shift was the increasing political and economic dominance of industrial capital, with an interest in cheap food as a means of keeping wages low. Another was the development of cereal production in, and transport routes in and from, the 'new world'. The combination of these two factors led (at least in Britain) to the lowering of tariff barriers, much reduced domestic cereal prices and lower profits in farming.[3] While, according to both Ricardian and Marxian rent theory, this should have had no significant effect on the profitability of farming, the difference being reflected purely in rent, and through the taking of land out of production until a new equilibrium was established, things do not seem to have happened this way – perhaps because the theory is based on too strict a distinction between non-productive land-owners and productive farmers. At any rate, this structural change of the mid-late nineteenth century seems in fact to have had two other effects. In the first place, it increased the impoverishment of small peasants, swelling the flow to the towns and into emigration to a flood. Secondly, it limited, if it did not halt, the movement of venture capital into farming and replaced it with the continuation of family farming. (In the strict sense of the term, one can only speak of capitalist farming as such when capitalist operators move into or stay in farming because the rate of profit offered there is (at least) equal to that in other sectors of the economy.) Transport costs seem less significant – or significant in another way – for North America, where the aboli-tion of British protection expanded markets and raised prices, while in continental Europe protective barriers were lowered much more slowly, if at all. But it does seem that the underlying factor – the dominance of industrial capital – operated similarly in all cases, to depress agricultural prices, especially for grains, halt the move of capital into agricultural production and generate sectoral shifts within agriculture.

This does not imply the stabilization of small-peasant farming. Indeed, for the most part it accelerated the exodus from farming of small peasants, through impoverishment and indebtedness. Moreover, many of the remaining family farms were larger in size, employed full-time labour and earned substantial incomes. The essential distinction is that pointed out by Chayanov (1965), that while capitalist farming is concerned with the rate of profit once all costs including labour have been deducted, the family farmer is concerned to maximize the combined income from labour and capital, which implies that the family will remain in farming so long as this provides a living. Another way to put it is that capitalists get out of farming into another branch when the profit rate falls. Family farmers are generally pushed out of farming when incomes fall to sub-subsistence levels or (more commonly) when indebtedness and forceclosure force them to leave.[4] This in itself contributes to the continuation of a situation in which rates of return to farming stay below those in other branches of the economy.

While capital moved out of agriculture as such, this was accompanied by the beginnings and expansion of a move into agribusiness – though a minuscule agribusiness by the standards of today. By the end of the century, the first agricultural tractors were being produced, and the production of equipment by blacksmiths had largely been replaced by industrial manufacturers. Organic fertilizers were beginning to be imported and chemical fertilizers had been developed, though their production and use were as yet minute. The development of dairy and other intensive livestock production, which accompanied the growth of towns and middle classes, was beginning to stimulate growth in the use of feed-grains, though by far the major proportion of these were still produced on-farm.[5] Apart from this, research institutes and agricultural colleges were beginning to spring up – especially in the USA, but also in Europe – together with state or co-operative banks for lending to agriculture. This set the scene for the beginnings of the third stage.

INTEGRATION AND MODERNIZATION

As with the other two stages, it is hard to date the beginnings of the third stage in agricultural development – the accelerating application of non-agricultural inputs and scientific techniques to agriculture, under the dual encouragement of industrial capital and the state. Some of the beginnings date back to the early years of the present century, if not before. Others can be traced to the 1920s, 1930s or the post-1945 period. As always, different aspects occurred in different countries and sub-sectors at widely different times. But one can pick out a number of relevant factors.

By the present century, and most especially after the Russian Revolution and the First World War, governments began to think more seriously than hitherto about the stabilization of peasant farming. Peasant farmers in situ tend to be among the more politically conservative social strata; dispossessed peasants can only too easily join the ranks of revolutionary parties. Even without the threat of revolution, peasants have often formed the basis of right-wing political parties, their value in this respect being often increased by the rural bias of electoral districts.[6].

Another factor was rising incomes and Engel's Law in both its economic and political manifestations. The first is that rising urban incomes reduce the proportion of income spent on food, while leading to a shift towards more expensive foods, notably fresh vegetables and livestock products, but also towards a greater degree of processing and packaging. The second is that, as incomes rise, urban populations become less politically concerned about the price of food, enhanced by the fact that, with the rise of the food industry, ex-farm prices of basic foodstuffs form a decreasing proportion of the final consumer price. The combined effect is to make protection and the subsidization of farm prices more politically acceptable, even while voting strength continues to move away from the agricultural sector, as its numbers decline.

Arguably more important than either of these was the emerging strength and integration of agribusiness and agri-state institutions, firms, banks, research institutes, marketing boards and the like, all largely motivated by the same vision of how agriculture should be developed, through increased investment in productivity and through state subventions to ensure that this should be profitable, at least for whatever size and scale of farming was considered 'economic'.[7]

This development resulted in the initiation of policies which were to lead to enormous increases in productivity and to a ratchet or treadmill process whereby this always destroyed the very increases in income which it promised to bring. The first aspect is the definition of the solution to the 'peasant' or small-farmer problem as increased productivity. At almost any historical period, this results in the definition of the smallest farms as unviable and concentrates on the next and higher levels. The next step is intervention through extension agencies (and salesmen of supplying firms) to increase productivity through increased use of purchased inputs and equipment, financed in many cases by state or co-operative banks set up for the purpose. But the bottom line is the fact that the demand for agricultural products is inelastic with respect to both income and prices. As incomes rise, a smaller proportion of them is spent on food (overwhelmingly the most important output of temperate agriculture), while price inelasticity means that if supply increases more than demand this will reduce prices to an extent that incomes actually fall.

The farmer is therefore caught in a trap. Since others are increasing productivity, he or she has to do so in order to keep up and survive. But this increases production, while labour-saving equipment substitutes for labour. This combination lowers farm incomes and pushes more people out of agriculture and, especially if the equipment involved is liable to economies of scale, starts the process all over again with a higher level of 'minimum viable farm'. This may be slowed by state subsidization, though, as shown below and in Chapter 7, subsidies on product prices provide minimal support to small and vulnerable farmers. Indeed, when all is said and done, this is not their purpose. That purpose, to increase the use of purchased inputs and equipment and to hasten the adoption of the most modern methods, is very well served by price subsidy – even if its effect is to introduce strong expansionary pressures which lead to production of surpluses – and further subsidies to get rid of them. This latter aspect of the current trend in agricultural development has been especially marked since 1945, with the USA leading the way in further acceleration of production growth and 'aggressive protection' to off-load the surpluses, but with the EEC most definitely following.

The majority of farms in the developed countries are still family-operated, though they do not account for the major proportion of output. But it is a fully commercialized family farming and one which is increasingly controlled from the outside through supply of inputs, equipment and credit, and through control over markets for produce. The dynamics of change come largely from the outside. Farming in the developed countries has been in a continuous process of technological change for many decades, and apparently at an accelerating rate. It is no longer accurate to speak of farms as economic units, they are merely a part (and one of declining importance) in a large food and agricultural system controlled and dominated from the outside.

For decades now, the impetus for change has come largely from corporations which develop and supply inputs, equipment, and increasingly whole systems, with products and services, backed up by extension, research and credit. The same is true on the other (down) side of farming. Production is increasingly bound up in one or other way by contractual specifications, and these change in line with the product-development, packaging, storage and appearance requirements of the food industry. Contract-farming is continually on the increase, and its specification of the production process is increasingly precise. It concerns not just what products and what grades should be produced, but the timing, size and sequence of deliveries and even the brand of inputs to be used. This again has profound effects on the production process. In order to meet the stringent specifications of purchasers, producers of, say, pigmeat must feed carefully designed rations, to avoid excessive fat, and must comply with strict hygiene regulations, affecting how the animals are kept and slaughtered. Pressure on product prices and rising land prices force farmers to increase scale and intensity continuously.

US agriculture

To give some idea of the extent of this development, by 1980 some two-thirds of the total final value of agricultural production in the USA was realized 'downstream' of agriculture itself, in processing, packaging, transport, storage and distribution. Of the third remaining to agriculture, 80 per cent went out again in expenditures on inputs, equipment, services and debt-service. This had risen from 50 per cent in 1970, and the proportion continues to rise. Within the agribusiness firms both up- and downstream, the degree of concentration is striking. Seeds, fertilizers, agro-chemicals, animal feeding-stuffs and agricultural machinery are all dominated by huge multinational corporations, the oil majors and major pharmaceutical corporations being prominent. The same is true downstream, where every major branch has four or less firms accounting for 55–90 per cent of the market. Even this fails to describe the full degree of concentration since almost the total production of items like broilers, fluid milk, sugar, citrus fruits and vegetables for processing are integrated within or produced under contract to processing firms. Producers of cereals, tobacco, oilseeds and cotton face 'free markets' dominated by huge corporations like Cargill and Ralston Purina. The modern farm is thus less an economic unit than a part of a larger unit dominated by agribusiness (George, 1979; Cathie, 1985).

Similar processes have occurred within farming. With an enormous substitution of capital for labour, the rural population of the USA had fallen by 1985 to only 40 per cent of its 1945 level. Agriculture is one of the most capital-intensive industries in the USA with more than twice the machinery and equipment per person employed in manufacturing. This has led to farm debt of over $200 billion, making US agriculture extremely vulnerable to changes in the interest rate (Thompson, 1986). Concentration within farming has been as rapid. In the late 1970s, some 10 per cent of farms accounted for two-thirds of production. Since then the process has accelerated enormously and by the early 1980s 'superfarms' with income over $0.5 mn, comprising no more than 1 per cent of the total, accounted for two-thirds of net farm income. Of the remaining farms, 74 per cent were operated by 'part-time farmers' with another source of income outside agriculture. The other 26 per cent were generally the poorest and most vulnerable to bankruptcy. (George, 1979; Cathie, 1985).

That this vulnerability is not just theoretical has been shown by a major surge of farm bankruptcies in the mid-1980s, hastening migration out of farming and tearing rural communities apart. And this has happened during a period when the USA was spending $25–30 billion per annum (more than the EEC) to protect its agriculture (but evidently not most of its farmers). Much of this has been spent on an attempt to secure or capture markets against the impact of Reaganite defence and macroeconomic policies. Defence expenditures have been responsible for much of the budgetary deficit which has kept interest rates and the dollar high, both highly damaging to a heavily indebted and export-oriented agriculture (Thompson, 1986). Though US government sources blame the problems of US agriculture on EEC protection, Paarlberg (1986) gives this only one-fifth the impact of exchange-rate factors. There is a certain irony here, in that the USA, always quick to point at indebtedness and over-valued exchange rates in relation to Third World economic problems is, as the world's major debtor nation, itself the prime example of this problem. A further irony is that US defence policies should be the major cause of this – a classic case of shooting one-self in the foot.[8]

Similar factors are operating in the EEC, though, as described in Chapter 7, they have not proceeded so far. All the same, well over half of total value accrues down-stream, while inputs, equipment and debt service account for 70–80 per cent of gross farm income. Here again large corporations are dominant at all levels and the impor-tance of contracting has increased. In the mid-1970s, it was estimated that 25 per cent of farms received 75 per cent of net farm income. With concentration proceeding here also, it is likely that the top 10 per cent now account for well over 50 per cent of income. Again, the farm population is less than half what it was in 1945.

Some related structural changes

The development of agriculture since the turn of the century has been marked by an exponential increase in the use of purchased inputs, the replacement of human and animal energy by machinery and fossil fuels, the increasing predominance of research, and a marked increase in specialization and precision. The major physical inputs have been farm machinery, fuel, fertilizers, pesticides and other agro-chemicals, seeds and animal feeds.

ENERGY

One of the most significant developments has been a drastically increased energy input into agriculture. Most forms of modern agriculture expend considerably more (fossil) energy into agricultural production than is produced in output – where the 'least developed' forms of agriculture extract much more energy than they put in (in the form of human labour). Tarrant (1980; 3) asserts that there is nothing necessarily wrong with a negative energy balance in farming. Fossil fuels are used for other non-regenerative processes and might as well be used for producing food. This would be a more convincing argument if a considerable proportion of the total product was not of surpluses which have to be dumped at below production costs. It should not, in any case, be taken to imply that developing countries should aim at fertilizer and mechanization levels equivalent to those in the USA and Europe. There is wide dis-agreement about how long this would take to denude the earth of its fossil fuels, but it is under most calculations much too short for comfort.

Mechanization is the most obvious but by no means the only form this takes. The internal combustion tractor was first introduced at about the turn of the century and by the 1930s was already significantly replacing animal traction. Developments since

1945 have taken this much further, while introducing new and specialized forms of machinery. *Grain-drying* uses about twice as much energy in maize farming in the USA as do all field operations combined (Doering, 1982: 12). *Supplementary irrigation* has developed very rapidly in northern European farming since the mid-1970s, and as a quite direct result of EEC price support. But the major area of innovation in new machinery has been in *livestock production*. Today's factory livestock farms are highly mechanized with automatic feeding, watering, milking, waste disposal and control of lighting (to lengthen the 'day' period and speed up eating/laying etc). In the most 'advanced' cases, livestock production is entirely transformed into assembly-line production, in which the stock never see daylight or leave their cramped quarters over lifespans reduced to a few months or even weeks.

Artificial fertilizers were first discovered in the later part of the nineteenth century, though before that organic fertilizers, like Chilean nitrates (accumulated bird droppings), had been used to supplement manure. At first their use was limited by high price and the abundance and cheapness of animal manure. But while nitrogen fertilizer prices have fallen to about one quarter of the 1900 level (in constant prices), the costs of spreading manure have increased along with labour and fuel costs. Growth in the use of artificial fertilizer has accelerated rapidly since 1945. For West Germany, Weber and Gebauer (1986: Table 4) show that it took 80 years for nitrogen fertilizer use to climb from 0.7 to 43.7 kg/ha, but only 20 years before this had tripled to 126.6 kg/ha. Similar growth patterns show for other countries. Manure still accounted for about one third of available nitrogen in Germany in 1980, but this probably overstates its use. In most of northern Europe and the USA, manure is too expensive to apply and constitutes a problem of waste disposal, adding to the more serious problem of fertilizer nitrates percolating into groundwater. Fertilizer also constitutes a very significant use of energy, since nitrogen fertilizer manufacture is largely based on the extraction of atmospheric nitrogen by application of heat, normally in relation to oil-refining. Nitrogen fertilizer accounts for about half the nutrients applied on a world-wide basis but some 85 per cent of the energy used in manufacture. Doering (1982: 12) shows that fertilizer accounted for four times the energy of combined field operations in US maize farming, though for all crops and all operations in developed country farming, fertilizer accounts for about one-third of energy used, compared with two-thirds for operation and manufacture of machinery (FAOd, 1976: 98). Fertilizer use not only increases yields directly but also allows higher plant densities and the use of land which might otherwise not be profitable to cultivate.

BIOLOGICAL AND CHEMICAL INPUTS

Another major factor has been *plant breeding*, the classic case being the rapid spread of hybrid maize in and from the USA, which has also had a very significant effect on fertilizer use, since high-yielding maize and other crops use much higher levels of fertilizer than the varieties they replace. While increased yields have been the major aim of most breeding programmes, others are also important. Hybrid maize not only yields more than open-pollinated types, but is far more homogeneous, to the extent of having all the cobs at the same height. This reduces the cost of combined harvesting, by allowing more precise setting of the harvester. For horticultural crops, breeding has been much concerned with the development of homogeneous varieties which keep well and can withstand transport, while being visually attractive. Unfortunately for consumers, this tends to mean a thick bright-coloured skin and reduced flavour. The whole area of breeding is currently undergoing a major revolution with new *bio-technological techniques,* capable of producing even greater yield increases and much higher degrees of homogeneity. It is reported that they will be able to produce new crops imitating the taste of others, thus competing tropical beverages, for example, out of their markets – though how realistic this is, and over what timespan it can be expected, is not clear. Mooney (1979, 1983) has shown some of the disturbing aspects of current trends in plant breeding, including the heavy domination by a few multinational corporations (often oil majors) and their interests in breeding for high fertilizer and chemical use, since they also produce these. Another serious problem has been the narrowing of the genetic base for many important commercial crops, with consequent increased risk from pest or disease attack (see also Ruivenkamp, 1986, Hobbelink, 1987).

The development and expanded use of *agro-chemicals* (pesticides, fungicides and herbicides) has been another major feature of agricultural growth and productivity, especially since 1945 (some having been developed as by-products of chemical warfare research during the War). This is closely related to developments in seeds, fertilizer use and mechanization. More homogeneous crops increase vulnerability to disease or insects and their increased yields imply greater potential losses to protect against. Machinery lowers costs of application, as well as dangers to the operative. With rising labour costs, it is often cheaper to spray herbicide than to weed crops, especially since use of fertilizers allows higher plant densities, leaving less space for machinery to pass. Again there are a number of problems, since many agro-chemicals are highly poisonous. One series of problems relates to operator safety. Another to the dangers of residues finding their way into the food chain or the groundwater. Another is that, in many cases, increasing doses are required with time to achieve the same effects, thus increasing the costs and risks of toxicity. For these reasons, a number of the more dangerous chemicals have been banned for use in the USA and Europe, while there has been a clear reduction in the amounts of pesticides used in the USA, as a result of the development of 'integrated pest management'. This involves combining a number of means of combating diseases and pests with the minimum possible chemical application.[9] Little of this seems to have penetrated to the formulators of agricultural modernization projects in Africa, where, in addition, most of the banned chemicals are still available.

SPECIALIZATION AND SEGREGATION

Another series of changes relate to a process, the reverse of what occurred in the 'first agricultural revolution'. Where that integrated the farm internally, and notably crops and livestock, the interpenetration of purchased inputs and increasing specialization undoes much of this. With manure replaced by purchased fertilizers and animal draught by machinery, the contribution of livestock to crop farming is much reduced, while the increasing dominance of purchased animal feeding-stuffs means the same in the other direction. Since the capital costs of farming have increased enormously, and most especially for animal production, the tendency has been towards more specialized forms of farming, depending upon either crops or livestock and reducing the number of products sold.[10] While economically efficient in the sense of minimizing costs, this wastes resources because it costs too much in labour or machinery time to use them. Manure is too expensive to spread and piles up polluting groundwater. Straw and other crop-residues are burned rather than ploughed under, to save time. Fertilizer is applied as a blanket dressing, allowing nitrates to percolate into groundwater. Rotations give way to monocropping, since there is no use for the products when livestock are fed on purchased rations. Such practices involve waste of resources and the danger of food and water pollution in Europe and the USA. They are even less suitable to the fragile environment and low-income populations of tropical Africa.

PRECISION AND SYSTEM-DEPENDENCE

The increasing use of machinery, fertilizer, bought seeds and chemicals requires an increasing precision in their timing, supply and use. Industrial processing requires standardized products and more precise timing of deliveries. Agricultural production is increasingly dependent upon complex systems for input supplies and increasingly called upon to adjust to the requirements of agro-industrial processors. This gives an advantage to large farms in the more developed areas. The companies which determine the shape of agricultural production are interested in large compact markets and in bulk supplies, to reduce costs of collection and delivery. They provide better and cheaper services for large farms and may even refuse to deliver supplies to, or collect produce from, small farmers. But if larger farmers have an advantage over smaller in the developed countries, all are more favourably placed than most Third World farmers, while the disparity between large and small is even greater there. Modern farming systems rely on precision, not only from farmers and firms, but from the general infrastructure of the country. If the electricity service or telephone system is unreliable, if spare parts are unobtainable, or if the roads are impassable at certain times of year, the systems which increase productivity in the developed countries may reduce it in the Third World. This applies with particular force to peasant production, where services are worst and costs of transport highest.

MARKET SEGREGATION

As incomes rise, people spend a decreasing proportion of total income on food, but become more selective about the foodstuffs they buy. So price differentials between preferred and non-preferred items widen. Feeding grains and oilseed cake to livestock only becomes profitable when the disparity between prices of meat and dairy products and those of grains exceeds the conversion ratio of grain to meat or milk.[11] By 1981, some 600 million tons of cereals, the equivalent of more than 36 per cent of total world grain production, was used for animal feed throughout the world.[12]

This development is related to an emerging set of international 'standard diets' for the middle-classes of developed and some developing countries. Meat consumption is increasing and to an increasing extent the meat is produced with purchased feed based on cereals or cereal substitutes and soya bean meal. An increasing proportion of the world's middle classes consume wheat products, dairy products, fresh fruits and vegetables of a number of standard varieties, further homogenized by the use of chemical additives. With advances in storage techniques, this has meant that availability of products is spread more evenly over the year. But it also increases the amount of basic foodstuffs directed to improving the diets of the rich. Its effects on the millions of people whose diet is primarily composed of 'feed-grains' (maize, sorghum, millets) or 'cereal substitutes' (cassava, other roots) are by no means obviously positive.

Western agriculture has achieved prodigious feats of increased production, at considerable cost. Agriculture has become a highly capital-intensive and energy-using industry and one whose use of fertilizers and poisonous pesticides increasingly affects groundwater and foodstuffs. At the same time, agriculture as such has diminished enormously in relative importance, both in terms of employment and of its contribution to the final value of output. Hedged about on all sides by agribusiness, threatened by market price trends and helped backed-handedly by state protection, the Western farmer, and especially the smaller one, is on a treadmill which steadily hoists productivity and surplus production, leaving him no better off. While this expansive tendency of capital increases state spending on Western agriculture without benefiting most farmers, its effect on the Third World and Africa seems rather more negative.

International trade in agricultural products

Prior to the industrial revolution, with the exception of a few areas like the Netherlands and some Italian city states, Europe (and the rest of the world) was overwhelmingly self-sufficient in basic foodstuffs. International trade in grains occurred only in the event of harvest failure and never exceeded 1 per cent of total consumption. International trade in foodstuffs consisted very largely of the import, at high prices and in small quantities, of tropical luxury products like coffee, cocoa and spices. The first impact of industrialization was to increase imports of industrial raw materials like cotton, jute and oilseeds. Imports of foodstuffs grew later. Even by the end of the nineteenth century, only Britain was heavily (80 per cent) dependent upon imported cereals. France was self-sufficient throughout the century and Germany produced 90 per cent of its requirements. Even small and heavily populated Belgium was self-sufficient until 1870 (Bairoch, 1973).

With European colonial control and the rising power of the USA, a pattern of trade emerged in the second half of the nineteenth century, which did not change markedly until the mid-twentieth century. Europe was by far the major focus of agricultural imports – foodstuffs, tropical beverages and industrial materials – though the USA was also a major importer of the latter two. Of the foodstuffs, cereals came primarily from the USA and other 'new world' countries, while most of the remainder came from tropical countries, usually colonial possessions. Europe continued to protect its own agricultural sectors from temperate product imports, but had no need to protect against tropical products, whose cheapness assisted industrial competitiveness. Most colonial powers took steps to prevent the development of processing in the colonies, either directly by forbidding it, or with tariffs. Several European countries protected domestic beet sugar production, though also importing some cane sugar.

The protection of domestic agricultural sectors increased during the depression of the 1930s, but remained essentially defensive – aimed at maintaining local markets, not penetrating foreign ones. In spite of protection, Western Europe was *the* major grain-importing region, virtually all other major regions of the world being net exporters. As a result of the depression and the dustbowl, US dominance of cereal exports declined and during the 1930s developing countries (mainly Argentina) were net exporters (Tarrant, 1980:222).

Since 1945, some remarkable changes have occurred in this pattern, as a result firstly of much accelerated technical advance and increased productivity, and secondly (and closely related) of much more aggressive protective policies, aimed at capturing foreign markets. The War itself did much to speed the change. The unusable surpluses of the 1930s were replaced by shortages, amounting in some cases almost to famine.[13] European wartime policies for agriculture stressed self-sufficiency in crop production and this continued to be important as post-war high prices were succeeded by those of the Korean War boom, and given the level of indebtedness accumulated during the War.

The War and lend-lease had done much to restore US agriculture to health, and the Marshall Plan continued this.[14] Cathie (1985:17) shows how US government policy towards agriculture changed after 1940 from a 'traditional protectionist mould, designed to protect a collapsing domestic sector' to the 'post-war "new" protectionism (which) promoted the growth of export markets'. When the Marshall Plan ended, there was a major gap between increased output and export market availability, soon to be filled by Public Law 480 and food aid. This maintained sales and growth of markets, until the take-off in US commercial agricultural exports from about 1972 (see Chapter 8). Somewhat after the USA, and with the formation of the EEC, a similar change occurred in European agricultural protection. While ostensibly aimed at preserving the internal market, protection has increased production so much that the EEC now has surpluses of most important agricultural commodities and export subsidies to off-load them.

Still more radical changes have occurred on the other side of international trade – exports from the Third World. The initial impact of industrialization, and later rising incomes in the West, was an enormous increase in trade. Coffee, tea and cocoa changed from being luxury products consumed by a small minority, to mass consumption goods for an increasing urban population. This, with rising incomes and growing industrial food processing, led to expansion in the world market for sugar. Trade in cotton was stimulated by increased incomes and demand for clothing, along with the expansion of the textile industry. Demand for other fibre crops, like sisal and jute, increased as the growth in international sea trade expanded requirements for sacks and rope, together with other industrial uses. Demand for vegetable oils expanded rapidly, both for edible purposes (margarine etc.) and for a variety of non-food (soap) and industrial uses (i.e. in lubricants and paints). Demand for rubber expanded rapidly, along with the motor industry.

But this very expansion set in motion the search for cheaper alternatives, which became more urgent in times of war, when sea lanes were vulnerable to enemy action. While the First World War generated some subsitution, that of the Second was far more important. Not only were there major technological advances, but in the 1950s fuel prices fell. Synthetic petrochemical-based fibres started to replace natural fibres and have continued to do so until now. Not only are synthetic fibres cheaper, but their much higher breaking strength allows use of larger-scale and faster-operating machinery, reducing labour and other costs (Clairemonte and Cavanagh, 1981). Bulk transport reduces the demand for sacks. Mineral oils have driven vegetable oils from many of their industrial uses, while petrochemical-based detergents have replaced soaps for many purposes. Synthetic rubber now accounts for some two-thirds of world utilization. This round of substitution primarily affected industrial raw materials.

In the first instance, the search for substitutes had been motivated by demand outstripping the growth of supplies, but the development of synthetics rapidly began to compete and to set upper limits on prices. Prices of Third World exports have generally declined in real terms, since the Korean War high-point. So rapid has been the cheapening of synthetic production that even the oil-price rise of 1973 hardly altered the picture. The slow-down in world economic activity and sharp rise in Western protection since the late 1970s have still further worsened the situation, with real prices of many tropical export products falling drastically. Coffee, tea and cocoa are semi-luxury products, demand for which is severely affected by falling real income levels. Western protection and the development of temperate agricultural production have hit tropical oilseeds in two separate ways. EEC butter surpluses compete and reduce prices, while the enormous expansion of soya bean production for animal feed produces huge quantities of soya bean oil (as a 'by-product') which has the same effect. More recent developments in bio-technology threaten even more devastating blows to many of the remaining products. Sugar is the major example to date, threatened both by EEC protection of beet sugar and the development of high fructose corn syrup (HFCS) in the USA. Palm oil is another example, since new varieties give much higher yields but are suited to production on plantation scale. The rapid growth of such plantations in Malaysia and Indonesia has already pushed marginalised African peasant producers to some extent and this process threatens to continue. It is claimed, however, that future developments threaten competition with coffee, tea and cocoa, though during what time period is not clear. One of the very few areas of growth in tropical primary exports – off-season fresh fruits and vegetables – seems threatened by similar processes – not to mention protection by EEC producers and preference for politically more important Mediterranean producers (notably Israel).

While Third World exports have been slowed by such factors, rapid urbanization and the expansion of urban middle classes has expanded the demand for food and especially 'temperate' foodstuffs. Thus one major factor, marking off the post-war period, has been that much of the new growth in markets for temperate agricultural products has come from the Third World. Twenty years ago, it was commonly accepted that poor countries 'specialized' in agricultural and food production, and the rich in industry. Some referred to this as the result of comparative advantage and resource endowment, others as exploitation and underdevelopment, but there was agreement on the 'fact'.[15] It was not entirely true even then, but in the intervening period the most significant growth in agricultural (especially food) production has occurred in the developed countries and of imports in the Third World. In 1982 and 1983, the Third World as a whole actually ran a deficit with the developed countries in agricultural trade.

Another important trend of the past thirty years has been the emergence of food as a major factor in international political bargaining and control. Only countries which are poor and powerless can be directly 'held to ransom' by withholding food supplies. The US embargo on grain sales to the USSR after the invasion of Afghanistan was a predictable (and predicted) failure, vulnerable to the US farm lobby and serving primarily to boost export earnings for other grain exporters and grain companies. But the strength of the USA in world affairs is significantly enhanced by the fact that it is the world's major agricultural exporter and that of the USSR correspondingly diminished by the fact that it is the world's largest grain importer. There are, moreover, many instances where 'food power' has been used against smaller adversaries or to control allies. The US withholding of food and credit helped to generate the coup which ousted the Allende regime in Chile. Egypt's enormous food aid receipts mean that its political leaders are virtually bound to support the USA and Israel, regardless of what the population thinks. This has been particularly characteristic of US food policy. With all its faults, EEC food aid has not generally been used or withheld so openly for political ends (see Chapter 8).

The other side of the coin is the increased use of political power to secure commercial advantage in agricultural trade. Chapter 8 provides examples of the uses of food aid for this purpose. Morgan (1979) provides many others. Recent years have seen increases both in non-tariff protection against imports and in aggressive policies to promote agricultural exports. The USA is the primary exponent of this type of policy, with the EEC (or its markets) the prime target and EEC protection the excuse most often used. For example, in early 1983, after some months of sabre-rattling and complaints about CAP export refunds, the USA sold one million tons of wheat flour to Egypt (a traditional French market) at below the price of unmilled wheat, and with the condition that any further wheat orders within the next 14 months should go to the USA (Petit, 1985:Ch. 6).[16] The issue flared up again in 1985, with the adoption by the Reagan Administration of a policy with the Ramboesque name of BICEP (Bonus Incentive Commodity Export Programme). This involved offering US exporters \$2 bn worth of agricultural stocks for sale at subsidized prices, starting with 1 mn tons of wheat to Algeria, where US exports have been pushed out by those of the EEC in recent years.[17] This trade war appears to have been concentrated upon the Mediterranean and Middle East, where imports of agricultural products have expanded most rapidly, and where the interplay of political and economic factors is especially close.

None of this is to deny the force or impact of EEC protection, which is discussed in detail below. The difference is as much of style as substance. On the other hand, the style does indicate aspects of the substance. In this particular trade war, the USA threatens, bullies and flexes its BICEPs. The EEC, as the admitted junior partner in the Western alliance, maintains a lower, more bureaucratic profile and argues about the unavoidability of its actions.

Trends since 1970

It was already apparent by the end of the 1960s that rates of industrial growth and profit were beginning to slow down in the developed countries. But both the tendency and the awareness of it were accelerated by the OPEC-inspired oil price rise of 1973.

This led to a rapid and massive accumulation of 'petrodollars' in the OPEC countries, which were then invested with Western banks, which in turn needed outlets for them. This, combined with the downturn in profitability in Western countries, led to a burst of investment in and lending to the Third World, much of it concentrated on 'middle-income' and 'new industrializing' countries (MICs and NICs) and on the development of 'export-oriented' industry. Froebel, Heinrichs and Kreye (1977) see this as part of a secular trend, resulting from a number of linked factors. On the one hand, low unemployment and strong union organization in the 'North' were pushing up labour costs and cutting into profits. On the other, improvements in transport and communications, and developments in the control of machine and labour processes, provided scope for making use of the cheap and disciplined labour forces of certain Third World countries. This was seen as resulting in a secular shift in the international pattern of investment or 'division of labour' in favour of Third World industrialization. It was also proclaimed by some as pronouncing the death of dependency theory and the notion that centre capital 'blocks' industrial investment in the Third World. Certainly it demonstrated one, fairly obvious, thing clearly – that international capital would not and does not 'block' Third World industrialization simply for its own sake. If there are advantages to be gained from investing in industrialization in the Third World, then they will be taken. But whether some of the other implications claimed are realistic is much more doubtful. This process has always been confined to a relatively small number of countries, and in some cases small parts of countries – tax-free export zones. It is also true that, in even fewer cases, it has led to real changes in economic structure and placing in the international hierarchy, even to some shifts in the world economic balance of power. But while the dependency formulation is shown to have been too rigid, the real difference has not been great, where the poorer Third World countries are concerned. That the 'four ASEAN tigers' (Korea, Hong Kong, Taiwan and Singapore), with parts of Mexico, Brazil and a number of other 'off-shore' havens and 'free-zones', have expanded industrial production means little to tropical

Africa, and certainly does not increase its chances of doing the same.

More recent events and analyses make such a development seem even less likely. Junne (1986) and others have shown that most of the conditions leading to the burst of investment in the 1970s have largely disappeared or been reversed. Further developments in the labour process and automation have reduced the importance of labour costs, while high interest rates have made reduction of inventory costs paramount. This tends to favour the siting of production capacity near to markets to increase flexibility. Most of the other Third World advantages have been affected by indebtedness, by limited availability of capital, by reduction in the subsidies and tax preferences which were part of the initial attraction, by increased political instability and by increased 'Northern' protection. At the same time, product life-cycles get progressively shorter as technological advance hastens obsolescence and rapid changes in fashion attempt to keep spending up. This again favours flexibility and production of smaller batches, which has been made cheaper by technological advance and gives a further advantage to production in the 'North' where the markets are. While Third World industrial exports continued to increase as a proportion of the total up to 1983, the prospects for any further advance seem much more limited. There is even less chance of export industrialization in low-income countries like those of tropical Africa.[18]

The impact of the period of easy money in accelerating the processes leading to crisis in Africa, has already been discussed in Chapter 3. Its impact on food imports has been mentioned and will be considered again below. In terms of impact on world trade in agricultural products, the impact of first easy money in the 1970s and then tight money in the 1980s has been to accelerate the underlying trends. Increasing oil prices hastened the search for innovations in saving fuel and materials, while the shift to industrialization in the NICs contributed to continuing low growth in production, incomes and imports in the West and consequent increased protection. High interest rates hastened the downturn in incomes and employment, reducing imports of Third World consumer goods and raw materials, leading, in turn, to severe deterioration in the terms of trade facing Third World countries, including those of Africa. At the same time, US policies led to a massive rise in cereal prices in the early 1970s, followed by increased production. This, together with steadily growing EEC cereal surpluses resulting from subsidies, led to world surpluses so massive that increasing food imports from Africa and other parts of the Third World were insignificant in comparison - generating a steady decline in world cereal prices.

Twentieth-century food crises

According to Johnson (1975), there have been four periods of general world food crisis in the present century (five if one includes that of 1973–5), during each of which prominent observers gave voice to Malthusian prophecies about the incapacity of world cereal supply to keep up with population. The first was at the turn of the century and related to the expectation that North American supplies would be increasingly used for domestic purposes, leaving the UK short of food. Again in the 1920s, it was thought that US grain supply could not keep up with growing demand. The next came in the late 1940s and is related by Johnson to population increases in the USA, though post-war and Korean War shortages would also have been important. The next came in the mid-1960s, relevant factors being rising world demand, the entry of the USSR and China as major purchasers, and poor harvests for two years running in India, at that time a major net importer of cereals. Finally, the shortage of 1973–5 gave rise to a flood of books and articles pressing the same Malthusian point. In each case, the 'crisis' period was relatively short and all of these crises have been interspersed by much longer periods of over-supply.

It is instructive to compare the two most recent food crises in Africa.[19] Two things show very clearly. Firstly, African problems have virtually no impact on world prices; they are simply not large enough to count. Secondly, the level of aggregate food production has only the most indirect effect on the current situation in Africa, though its disposition, as a result of policies, has significant effects on prices and imports.

THE EARLY 1970s

As early as 1970, some Sahelian countries had already experienced one or more serious harvest failures and problems were beginning to arise on the margins between cropping and pastoral areas. At this time, there was no world crisis. Talk on world markets was of overproduction, high stocks and low prices. As from 1971, the Nixon Administration in the USA was determined to reduce stocks and food aid, putting US agriculture on a more competitive basis and increasing commercial exports.

Problems in the Sahel hardly merited a footnote in 1972, though a world crisis began to emerge in this year. Devaluation of the dollar in 1971 had improved cereal export prices in the USA and led to some drawing-down of stocks, added to which direct curbs on production reduced US grain production by another 8 million tons. 1972 was a poor harvest year in many parts of the world, leading to 'an unusual scarcity of wheat (which) coincided with a rice shortage in the Far East' (FAOd, 1973: 1), even though, as the same source recorded, 'world cereal production was the second highest ever'. Among other factors was a drought in the USSR. Combined world production of all cereals fell by about 40 million tons.

But demand factors were at least as important. The Western world was emerging from a slump in 1969–71; incomes and consumption were rising. Incomes were also rising in the USSR and China. More importantly, the USSR was making every effort to increase and stabilize meat consumption and, to this end, had changed policy radically. Previously cattle numbers had been adjusted to cereal supply, with large-scale slaughterings when harvests were poor. Under the new policy, cereal supply was to be adjusted through imports. As a result of these combined factors, it is estimated that world demand for cereals increased by 1 per cent in 1972, against 0.7 per cent during the 1960s and 0.55 per cent in the slump year of 1970 (Tarrant, 1980).

The impact was further increased by an already growing shortage of livestock feeds. The unpredictable but often devastating ocean current El Niño led to a failure of the Peruvian anchovy harvest, which fell from 19 to 3.5 per cent of a reduced world fish catch. This is used primarily as a high-protein livestock feed, and the failure increased demand for and prices of soya beans, since soya bean cake is another high protein animal food. But soya beans compete with maize for land in the USA, thus pushing maize prices up too.

In addition, the world *fertilizer* industry was going through a cyclical period of low production and high prices. During the late 1960s, the industry had been beset by over-production and excess capacity, as a result of scale-related advances in technology and over-investment in large factories. With prices too low for profitability, investment fell, so that supplies were short and prices rising by 1972. Demand was increasing rapidly, especially in parts of the Third World where 'Green Revolution' High-Yielding Varieties were catching on. The situation was given a further twist by the 1973 rise in oil prices, but cyclical factors internal to the fertilizer industry are thought to have been more important. The effect of this shortage was felt mainly in the Third World, since companies in the 'North' supplied home markets first, despite the attempted OPEC embargo.

It was in this situation that the US-USSR grain deal of mid-1972 unleashed an explosive increase in the price of wheat, which doubled over the next few months and reached a peak in February 1974, at almost four times the August 1972 level (Tarrant, 1980: Fig. 50, p. 281).[20] The sale of almost 20 million tons of grain in the much reduced market of 1972 had an immense impact on prices, and meant a further drawing-down of US cereal stocks, bringing them as a proportion of world consumption well under the 17 per cent 'safety margin', where they stayed until after 1976[21] (Tarrant, 1980, Morgan, 1979). So prices had already started to rise when the African crisis came to the surface in 1973 and 1974. The world harvest was better in 1973, but with demand strong, this did not prevent a further fall in stocks and rise in prices, not helped by a further poor harvest in 1974. It was only with a record world harvest in 1975 that things began to return to normal, and stocks only reached the FAO-defined minimum adequate level in 1976. African problems were of too small size to affect world markets, though world market manipulation greatly worsened the situation of African countries with deficits, aggravated, as shown in Chapter 8, by trends in food aid.

THE CRISIS OF THE EARLY 1980s

Expressions of worldwide neo-Malthusianism have been less common during the most recent African crisis – scarcely surprisingly, since the world market was in surplus throughout, except for 1980–81. From 1982, world cereal stocks increased steadily, with the sole exception of 1984, since a 1983 acreage restriction programme in the USA reduced coarse grain production (as intended) by 125 million tons. Even this only reduced stocks for one year and increased prices slightly. In this context, the few million extra tons imported into Africa made little difference.

Overproduction had to do with both supply and demand factors. Agricultural policies in the EEC and USA kept production rising, while the severe slowdown in world economic activity after 1979 cut into demand. Reduced demand for, and prices of, meat had negative effects on prices of feed-grains and soya beans. Falling real incomes and rising indebtedness were leading to cuts in food imports in the Third World. But, as shown by Gill (1986) and in Chapter 4, even with world cereal stocks at around 300 million tons, it tooks months before even a few tens of thousands were shipped to Ethiopia.

Tropical African imports of cereals seem to have stagnated since 1982 *except for* food-aid shipments to cope with disasters. The reasons for this were balance-of-payments deficits, and foreign-exchange shortages, compounded by IMF/World Bank conditionality. Available figures for cereal trade suggest that increases in food imports during the 'crisis' period 1982/3–1985/6 were highly concentrated.[22] They seem to have increased by some 7–10 million tons, but this includes the Maghreb and South Africa (which was a net importer for the two years), and significantly exaggerates the real increase. A special FAO report on the African drought situation of April 1986[23] covers 45 countries of sub-Saharan Africa and compares imports in 1984/5, the most serious famine year, with the average of the five previous years (Table 1). This shows an increase in imports of about 2.5 million tons, all of which can be accounted for by increases in shipments to the main famine areas[24], since elsewhere the trend was downwards. Recent reports foresee a much lower growth of African food imports than for some years. Preliminary figures for 1985/6 (for Africa as a whole) show a drop of 6 million tons over the previous years, while forecasts for 1986/7 show no increase over 1985/6 (FAOe No. 9, October 1986). If the sharply rising imports of the 1970s have been exaggerated by official agencies as an indication of real shortfall, there is no evidence that this most recent trend shows any significant improvement.

Notes on some specific food crops

Before the concluding section of this chapter, the following summarizes briefly specific processes and trends applying to some particular major food commodities.

Cereals. Since the early years of the present century, world trade in cereals has been dominated by a small number of exporting countries. Industrialization and urbanization in Europe led to a major increase in world grain trade, with Western Europe as the focus. Initially, Russia and Poland were the major exporters, but 'new' grain-exporting countries came rapidly to dominate. These were, and are, the great plain and prairie lands of the world in North and South America, Oceania and South Africa, where previous inhabitants were pushed out, or wiped out, by white colonists, who then set up large-scale production on cheap high-quality land.

Of these countries, the USA is by far the most important, accounting for between 50 and 60 per cent of world cereal exports from 1970 until 1982, since when, with decreased competitiveness, it declined to 45 per cent but is now rising again. Next come Canada, Argentina and Australia, accounting, between them, for about 25–30 per cent. Other than South Africa, which is normally a major maize exporter, and Thailand as a major rice exporter, only the EEC regularly exports over 2 million tons of cereals. And the EEC has only become a net exporter of all cereals since 1980. In the mid-1950s, the countries making up the present EEC accounted for 36 per cent of total world imports. As recently as the mid-1970s, they still accounted for almost 10 per cent of net world imports. Since then, entirely as a result of protection and export subsidies, the EEC has become a net exporter, responsible for almost 10 per cent of the world total. Currently Japan, the USSR and China are the world's major cereal importers, followed by the richer and more industrialized Third World countries (Brazil, Mexico, South Korea), the major exception being Egypt, a low-income country but the world's largest food-aid recipient, in addition to other 'special offers' from the USA (see above).

Lying behind this, there has also been a major change in the composition of cereals produced and traded, the main distinction being between wheat and rice as the major human food cereals, at a world level, and maize, sorghum and barley as major feed-grains. From the 1950s to the 1980s, production of wheat and rice has grown rather faster than that of feed-grains, a significant factor having been rapid growth in India and China. But over the same period, trade in coarse grains, primarily exports from the USA, grew more rapidly than that of wheat. Both are far more important in world trade than rice. While the production of rice is about 80 per cent that of wheat, only about a tenth as much is traded internationally. This reflects the fact that most of the world's rice is produced and consumed in Asian countries.

During the period from 1972/3 until 1982/3, African cereal imports increased by roughly three times, to a level roughly equal to that of Japan. But three-quarters of this went to the five Maghreb states, with Nigeria taking another 8–10 per cent. By far the largest proportion of African cereal imports were composed of wheat, indicating the importance of the growing urban middle class. In Africa the use of coarse grains for livestock feeding has not proceeded very far, but changes in taste from maize and millets to wheat have been quite rapid.

Cereal substitutes. A recent development has been the rapid growth of international trade in 'cereal substitutes' for use as animal feed. These include cassava, corn gluten, other maize products, citrus pellets and various other by-products from the manufacturing of starchy raw materials. By far the largest market for cereal substitutes is the EEC, partly because it is the largest market for animal feeds, but also for reasons connected to CAP protection. Under the Common Agricultural Policy, imports of cereals are subject to a variable levy which keeps their prices up to those of local products. Since local cereals are protected up to and over 50 per cent above world market prices, this significantly increases costs of animal production, especially as consumption of purchased feed has increased enormously. But cereal substitutes are not (yet) subject to protective levies and are thus very much cheaper. This hole in the protective fence seems likely to be stopped in the future. Some of it has already been patched, in the form of a 'voluntary agreement' forced upon Thailand to limit its exports of cassava. But since the main effect of this was to encourage imports of corn gluten and citrus pulp from the USA, further measures are being sought (and resisted strongly by the USA). A large part of the EEC export surplus of cereals derives from this. With imports of cereal substitutes much cheaper, local feedgrains are not needed and are dumped on the world market.

Oilseeds and animal feeding. Thirty years ago, the term 'oilseeds' was unambiguously appropriate. The edible or industrial oil deriving from this group of products was by far the most important final product and the cake a low-valued residual. Since then, there have been major changes in the world market. The development of mineral oils and new industrial processes has driven industrial vegetable oils out of many of their previous uses, as has the replacement of soap by detergents. But growth in demand for animal feeds has been very rapid. So, while in 1945 the most valuable oilseeds were those which produced large amounts of high quality oil, regardless of the quality of the cake, the current situation is the reverse and the most important oilseeds are now soya beans, with a low oil content (of medium quality oil) but with a cake of high protein quality not vulnerable to aflatoxin (a carcinogen found in groundnuts and some other crops when mouldy).

In 1955, soya beans and cake accounted (by volume) for about one third of world exports of oilseed cake (including the cake equivalent of oilseeds exported). By the late 1960s, this had grown to nearly half and by the late 1970s to over 65 per cent of cake and 80 per cent of oilseeds (FAOb, 1969:165 and FAOc, 1978:83). At the same time, total exports of oilseed cake had increased from about 8 million tons in 1955 to over 20 million in the late 1960s and 32 million tons by 1983, the real increase being even larger and masked by increased trade in raw oilseeds. In short, one of the relatively few potential growth areas for tropical exports was entirely pre-empted by soya beans from the USA and later Brazil. By far the major growth in concentrate-feeding has come from developed countries, notably the USA, Germany and Japan, even though the *rate* of growth has been more rapid in the Third World in recent years.

Another problem for oilseed producers is that the increase in animal production arising from increased feeding with oilseed cakes includes butter and lard which compete with the oils. This is also a problem for the EEC, since it has a large surplus of butter but keeps on importing oil in the oilseeds required for animal feeding (to produce more surplus butter!). The vegetable oil problem could be solved by importing oilseed cake, but this would not be acceptable to European oilseed processing firms, which are large and powerful and which have already achieved high rates of effective protection to keep out processed oilseed products. It would not, in any case, solve the butter surplus problem. Since the USA is the major world exporter of soya beans, increased protection would further sharpen the developing agricultural trade war between the USA and the EEC.

To the extent that this round of absurdities affects hunger problems in tropical Africa at all, it is through effects on the prices of oilseeds. These are grown as export crops, and downward pressure on prices reduces export proceeds. But groundnuts and other oilseeds are useful sources of both energy and protein for local consumption, and in a number of countries increasing local demand has already diverted groundnuts away from exports.

Animal production. Increases in purchased feed use relate not just to increases in animal production but to changes in its form. Pigs and poultry are more dependent on purchased feedstuffs than beef cattle or sheep and goats. Production of liquid milk for a year-round fresh milk market requires far more supplementary feeding than does production of a storable commodity like cheese, which can be produced seasonally. With rising costs of land, labour and machinery, yields have to be increased, and this requires both increased quantities and more precise specification of purchased feedstuffs.

Meat. Between the late 1950s and 1983, total world meat production (by volume) increased by about two and a half times. But cattle, sheep and goat meat production has less than doubled, while pigmeat production has tripled and poultry production increased by five times. As a result, pigs and poultry have grown from about 45 per cent of the total to almost 60 per cent. (FAOb, 1969: 45; e, 1984: 30). This trend has hardly affected most of Africa. There is a certain amount of intensive and semi-intensive poultry rearing, but usually not more than can be fed from milling by-products. Extensive beef-ranching is a greater problem, in that it often uses land which could better be cultivated or grazed by peasants and pastoralists.

Dairy products. Where the USA dominates the world market for cereals, the EEC is even more dominant in dairy products, with up to 70 per cent of world exports in some commodities. This is a relatively recent phenomenon, dating from the 1960s and 1970s for skimmed milk powder and butter respectively (the two major dairy products traded internationally). It is entirely due to EEC protection. As of the mid-1950s, the USA was by far the largest exporter of milk powder (almost all as food aid), followed by Australia and New Zealand, with the EEC-6 accounting for only 13 per cent of the total (FAOb, 1969:39). For butter, Oceania was by far the largest exporter (44 per cent of the total, ibid: 37) and the EEC accounted for only 9 per cent of exports. Future members, Denmark, Ireland and the UK (re-exports) accounted for an equal amount. Between 1965 and 1967, the EEC replaced the USA as major exporter of skim milk powder and by the end of the decade accounted for over 50 per cent of world exports. Although New Zealand and Australia are very much lower-cost producers than the EEC or USA, the EEC has used export subsidies to achieve and retain market leadership. For cheese and condensed and evaporated milk, the EEC countries have been world market leaders throughout the post-war period.

The issue of world trade in dairy products is complicated by the size of internal trade within the EEC (intra-trade). The EEC market itself is about three times the size of the 'world market' (the proportion of total milk production which is traded internationally is very small). In recent years, EEC intra-trade has comprised a very high proportion of total world trade. Thus, for skim milk powder, it accounted for about 40 per cent between 1975 and 1982, jumping to 66 per cent and 64 per cent in the next two years. For butter, it has been in the region of 35–40 per cent from 1979–84 and for cheese 50 per cent over the same period (FAOb, 1981:60-1, 1984:69, 1985:78, all in volume terms). While this is understandable for cheese, which is produced in a wide variety of types and flavours, butter is relatively homogeneous and SMP almost completely so. The only possible motive for intra-trade seems to be speculation on variations in the 'green currency' rate (see Chapter 7). The total value of EEC intra-trade in dairy products is over $2 billion.

Western Europe has been a dairy producer for centuries, but production increased significantly in the latter half of the nineteenth century with urbanization and the emergence of large markets. Dairy products had a high degree of 'natural protection' against imports from their bulk and perishability, and were well-suited to small-scale family farming. European dairy sectors were protected, but not for export production. Prior to 1940, world trade in dairy products was in any case limited by the weight and perishability of milk, butter and cheese, and the poor quality of processed milk (condensed, evaporated and dried). A major breakthrough was the recombination of milk powder and butter oil. Fresh milk can be broken down and all the water extracted, leaving products with minimum weight and optimal keeping qualities. These can be shipped cheaply to anywhere in the world and then recombined to produce a product which looks (but does not taste very much) like fresh milk. Factories for recombination can often be funded as part of aid programmes because of the effectiveness of the 'dietary improvement' lobby.

All this has minimal direct relevance to the food needs of the hungry in Africa or in the Third World in general, since dairy products, and especially recombined milk, are too expensive to be consumed by poor people. This has not prevented export promoters from making extravagant claims for the value of dairy products in improving the protein content of Third World diets. Thus dairy plants have been a popular form of aid investment in Africa, though by far the largest expansion in exports of both dairy products and processing plant has been to the richer NICs and MICs, to supply their burgeoning urban middle classes. With easy money and a number of aid projects, dairy imports into tropical Africa rose rapidly during the 1970s, but have stagnated during the 1980s.

Sugarcane is a crop for which world market prospects are unusually poor. Production of cane sugar for export was developed by a number of European countries in their tropical colonies, usually on the basis of slavery or indentured labour, since cane growing and cutting involves hard and unpleasant labour. Many of these countries are heavily dependent on sugar. But since the mid-1960s, production has increased more rapidly than demand, pushing prices downward. Only a minority of sugar is sold at the 'world price', the majority being traded at significantly higher prices under special agreements, like the EEC quota, the Soviet quota for Cuban sugar and US arrangements with certain countries. Prior to Britain's joining the EEC, the Commonwealth Sugar Agreement was the most important of these. Since its accession, part, but not all, of its CSA obligations have been assumed by the EEC.

The major long-term problems with sugar are many. Countries which depend on sugar exports often have few alternatives. But cane exporters are increasingly threatened by competition from other forms of sugar. Western Europe has been a major producer of beet sugar under protection since the early nineteenth century. But the development of the EEC sugar regime since 1968 has led to rapid growth in production, and transformed the EEC into one of the world's major exporters, with disastrous effects on world prices. There is also a tendency within the EEC not to renew cane-sugar refineries as they age, but rather to increase capacity for beet sugar. This seems likely to doom cane-sugar exporters even more surely than protection and low prices. Potentially still more serious has been the development in the USA of high-fructose corn-syrup (HFCF) from maize, which is currently capturing ever more of the industrial uses of sugar. A measure of the success is that the major soft-drink corporations (Coca-cola and Pepsi-cola) now use HFCF. This has led to the halving of US sugar imports, with catastrophic effects on suppliers like the Philippines. The long-term prospects for sugar exporters are thus exceptionally poor.

Implications and prospects for Africa

The development of agriculture in the developed countries can affect that of tropical Africa in a number of ways. Firstly, it can, and does, serve as a model for emulation. Many, if not most, agricultural development projects are concerned to 'modernize' agriculture on this model. But it also competes with African agriculture in some spheres, through its technical efficiency and the development of new products, and through subsidies and protection. Thirdly, processes similar to those involved in the development of 'Northern' agriculture are occurring in the 'South', notably the increasing dominance of agribusiness.

Those who find in advanced agriculture a model for the development of African agriculture are apt to look at the gains in productivity and forget everything else. This process is compounded when 'experts' proclaim that increasing productivity is the only way out of Africa's food crisis and must therefore be given overwhelming priority. If one looks at the development of agriculture in Europe and the USA in somewhat broader terms, its suitability for uncritical transfer as a model becomes much less certain.

This development has been a wildly contradictory phenomenon, most especially during recent decades. There has been unparallelled advance in technology and growth in productivity, a process which shows no signs of abating. Yet the results in terms of human welfare have been, at best, extremely meagre. The aggregate flow of people out of agriculture has declined, but only because there are so few left to migrate. Rates of exodus are no lower. Farm incomes in both Europe and USA are so low, except for a small rich minority, that less than half can live by farming alone in spite of the massive increases in productivity. Many of those who do not have second jobs are the very poorest, kept in farming by old age and lack of alternatives. Capital investment per person employed is higher than in manufacturing, but this has had no corresponding effect on wages, which remain lower than in other sectors. This is particularly marked at the 'sharp end' of this development, for example in the huge farms of the south-west USA, where millionaire farmers employ hispanic labour under some of the worst conditions to be found outside the Third World.

If this is one side of the massive increases in productivity, the other is over-production, the build-up of unsaleable stocks and subsidies to get rid of them. As should be abundantly clear by now, the existence of surplus food stocks in Europe or the USA does nothing for the hungry people of the Third World, and nor would shipping more of them to areas of hunger. People need income to buy food, not food which they cannot afford.

And what of the effects on the non-agricultural majority of the population? Prices of food and agricultural products have fallen in relation to real incomes in most Western countries, though some of this has been taken back in costs of packaging and presentation and more in taxes to subsidize agriculture. 'Convenience foods' have reduced cooking time and household labour, though in many cases at the expense of taste and nutritional value. Consumers in the industrialized countries can purchase a wider variety of different foodstuffs with less seasonal restriction than before, though some of this variety is imported and some is more apparent than real. While there may be more different products, standardization has reduced the variation within each (fresh fruits, sausages, frozen foods).

There have been enormous costs of storing and dumping excess produce on the world market, where it competes with products from poorer parts of the world. But these are only the measurable costs. The proportion of chemical additives, to give colour, synthetic taste and longer storage life to food, steadily increases, with control over the use of dangerous substances lagging behind their rate of introduction. Varietal selection and breeding strive after maximum yield and storage qualities. It is not nostalgic imagination that fruits and vegetables have become more tasteless. Increasing yield is associated with blander taste, while a thick oily skin improves storage life and resistance to handling. The proportion of plastic covering in the total weight of food seems continuously on the increase – as does its impenetrability. Prices of meat and vegetables are 'held down' by increasing the proportion of water in them, this requiring further chemical additions to prevent it dripping out before cooking. The cost of meat is further held down by factory production, involving horrific conditions for the livestock in question, during their very short lives.[25]

Vastly increased fertilizer use means more nitrates in the groundwater, with serious negative effects on human and animal health. In some of the worst affected coastal waters off Denmark, fish grow hideous sores and cannot be eaten.[26] Pesticides and other agro-chemicals are found in dangerous concentrations in both water and food products. One is told that control would 'cost too much' and make farming unprofitable. But without the heavy subsidies most agriculture in developed countries is already unprofitable. Having paid enormous sums in tax and higher prices, the mass of the population must pay still more in pollution to keep farming profitable for a very few farmers and allow rather more to hang on. But removal of subsidies, the 'free market' solution, would not help – rather the reverse. It would drive most of the remaining farmers out of business, leaving only the very largest behind. These would be forced by economic pressures to cut costs and labour forces, with even larger and more gruesome factory farms, with even less concern for the environment and with products whose excellent yield and storage qualities had taken most of the remaining taste from them.

If these are some of the effects of modern agriculture in the places for which it was developed, what then of its likely impact in tropical Africa? A number of factors likely to reduce technical efficiency have already been mentioned. Dependence on tightly organized systems and schedules becomes a disadvantage where the country's infrastructure cannot deliver inputs or collect produce on time, when it is a matter of luck whether one gets credit next year, or when foreign-exchange constraints may make certain vital inputs or spares completely unobtainable.

There are even worse problems with control over the use of chemicals. These are dangerous enough where the farmer can read and the instructions are written in his language, where protective clothing is available and (in theory) compulsory, where factories are safe(r)[27] and where the most dangerous chemicals are disallowed. The dangers are many times worse where none of these applies; where even the extension agent does not know how to use the product safely – and in any case would not find it necessary to tell the farmers; where DDT is used as a storage chemical for maize for lack of any safer chemical; where protective clothing is non-available and most work in torn shorts on shift and nothing else; where one cannot even learn from experience since 'they' bring a different chemical every time.

And behind all this is the double-edged effect of increasing productivity. Under ideal circumstances increasd productivity allows more to be produced with less work, increasing incomes and leisure. But here on earth it has yet to work that way. Productivity increases profits and saves labour – that is, it saves the employer the cost of hiring labour. Competition ensures that it cannot be used by the direct producer to take life more easily, but rather to produce more for the same income. Moreover, all historical experience indicates that increasing productivity under capitalism throws the smallest out of farming. Family farming may have survived until today in Europe and the USA, but not many of the families are left – and in Africa there is no 'new world' to migrate to, nor many urban jobs.

The 'experts' talk glibly about 'scale-neutral' techniques, which increase employment in agriculture, but nowhere are distribution and credit 'scale-neutral'. Under market conditions, the larger and richer farmers receive preferential treatment, because this makes 'economic sense'. In most tropical African countries they receive even more preferential treatment, because they have the money and influence to buy it – and because 'experts' find this more economically efficient. Even the most cursory survey of where 'modern agriculture' is advancing in tropical Africa shows quite clearly that it is in the best-watered areas, nearest to the main towns. The dry and marginal areas are where many of the problems lie, but they tend to be ignored because the technical solutions are not yet known (by experts) and because governments and aid agencies prefer to avoid the risk of failure.

If 'modern agriculture' is a problematic model for emulation, it is an ever-present competitor. In part this is because of its very real technical efficiency, but protection is also a significant factor. Industrial development in the post-war period hit Third World exports hard. Early examples of bio-technology, as in sugar, have had similar effects. And there is no reason to expect the process to stop now. Not only do Europe and America stand on the brink of further technological breakthroughs, but there is enormous pressure to find uses for thousands of tons of excess produce. The development of HFCS is clearly related to the existence of unsaleable maize surpluses in the USA, and even so, it has not managed to use enough to prevent stocks rising. In both Europe and the USA (not to mention Japan and other developed countries), the existence of large surpluses generates pressures either to find something which can be done with them or to find other crops/livestock with which something can be done. One approach is to find something which can substitute for imports, another to find new export possibilities – almost certainly in competition with Third World producers. Junne (1986) was cited above in opposition to the view that export-industrialization provides a 'way forward' for more than a handful of Third World countries. Another, still more depressing, conclusion from his research into bio-technology, is that there is no 'way back' to agricultural primary product exporting. Current short-term market prospects for Third World agricultural exports are poor enough. Long-term prospects are much worse. Some Third World countries – the richer, like Brazil or Malaysia – will be able to compete through adopting the most advanced methods, though this will not achieve much for the poorer sections of their populations. Most countries will lose export markets – and find their domestic markets protected only by poverty which leaves them unable to import much.

The above discussion of 'modern agriculture' as a model for emulation presupposes that there is a real choice which can be made on 'policy' grounds. While fair enough for purposes of illustration, this is not really the case. Choice of technique, choice of model for development are not neutral policy choices, they reflect political and

Trends in World Food Production & Trade 133

economic interests and power. The penetration of international agribusiness into African agriculture is already substantial, is continuing all the time, and is encouraged by technicist approaches to solving the hunger (or balance-of-payments) problem. Private firms, large or small, are concerned to make profits; so much is obvious both from observation and from the first chapter in every economics textbook. According to the textbooks, this leads to 'welfare maximization' (taking no account of distribution, of course) because free competition in the market forces all producers to be efficient. This part of the story is less easy to observe from the real world. Agriculture is often cited in economics textbooks as the prime example of free competition. Nothing could be further from the truth. There are few sectors in which control is so concentrated, from oligopoly and contracting on the world scene to monopoly marketing boards and enforced production for African peasants. Given the power of the forces pushing for agricultural modernization under the control of capital, it will take a good deal more than 'policy choices' to stop it.

NOTES

1. Agricultural protection did not, of course, start in 1945. Some of the European countries have been protecting their agricultural sectors without break since before the industrial revolution. As discussed below, to some extent one can date this change back to the early years of the present century. But there has been a major change since 1945.
2. Even with all these qualifications, I am still uncomfortably aware of its rough and ready nature. Most of what I have read on the development of agriculture during the past two hundred years lays great stress on the variability of the process in line with specific histories and social structures. One useful overview on this, with references, is Bairoch (1973).
3. On this see Djürfeldt (1981), who refers, however, to declining rent. In a comment (Raikes, 1982) I state reasons for preferring to think of it in terms of declining profit, although this is out of line with both Ricardian and Marxian theories of rent.
4. Since one is talking of a wide range of families, ranging from 'gentlemen farmers' to poor peasants, the definition of 'a living' will obviously vary widely. In the former case, this will often be a very comfortable living. What it will not be, under the conditions described, is a return to invested capital (including the value of land) comparable with the average rate of profit in the economy as a whole.
5. This is related to the growth of middle classes and incomes. Feeding cereals to livestock only becomes economically feasible when the ratio between the price of the livestock product and that of the products fed exceeds the conversion ratio. It makes no economic sense to feed ten kilos of grain to get one kilo of meat unless meat fetches over ten times as much as grain. This, in turn, only occurs with the emergence of substantial groups of consumers for whom income is high enough for the quality, rather than simply the sufficiency, of food to become a major consideration. This is discussed in Crotty (1980).
6. This was particularly true of second chambers, of which there were more in the inter-war period than now. The rural vote (not peasant) is still over-represented in the USA, through the method of election to the Senate.
7. The irony here, to be considered in more detail in Chapter 7, is that 'economic' farming is often only so because of subsidies.
8. The most recent information available to me dates from around the end of 1986. The more recent fall in the value of the dollar has improved the competitiveness of US farming on the international scene.
9. See Chapin and Wasserström (1981). For a broader discussion of the issues surrounding pesticides, see Bull (1982).
10. This process is discussed, in relation to Denmark, in Jensen and Reenberg (1980 pp. 55 ff.).
11. With livestock raised at low intensities this ratio is around 10:1. Intensive rearing, especially of pigs and poultry, can reduce this to 4:1 or even less. This is achieved by eliminating

maintenance and other 'waste' feeding. Movement is reduced to a minimum by the confines of the pen or cage. Growth is speeded and time to slaughter reduced by intensified feeding over long, electrically lit 'days', while the risks of disease are combatted with blanket use of drugs. Broiler chickens live as little as nine weeks from day-old-chick to carcase.

12. This is only a rough equivalence, since a large proportion of animal feed consists of oilseeds and even animal products, like skimmed milk powder. The latter is only possible because of heavy subsidies, which feed surplus milk to livestock so that they can produce more surplus meat and milk.

13. Holland did experience famine just after the War, though not on anything like the scale of Bengal, Ethiopia or Mozambique.

14. Both lend-lease and the Marshall Plan involved major transfers of credit to European countries to allow imports which their depleted foreign exchange would not otherwise have allowed.

15. As so often, there is some element of truth in both. The inferior 'resource endowment' for industry of the Third World is largely a matter of lack of infrastructure, trained manpower and experience. While (with the significant exception of some NICs) this is largely true, it can scarcely be denied that the very low levels of investment in infrastructure and education and the outright suppression of industries in colonial possessions were major contributors to that situation.

16. As Petit and others indicate, one major frustration of the USA is that it has been unable to use the regulations of the General Agreement on Tariffs and Trade (GATT), because of its own pressure for maximum imprecision in their formulation during the 1950s. At that time, the major issue in dumping of agricultural produce was US food aid.

17. *Telex Africa* 260, 11 June 1985 pp. 8–9.

18. G. Junne – seminar held at Centre for Development Research, Copenhagen, 29 October 1986. See also Junne (1986).

19. For the 1970s and previous (above) see Tarrant (1980) Ch. 8. For the 1980s, information from FAO (a,b,c,d,e – annuals) and Special Reports on the famine in Africa.

20. At the time, this deal was widely referred to in the US press as 'the Great Grain Robbery', implying a swindle pulled by the Russians through the secrecy of the negotiations. The Soviet negotiators were undoubtedly secretive, as all negotiators of such deals are, but it was the responsibility of the US Government, not of that of the USSR, to keep the American people informed. In fact, the Nixon Administration kept them misinformed, by failing to reveal the scope of the deals and by falsely denying that export subsidies would be applied to these sales. In any case, the losers were third-country and often Third World importers of grain.

21. FAO computes the minimum levels of world stocks required to prevent major price instability and severe risk of shortage variously at 17%, 18% and 20% of estimated annual world cereal consumption.

22. By end 1986, I had not been able to see FAO Trade Yearbooks after 1983. *Food Outlook* has more up-to-date figures, but does not provide a complete breakdown within Africa. A Special Report provided further incomplete figures. Since early 1987, I have not had access to up-to-date libraries for this sort of material.

23. 'Food Supply Situation and Crop Prospects in Sub-Saharan Africa – Special Report', FAO, Rome, 21 April 1986.

24. Ethiopia (1.09 mn tons), Sudan (0.93 mn), Kenya (0.42 mn). Other famine-hit countries did increase imports (Zimbabwe, Chad, Mali, Niger) but this was more than offset by reduced imports from countries feeling the balance-of-payments pinch (Zambia, Zaire, Ivory Coast, Senegal and Nigeria).

25. While finishing a previous draft of this book, a case hit the Danish papers, indicating what can arise from factory farming combined with an economic squeeze. A farmer, realizing that, whatever happened, he would soon be bankrupt, went on holiday abroad, leaving some 100 intensively reared pigs to starve and drown in their own excrement. When the police arrived, alerted by neighbours who had noticed the smell, almost all had to be shot to put them out of their misery.

26. I cite Danish cases, not because the situation is any worse than normal there – the reverse is

probably the case – but because I was living there for most of the time I was writing this book.

27. The examples of Seveso and Sandoz in Basel show that the claims of industry spokesmen about the 'impossibility' of such accidents are worth little. But in terms of human life, these hardly compare with the disaster in Bhopal.

6

The EEC & Africa

Outline of Major Issues

The following four chapters present material on EEC policies likely to affect the food situation in Africa. This chapter introduces the issues at a general level. The others take up particular issues in greater detail.

The EEC is a regional economic grouping of Western European states, whose main general purpose is to improve the economic strength of its members, small and medium-sized states, in a world dominated by the USA, Japan and other large economic units. Since its members are among the most highly industrialized countries in the world[1], the major gain in economic strength could be expected to derive from concentration of industrial capital and the economics of scale which can be reaped from a market of some 270 million.

In view of this, it is perhaps surprising that by far the largest item in the EEC budget is agricultural protection, which has taken, on average over the past ten years, almost 80 per cent of the total. True, this comes to only about 0.66 per cent of EEC GNP, but it is still more than the combined members spend on development aid and even then is only between one-third and two-fifths of the total cost of agricultural protection. The national governments also subsidize their agricultural sectors. More importantly, however, agricultural protection increases the prices paid for food and other agricultural products, and this hidden tax accounts for 55–65 per cent of the total cost to consumers and taxpayers (BAE, 1985: 99–100). Since agriculture, forestry and fisheries account for some 3.8 per cent of EEC GNP, it can be seen that protection accounts for about half of total value-added in EEC agriculture. Indeed, it is estimated to have accounted for some 60 per cent over the past decade (BAE, ibid.:4).

While it has other aspects, the EEC is fundamentally a customs union. That is, it operates a system of (relatively) free trade internally, presenting a common protective wall to the outside world. While this is particularly obvious in the case of agriculture, it is by no means restricted to that sector. Indeed, where the Third World is concerned, a good case could be made that industrial protection has more serious negative effects. One common characteristic of the pattern of EEC (and most other industrialized country) protection is that the level of tariffs or levies increases significantly with the degree of processing. Most tropical agricultural products are free or virtually free of tariffs in their raw form, but in processed form, they are subject to tariffs. In this way, the rate of 'effective protection' on processing is far higher than the nominal tariff, since it falls entirely on the value-added in processing.

There are other ways in which the real degree of protection can be very much higher than would appear from nominal tariff levels. One comprises so-called 'gentlemen's agreements' or 'voluntary restraints' in which prospective exporters are forced by threat of retaliation to reduce their levels of exports. Limits on Japanese car exports to

Europe and the textile Multifibre Arrangement are examples of this, as is the 'voluntary restraint' on Thailand's exports of cassava to the EEC (see Chapter 5). Another is the imposition of such strict phytosanitary or hygiene regulations that prospective exporters cannot meet them.[2]

While protection is central to the EEC, it is not its only facet. There are a number of relatively small funds for social expenditures, assistance to depressed regions and research. Between 2 and 4 per cent of the EEC budget is devoted to development assistance. In addition to this, a larger sum is allocated to the associated ACP (African, Caribbean and Pacific) countries under the Lomé Convention, which is funded outside the budget by contributions from the member states. Average commitments under the second Lomé Convention would have amounted to some 4–5 per cent of the budget, though disbursals would have been somewhat less, because of delays and inflation. EEC aid is linked to agricultural policy through the fact that up to 40 per cent of the total (thus the large majority of budgetary aid) takes the form of food aid. This is the highest proportion for any major aid donor and relates to the surpluses resulting from EEC agricultural policy.

The EEC and Africa

The EEC policies which can be expected to affect Africa can thus be divided into two major categories, trade and protection policies and those concerned specifically with development aid or investment. In the former case, one can make a further division between policies affecting Africa's exports and those which affect imports. In the latter case, it is worth distinguishing between food aid and other aid. In addition, as part of the Lomé Convention, there are some 'hybrid' policies with both trade and aid implications, like Stabex and Sysmin, whose proclaimed intention is to stabilize export proceeds for ACP countries.

EEC PROTECTION

The general effect of economic protection is to increase prices of the protected items within the area to be protected (in this case the EEC) and to lower them elsewhere, other things being equal. This latter effect arises because protection, by increasing prices within the protected area, usually also increases the level of production within the affected area, thus reducing its imports from (or increasing its exports to) other parts of the world. Moreover, increasing internal prices has the effect of reducing demand, again reducing imports or increasing exports. By and large, the formation of a customs union tends to increase the general level of protection, even if the level of tariffs is not raised. This is because the larger the area to be protected, the more varied its pattern of production is likely to be and the more trade is likely to be diverted from third countries. Theoretically this need not be the case, but in practice it usually is.

These effects can easily be seen in the history of EEC agriculture. Before the formation of the EEC, the current members were, as a whole, major net importers of temperate agricultural products.[3] All had previously protected their agricultural sectors, but not in ways, or to the extent, that surpluses were generated. With the formation of the EEC there occurred both a considerable diversion of trade from third countries and a steady increase in production. Over the period since 1960, this has grown at about 2 per cent per annum, compared with a growth in demand for the products in question of about 0.5 per cent per annum. While the high growth of production is clearly a direct effect of EEC protection, the slower growth of demand is less so. In part, it can be attributed to price increases, but the major factors have been high income levels and low population growth. At high income levels, the proportion of any increase spent on food is small and consumption does not respond strongly to price change. The amount of this which reaches the agricultural sector is even smaller, since well over half of the total consumer price is accounted for by packaging, processing and presentation.[4]

Third countries lost more export markets when the UK, then one of the world's most important food markets, joined the EEC. Again, this involved both trade diversion (specifically from Commonwealth agricultural exporters) and growth in local production. This coincided with changes in the method of protecting dairy products and sugar, which still further increased growth in production. At present the EEC is, by some way, the world's major exporter of dairy products, beef and veal, and sugar, and among the top few exporters of cereals. Without exception, the agricultural exports of the EEC are heavily subsidized, these subsidies having recently become the most expensive item in the Common Agricultural Policy.

At the same time, EEC protection of its livestock sector has significantly *increased* the market for livestock feeds, imports of which have increased in line with the surplus of animal products. There are two further complicating factors here. First, the oil in oilseeds, which are imported for their protein-rich animal feeding cake, competes with EEC protected products (butter and olive oil). This strengthens a preference for soya beans (which have a low oil content, a high quantity and quality of oilcake and are temperate in origin) over groundnuts and other tropical oilseeds. Secondly, while the EEC is an exporter of feed grains, it is a major importer of cereal substitutes. The level of protection on feed grains (local or imported) makes them too expensive to feed to livestock in the EEC. So EEC taxpayers subsidize their export. But cereal substitutable materials (roots like cassava and grain or fruit residues) are not covered by CAP protection and can thus be imported cheaply.[5]

EEC PROTECTION AND AFRICA

In considering the impact of EEC protection on the economies of African countries, there are basically three questions to answer:

- What have been the effects of EEC industrial protection on tropical African industrial exports?
- What are the effects of agricultural protection on countries exporting 'CAP-products' or similar?
- What are the effects of agricultural protection on African net importers of 'CAP-products' (mainly temperate foodstuffs)?

The general answers seem obvious enough; EEC protection harms exporters, whether of agricultural or industrial products, by reducing the prices they receive or by limiting their access to markets. Conversely, it helps food importers by reducing the prices they must pay for their imports. The reality is rather more complicated.

EEC INDUSTRIAL PROTECTION

Industrial protection 'works both ways' for many African countries, since, as members of the ACP group, they receive a certain degree of 'preference' over other exporters, in the form of lower tariffs and improved market access. While EEC protection has negative effects on all external exporters and all Third World exporters taken together, African exporters might actually gain through improved market access.

Moss and Ravenhill (1983) show, however, that this was not sufficient to prevent their share of the EEC market falling in the period between 1973 and 1979. They also show that by far the largest categories of 'manufactured goods' exported by the ACP to the EEC consisted of aluminium oxide and uranium, which are in fact mineral ores processed solely for convenience in transport (ibid. Table 10.10). This can be read to indicate the insignificant level of preference, the negative impact of protection, the lack of competitiveness of African industry, or any combination of the three.

Probably the most important aspect of industrial protection for the African countries is the high 'effective protection' on processed agricultural commodities. That is, by increasing the level of protection with each stage of processing, each addition to the level of protection falls entirely on the value added in processing. However, since this is an established practice among industrialized countries, it is not clear that the EEC, as such, bears the responsibility. It is highly likely that the individual states would operate similarly even if the EEC did not exist. Certainly they were doing so before the EEC was formed.

To summarize, the effect of the EEC on African exports of manufactures is far from clear. The general effect of protection is to reduce prices and market access. ACP preference tends to improve access, however marginally. Given that growth in manufacturing for export has been far greater and generally far more competitive in other parts of the Third World, EEC protection is probably not the most important factor. Moreover, given the sharply rising trend of industrial protection (much of it in the form of 'voluntary restraint') over the past decade in nearly all OECD countries, it is far from clear how much of this can be attributed to the EEC as such. Similarly one cannot compare levels of protection over time, to see if they were lower before the EEC was formed or particular members joined. Levels of protection have been increasing throughout the developed countries since the mid-1970s, under the dual pressure of increasing Third World industrial exports and declining economies at home. The EEC has certainly taken part in this, but no more than others.

EEC AGRICULTURAL PROTECTION

Where agriculture is concerned, there is no doubt that the existence of the EEC has significantly affected the level of both protection and production. EEC subsidized exports of cereals, beef and veal, dairy products and sugar clearly have a depressing effect on world prices, as, to a lesser extent, does increased self-sufficiency in other products. Just as important, though less easy to measure, EEC protection increases the instability of world markets and prices. By stabilizing its own large internal market, the EEC passes on variations in the relationship between supply and demand to the world market.

The precise effects of this protection on African countries are less easy to measure or assess, however, than would seem the case at first sight. Firstly, there are wide disagreements on the price effects of CAP protection. But there are other complicating factors. Some ACP countries receive quotas for protected EEC markets, this being the case for some producers of sugar and beef. The effects will be vastly different for different countries, some being more affected via their exports, others through reduced prices of imported food. Most tropical African agricultural exports are relatively little affected by CAP protection, for the obvious reason that they are not competing products. It is only where there is European produce to protect (beef, sugar, tobacco) that protection against African exports becomes significant. Even here though, the situation is complicated by EEC quotas for sugar and beef, under which ACP, including African, countries can export to the EEC at prices near to the high domestic ones. Oilseeds are a rather odd case. As noted above, the oil competes with EEC butter, lard and (to a lesser extent) olive oil, but the EEC does not protect the seeds heavily since the oilcake is important for its animal production. It is true that EEC imports have increasingly been composed of soya beans from the USA and Brazil, because of their low oil and high cake content. But internal price and other factors within Africa seem as important a reason for the decline in exports of groundnuts from Africa.[6]

As is the case for industrial protection, it is difficult to distinguish the effects of EEC protection from those of protection by other industrialized countries. The USA, Japan and non-EEC countries within Western Europe also protect their agriculture and, with the exception of the USA, at rather higher rates than the EEC. Moreover, a significant proportion of agricultural protection within the EEC is carried on by its member governments independently of the EEC. Yet one cannot conclude that EEC protection is unimportant, simply because its effects cannot easily be measured. The Common Agricultural Policy has turned an area of net agricultural imports into one of the world's largest exporters of temperate agricultural produce and this at least certainly does some have some effect.

EEC PROTECTION AND AFRICAN EXPORTS

The question of the impact of EEC protection on African exports is considered in more detail in Chapter 7, but it is worth posing one of the questions it raises. EEC protection affects African exporting countries, through reducing foreign-exchange receipts. While the negative economic impact of this is clear, much less so is the impact on food production and availability.

If one could assume that African states were economic actors in a market, and *if* one could assume that they passed on all variations in world market prices to local producers, then one could make two predictions:

- that reduced export proceeds would be reflected, other things being equal, in reduced food imports and food availability (assuming, as is generally the case, that African states are net food importers).
- that producers would shift out of products whose prices were reduced by protection and towards others whose price was unaffected (other things being equal again).

In reality neither of these presumptions can be made. As already discussed in Chapter 3, African states do not generally act as 'rational consumers'. They tend to import food when it becomes necessary on human or political grounds and seldom (except for food aid) on the most favourable terms. Food prices tend not to be closely related to import prices and the relation between food imports and food availability to those in need is unclear and ambiguous. Regarding the second 'conclusion', it is rather seldom that producer prices in African countries reflect export prices with any accuracy, since a variety of internal taxes and other charges intervene. If anything, increases in export price are often taken as a chance to increase export taxes, while export-price reductions necessitate their reduction.

But even if one could make simple predictions about the impact of protection on relative prices and production patterns, as shown in Chapter 4, this still tells one relatively little about how the food supply or security of vulnerable sections of the population is affected. There is no special reason to suppose their interests to be identical to, or even congruent with, those of the state.

EEC AGRICULTURAL PROTECTION AND AFRICAN IMPORTS

Here the main issue is the impact of CAP protection on world prices for cereals and dairy products, and the impact of these on African importing countries. Most tropical African countries are net importers of both these commodity groups, so there is no significant complicating effect from restrictions to potential African exports.[7]

The issue will be considered in detail in Chapter 7. Here I shall just indicate some of the questions to be asked. The first concerns the effect CAP protection has had on world prices for cereals and (to a lesser degree) dairy products. Econometric studies have reached widely varying conclusions on this question.

An even more difficult question is how this affects African food imports. There is no doubt that they increased very rapidly during the period when CAP protection had its most significant impact on world prices. But it is far less clear how much of this was due to CAP protection. African cereal imports have been determined by a variety of factors, including the gap between urban requirements and procurements from the rural areas, the political pressures which this generates, and the constraints imposed by shortage of foreign exchange. The first of these is affected in part by the price and other agricultural policies of the state. The availability of cheap imports may be *among* the factors affecting this (if one can import cheaply, there is no need to improve local prices), but it would be far-fetched to suppose that it was the only or even a major cause. Indeed, as soon as one considers concrete examples, other factors become clear. Moreover, the major growth of food imports occurred during the period 1973–80, when access to credit made foreign-exchange constraints less binding than now, making it easier to turn to imports to make up for the inadequacies of domestic procurement systems. During the past few years, imports of cereals into Africa appear to have declined (except for areas of specific crisis and famine), as shown in the previous chapter. EEC protection has not declined in the meantime. Clearly foreign-exchange constraints have been a major factor, probably significantly increased by IMF conditionality.

This raises another issue. The absolute maximum effect estimated for EEC protection on world cereal prices is between 15 and 30 per cent. Yet during recent years, many African governments have kept their exchange rates 'overvalued' by up to several hundred per cent. Would this not be a much more potent cause of increased imports? The answer here is both yes and no. There seems little doubt that exchange-rate overvaluation, by reducing the *apparent* cost of foreign exchange, has increased import levels. But it does not affect the hard-currency price of imports, and with foreign-exchange shortages widespread, this would have been a major factor determining food imports, though there is little doubt that exchange-rate overvaluation has contributed to 'cheap food' policies and so to demand. In sum, it seems likely that exchange-rate overvaluation would have been a contributory though minor factor. But here again one is back to state policy, for overvaluation of exchange rates leads to shortages and to rationing of what there is by the state. It may well be that certain state agencies have based their calculations about whether to import or not on the official exchange rate, which would invariably favour increased imports.

Again, there are further questions to be asked about the impact on food availability. *Other things being equal*, food imports would increase aggregate availability, though this tells one nothing about the distribution of imported food. The large proportion which has been made up of wheat and dairy products is probably mostly consumed in the urban areas, and to a significant extent by the middle- and upper-income groups. Even in such cases, it could reduce the prices of alternative foodstuffs, though price controls and market segregation tend to limit this effect.

Cheap food imports will tend to affect production and procurement in different ways. In the first place, they may (but may not) encourage governments to reduce official producer prices (or fail to increase them in line with inflation), thus tending to reduce procurements through the official system. But this is as likely, for reasons discussed in previous chapters, to involve diversion to black markets as reduced production. The effect of food imports will in general be to reduce black-market food prices too. But the price effect and its impact will depend on the amounts involved and the sequence of deliveries. Its effect on production decisions will depend on a myriad of specific factors. What time of year do import deliveries arrive? Are local stocks with producers or traders when prices fall? Are seeds, tools, fertilizer etc. available and at what prices? Once again, detailed study of local situations will show that in some cases cheap imports (and *a fortiori* food aid) do affect production, while in others they are not among the significant factors.

STABEX

Stabex is one of the mechanisms of the Lomé Convention. It is a trade-aid 'hybrid', whose purpose is to stabilize the export proceeds of the ACP countries or, more accurately, to offset the effects of downward fluctuations. The means to achieve this is a system of subsidized loans (grants for the poorest and landlocked countries), which are given automatically if export proceeds for the affected commodities fall below a given proportion of the average over the three previous years. There is no specification that the reduction in export proceeds should arise from price variation. Indeed, a large proportion of Stabex transfers up to 1980 were on account of declines in production rather than prices. Since decisions are made on the basis of individual products, countries can (and do) receive Stabex transfers even when overall export proceeds have risen. Further details on the operation of Stabex can be found in Hewitt (1983) or Raikes (1984a).

Stabex has been a very popular policy among ACP countries, at least among the small minority which have received the bulk of the transfers. It provides loans on easy terms or grants, with little effective tying as to end-use, at least until recently. It has been specified that the loans should be used either within the sector affected or for 'diversification' from it. But the latter can be taken to imply almost any economic activity, while the transfer of foreign exchange directly to recipient treasuries has hitherto made it so 'fungible' that it can be used for virtually any purpose (for example by Idi Amin, to purchase a fleet of Mercedes Benz cars). Some of the worst uses have prompted the EEC Commission to attempt to tighten the conditions, thereby probably lessening its usefulness to more responsible governments with legitimate uses for untied foreign exchange. It will be particularly disadvantageous if this (as seems likely) implies that Stabex must be used for 'projects', since what many countries desperately need is free foreign exchange for the purchase abroad of small but vital items which do not fit under the project heading, like spare parts, tools, nuts and bolts.

One thing which Stabex seems clearly *not* to have done, is to stabilize aggregate export proceeds (Hewitt, 1983). Although Stabex transfers are more rapidly disbursed than most EEC aid, they still arrive up to a year after the decline in respect of which they are given. The selection of items for inclusion is also incomplete, notably excluding sugar, beef and tobacco (CAP products). Basing transfers on individual products has also led to a small number of favoured countries (notably Senegal, Ivory Coast and Sudan) receiving a very large proportion of total transfers and sometimes in cases where overall export proceeds are above average. Moreover, the funds allocated are insufficient to cope with any serious decline in export prices, as occurred in 1982 and 1983, with the result that not all cases qualifying received transfers. In short, Stabex transfers are so arbitrarily distributed with respect to aggregate export-proceeds shortfalls that they can best be characterized as windfall gains – useful to those who receive them but unpredictable and bearing little relationship to needs. There is no evidence that they affect food production or security one way or the other. Nor, to cite another common criticism, is there much evidence that they have had much effect in further locking countries into independence on export crops. Most of the countries in question are so locked in by broader circumstances and overall policy choices that the chance of getting a Stabex transfer is simply not a significant factor.

An analogous policy for minerals (Sysmin) suffers from the same problems as Stabex, but is far less popular with the ACP countries, as it is seen to reflect the interests of mineral firms in Europe more than those of exporting countries. To date, it has resulted in relatively few, but relatively large, transfers, with a focus on the Zambian and Zairean copper sectors. Its impact, if any, on food matters is inextricable from the policies of recipient states.

EEC aid and investment

The other, and perhaps more obvious, portion of EEC policy which affects Africa is aid. Under the Lomé Convention, the EEC operates a medium-sized aid programme which, because it is very heavily concentrated on Africa, is one of the larger programmes in the continent. The Lomé Convention is a renewable five-year contractual agreement, stipulating trading conditions (ACP preferences, Stabex, Sysmin, aids to industry) and a framework for grant aid, together with loans from the European Investment Bank (EIB). By far the largest category of financial transfers is grant aid from the European Development Fund (EDF), which is used either directly for development projects or for subsidizing loans from the EIB.

Relevant to assessment of EDF aid is the fact that it is *not additive*. It does not result in the aggregate transfer of extra funds from the member states. The latter include the portion of their aid channelled through the EEC in their total aid transfers and there is no evidence that the existence of the EEC aid programme increases the total. It does, however, shift both the sectoral and the geographical balance. Almost 40 per cent of total EEC aid is composed of food aid, a much higher proportion than for the member states, so it may be that the existence of the EEC increases the proportion of food aid in combined (national and EEC) aid. This is not, however, certainly the case, since individual members probably reduce their own food-aid transfers correspondingly. This is because surplus agricultural producers, who would otherwise be the major food-aid donors, find it preferable to sell their surpluses into intervention, from whence other members share the cost of shipping it as food aid.

Food aid is poorer value to the recipient than a corresponding amount of money.

Apart from the greater flexibility of money, food aid tends to be valued at what it would have cost in the (normally protected) markets of the donor. In the case of the EEC, this means that the recipients often receive less than half the value marked in the EEC records (EC Court of Auditors, 1980). Correspondingly, EEC members increase their standing in the 'league tables' of aid as a proportion of GNP rather cheaply. Recipients would clearly be losing if the alternative to food aid was aid in liquid form. But in most cases it is not, or appears not to be.[8]

Other issues surrounding food aid concern its distribution, the considerable delays in arrival, its tendency to encourage new and foreign-exchange costly eating habits, and its effects on local production and prices. Only a relatively small proportion of total food aid takes the form of disaster relief, by far the major proportion being transferred to governments and used for increasing urban food supplies. Of all forms of food aid, dairy aid (skimmed milk powder, butter oil) has been most criticized. This is the major form (in value terms) taken by EEC food aid.

The other shift which arises as a result of the diversion of national aid funds to the EEC, is an increase in the proportion going to tropical Africa. Over half of EEC aid is delivered under the Lomé Convention, by far the largest proportion of which goes to tropical Africa. A significant proportion of EEC budget aid also goes to Africa, making the proportion of total EEC aid devoted to Africa higher than for most member states.

An important question about EEC aid is whether it is better or worse than aid from other sources, notably the member states. This is considered in Chapter 9. To the extent that any general conclusion can be drawn at all, it is that, in line with general opinion, EEC aid is probably somewhat worse organized and implemented than is the case for most member states. But the grounds for this generalization are relatively weak. More important, in my opinion, is the fact that the variation within aid programmes is at least as great as that between them, that the variations between recipient countries are at least as large, and that the problems of development aid seem common to the majority of donors. Certain improvements could be made to both EEC and other donor aid, through bureaucratic reorganizations and tightening up of procedures. But this sort of change seems often to miss the point. So much time is spent making sure (or trying to) that funds are not misappropriated, that projects are not delayed, or that definable 'project goals' are achieved, that the purpose behind the whole slips out of focus. This problem is taken up in Chapters 9 and 10.

The purpose of this short chapter has been to introduce some of the salient features of the EEC and its impact on Africa. With the exception of items like Stabex, which are not highly relevant to the present book, all are considered in more detail in the following chapters.

NOTES

1. In 1983, only 7.6% of employment and 3.8% of GNP in the EEC-10 came from agriculture (Comm. 1985a: 195).
2. This is not to argue that such standards are bad things in themselves. It seems, though, that they are more often imposed in response to producers' requirements for economic protection than to people's needs for health or environmental protection. An analogous case from industry concerns motor vehicle safety regulations, especially in the USA. These were fought for several years by the motor industry in every way possible, including character assassination of the safety crusaders. Then it occurred to the manufacturers that this could be used as a means to keep foreign vehicles out of the market, at least temporarily, and what had been impossibly expensive was suddenly able to be achieved.
3. Even in the mid-1950s, the original six EEC members were fully self-sufficient in sugar,

potatoes, pigmeat and dairy products, almost 95% self-sufficient in beef and veal, and 85% for cereals. (Fennell, 1979: 104, cited in BAE, 1985: 83). But the UK, which joined in 1973, was still at that time the world's major food importer, and specifically for those products for which the EEC-6 were approaching self-sufficiency (cereals, sugar, dairy products and beef).

4. Furthermore, the increase in non-agricultural incomes since the 1950s has been such that the proportion of total income spent on food has declined, as have the number of working hours required to purchase any given 'basket' of food. I have no figures for the proportion of total value of food products which gets to the farmer. It may be assumed that the trend is similar to that in the USA where it has declined from 16.8% in 1950 to 7.9% in 1983, (Hansen et al., n.d.: 2). It seems likely, however, to be rather higher, partly as a result of protection.

5. But see Chapter 5 and below Chapter 7, where it is shown that non-tariff protection *is* imposed on Thailand, which cannot retaliate, though *it is not* (yet) imposed on the USA, the other major source of cereal substitutes, for fear of retaliation.

6. The major decline in African oilseed production has been of groundnuts. Part of this can be attributed to declining relative prices – arising both from the advantages of soya beans in cattle feed and to increasing concern over aflatoxin in groundnuts (see Chapter 5). But soil exhaustion and increased incidence of pests and diseases (notably rosette disease) has been another problem in some producing countries (Senegal in particular), strongly related to long-term monocropping. Another has been an increase in local markets and prices for groundnuts for human consumption, this being closely related to African urbanization. All told, it cannot be said that EEC protection is likely to have been a major factor in the decline in raw groundnut exports. Where processed products are concerned, because of effective protection on processing, the story is different.

7. There have been possibilities of exporting cereal substitutes, notably cassava. Some years ago, a large EEC project for the production of cassava in Ivory Coast for export to the EEC was cancelled when it became clear that this would soon become subject to protection.

8. It is extremely difficult to get information on the way in which donor agencies actually decide on how much food aid (or come to that, other aid) to give and how to allocate it between different countries. It seems rather likely, though, that a large consignment of food aid might be taken as the occasion to cut back on some other form of aid. Some of the factors determining the allocation of US food aid are considered in Chapter 8, where the point is made that EEC food aid is rather less easy to assess. In part, this relates to the more obviously political basis of much US food aid. But it also relates to the multitude of bureaucratic processes through which EEC food aid passes, the prevalence of bland and impenetrable language in many of the papers, and the far more limited involvement of parliaments. In any case, as shown in Chapter 9, a wide variety of factors affect donor allocation of aid, including the general feeling in the 'donor community' about the recipient country. These tend to lag somewhat behind the event, and may increase variations in availability of finance. In the aftermath of the financial tightening up in 1979, a number of donors suddenly 'noticed' the problems with use of aid, and reduced transfers – thus aggravating the problems caused (or brought to the surface) by high interest rates and shortage of loan finance.

7
The EEC & the Common Agricultural Policy

The Common Agricultural Policy (CAP) is the financial heart of the EEC, taking about 70 per cent of the total budget. It is also one of the most widely criticized of all EEC policies. Most of this criticism concerns its cost, and effects on European agriculture, but it is also criticized by some people for its negative effect upon Third World countries. Others assert precisely the opposite and go so far as to propose CAP-style policies as a solution to the problems of tropical African agriculture. In general, those who are critical of the effects of the CAP within Europe also tend to claim harmful effects elsewhere and vice versa.

The case is complicated by two things. Firstly, the CAP has different effects for different commodities and countries. Secondly, criticisms of the CAP come from two rather different viewpoints. One of these is a strictly market-oriented economic argument, the gist of which is that protection is always a bad thing. The other is more concerned with food security, and sees this as being endangered by dependence on external sources. The CAP is seen as increasing this dependence.

The core of the factual argument is generally agreed and very simple. The CAP, by increasing European agricultural prices, increases production of the products covered. This either reduces European imports or increases exports. In either case, it exerts a downward pressure on world market prices of the products in question. Where CAP products compete with exports from the Third World (sugar, beef, tobacco), Third World exporters lose from the reduction in world prices.

The complications arise when Third World countries are net importers of CAP-affected products (cereals, dairy products). Here the immediate effect of the CAP is to save them foreign exchange by reducing world prices. The question which then arises is whether this short-term advantage is offset by the dynamic effects. In one version, it has a harmful effect by distorting the pattern of production away from that which would be generated by the free market. In the other, the problem is rather that it generates dependence. In both cases, discussion is further complicated by wide disagreement about the size of the price effects of the CAP.

Before looking at the arguments and attempting to draw conclusions, the chapter begins with a discussion of what the CAP is and how it affects the interests of European farmers and consumers and those of other agricultural exporting countries. There are several reasons, for this. European agriculture has mostly been heavily protected for well over a century and the CAP itself has lasted and grown for about 25 years, in spite of constant criticism. It is clearly supported by significant interest groups and is not going to disappear in a hurry. Proposals for mitigating its ill effects must therefore take account both of its structure and mode of operation and of the political forces which are ranged for and against it. This is particularly important because all available evidence indicates that African food security is *not* among the factors likely to affect decisions about the future of the CAP. Related to this (albeit tangential to African interests) is the point, which will be argued below, that the vast majority of the European population, including most farmers, have an interest in major changes to the CAP, which currently benefits a small but powerful minority at the expense of the rest. This relates not simply to higher taxes and food prices in the present, but also to the pattern of farming and land use which will emerge in Europe in the future, raising a number of fundamental questions about employment and liveli-hood, environment and pollution. Current discussions in Brussels about the future of the CAP ignore most of this, concerning themselves almost solely with the budget on the one hand and competition with the USA on the other.

The Common Agricultural Policy

The CAP is the collective name given to an enormous number of different policies and regulations for the protection of EEC agriculture. These consist of instruments for fixing prices, purchasing and disposing of surpluses, and protection from external competition, which differ widely as between the hundreds of different products, varieties and grades concerned. In addition, they involve instruments for fixing agri-cultural exchange rates (the so-called 'green currency') for trade between member states and then for adjusting for the anomalies which this brings about.[1] There are also a variety of other instruments and policies for assisting 'structural change', encouraging the use of approved practices, defining and limiting the varieties of seeds or other planting material which may be used[2], and for subsidizing 'hill-farmers'. All this comes under the aegis of a body known in English as the European Agricultural Guidance and Guarantee Fund (EAGGF), but more conveniently referred to by its French initials as FEOGA. Of the total expenditure under FEOGA, guidance has taken between 3.5 and 5 per cent over the past ten years, guarantees (that is price-support) the other 95–96.5 per cent (BAE, 1985:54).

In the beginning, the majority of the funds disposed of by FEOGA came from the tariffs and levies imposed on agricultural imports. But as the EEC has approached and then passed self-sufficiency in an increasing number of products, revenues have dwindled while costs have risen. In 1971, as part of a new system, effectively provid-ing the EEC with its own tax-base, these sources of revenue were expanded by the addition of a Value Added Tax of up to 1 per cent, this taking full effect in 1979.[3] Since then, further increase in costs has made it necessary to increase this VAT to 1.4 per cent (as from 1 January 1986), with the prospect of a further increase to 1.6 per cent in 1988. In 1984, VAT contributed 58 per cent of total EEC budget revenue (ibid:58).

In 1984, the total cost of the CAP to the EEC budget was 19 billion ECU. Since 1974, it has never taken less than 65 per cent of the total EEC budget, and on average about 75 per cent (ibid:54). A large and increasing proportion of this goes to subsidize the export of surplus agricultural produce, whose existence in the first place is a direct result of the CAP itself. It has been estimated by OECD that the total cost of EEC agricultural protection in 1984 amounted, on average, to nearly 1,000 ECU per non-farm family of four (ibid:4), of which one-third was in the form of tax and the other two-thirds in higher prices for food and agricultural products. At that time, the ECU was slightly less in value than the US dollar, though now it would be rather more. So, allowing for variations of income within the EEC, it would be a reasonable estimate that an average European non-farm family of four would be contributing some $1,000 per annum to CAP protection, mostly in the form of high food prices, but some in VAT contributions.[4] This is the cost of agricultural protection, and *not* the cost of EEC membership. Without the EEC, some members would be paying more for the same level of agricultural protection, others less.

At the same time, the number of farmers in the EEC has declined secularly since the CAP was initiated, and continues to do so. It is true that the rate of movement out of agriculture has declined since the early 1970s, but even EEC Commission sources admit that this is almost entirely the result of rising urban unemployment and thus nowhere to go, rather than the CAP's success in maintaining rural livelihoods (Comm., 1985a:ii). Agricultural employment in the EEC (10) has declined by 59 per cent since 1960 and by 24 per cent since 1973 (Comm., 1985a: Table 7), giving annual rates of decline of about 3 per cent per annum up to 1973 and about 2 per cent since then.

Yet EEC sources, not uncommonly, refer to the CAP as a success. The 1985 Commission 'Green Paper', discussing proposals for reform, starts: 'The common agricultural policy has sustained the development of Community agriculture over more than twenty years, with results that are substantial and positive' (Comm., 1985b:i). A bit further on it is claimed that 'the economic objectives (of the CAP) have in many respects been well achieved' (ibid:1). What are these?

There is no doubt whatever that the CAP has led to enormous increases in agricultural production and productivity and, since this was what it was intended to do, this is generally conceded as being a success. But looked at in another way, this is precisely the problem. The problem with European agriculture has never been low productivity (not, at any rate, during the period since the EEC was formed). It has been (and remains) high costs. Prior to the inception of the CAP, all the current member countries (with the possible exception of Greece) subsidized their agricultural sectors, but the cost of this was limited for most of them by the fact that they were net importers of temperate agricultural products from cheaper producers. One effect of the enormous productivity increases achieved under the CAP has been to produce large surpluses of most major EEC agricultural products, so that the cost of dumping these on the world market has to be added to that of protecting the internal market. Another effect has been to reduce the labour requirements for producing this much increased output, leading to a decline in agricultural employment and population. Thus in terms of either providing cheap food or maintaining farm livelihoods, the CAP has been anything but a success.

The evidence on farm incomes is more difficult to summarize, but in the final analysis no more encouraging. Farm prices have declined in relation to the cost of farm inputs by some 10 per cent since 1970, though probably to a lesser degree than in other developed agricultural producing regions (BAE, 1985: pp. 50–1). Average farm income per labour unit (inflation-adjusted) remained virtually constant between 1973 and 1983, though other sources claim a slight fall. This is generally claimed as a success for the CAP, on the assumption that it would have fallen considerably without it.

But averages are more than usually misleading in this case. The total farm population has fallen to about 40 per cent of its 1960 level and by about 24 per cent since 1973. Farm size has correspondingly increased very significantly. The distribution of activity and income within European agriculture is enormously unequal and becoming steadily more so. For the EEC as a whole, per capita incomes of the richest 25 per cent of farmers were some twenty times those of the poorest 25 per cent. It is true that the variations were smaller within the different countries because of the very substantial disparities between them. None the less, they were still very substantial (Comm., 1985a:126).

In 1975, it was estimated that the largest 25 per cent of farms (in terms of production) accounted for 73 per cent of total output (Caspari, 1983:87–8), and the degree of concentration has increased significantly since then. Between 1975 and 1983, the number of farms in different size categories changed as shown in Table 7.1, in the seven countries for which figures were available.

Table 7.1: *Trends in EEC farm size 1975–83*

Size (ha)	Number of farms ('000)		Change 1975–83(%)
	1975	1983	
1 – 4.9	682	527	−22.8
5 – 9.9	478	367	−23.1
10– 19.9	640	507	−20.7
20– 49.9	703	665	−5.5
50 +	266	292	+ 9.7

Note: date are not available for Ireland and Greece, which seem to show a similar pattern, or for Italy which (possibly due to statistical anomaly) shows increases in all categories of farms, but with figures only up to 1977.
Source: Worked out from Comm., 1985a: Table 54, pp. 292–3.

Both the degree of, and the increase in, concentration were even greater for livestock holdings. In all cases, smaller than average farms declined between 1975 and 1983, while larger ones grew. By 1983, 21 per cent of livestock farms controlled 64 per cent of cattle, 15 per cent of dairy farms had 52 per cent of cows and 5.2 per cent of pig farms, with more than 200 head, controlled almost 70 per cent of all pigs (Comm., 1985: Tables 56,57,58). Given this increase in farm size, together with the increase in productivity, it would be surprising if farm incomes had not been maintained *on average*. At the same time, it seems quite clear that real incomes for the majority of farm families have declined significantly.

The CAP has done very little for the Community's smallest farmers, many of them in the southern parts of the EEC, who are chained to farming at grossly inadequate incomes by the lack of alternative opportunities and by increasing age. The top 25 per cent of farms receive about 75 per cent of total CAP support, an average of 9,700 ECU per farm per annum. The remainder receive an average of 1,100 ECU per annum, with the smallest getting at most a few hundred ECU apiece (BAE, 1985:5).

None of this is in the least surprising, given that the major instruments of the CAP have been price support and encouragement to technical progress through investment and increased use of purchased inputs. Price support assists producers in direct proportion to their level of output, give or take variations for differential levels of protection between crops. Moreover, since many of the technical innovations which the CAP has so proudly initiated are subject to economies of scale, the margin between gross sales income and farm costs would tend to be greater for larger farms. The proportion of net income earned by the largest farms would thus be even greater than indicated by their proportion of total production. Given this and their generally superior access to credit, it is scarcely surprising that the concentration of European agriculture has increased and with it the disparity between the few rich and the many poor.

Indeed, the effect of technical innovation has been even more direct. Not only does new equipment steadily increase the minimum economic scale of production, but in many cases farmers are forced to innovate by agro-industrial processing and marketing firms which are reluctant to collect small amounts and stipulate the use of scale-economic equipment like milk-coolers.

Of course, all this is not only the result of the CAP. Agricultural modernization is a worldwide phenomenon which has proceeded most rapidly in the USA. Nor is it necessarily a negative phenomenon in itself. Under different socio-economic conditions it could contribute to increasing the income and decreasing the labour burden of small farmers rather than generating surpluses and reducing employment. But under existing conditions, its primary effect is the latter. Moreover, certain of its environmental effects can be considered harmful under any circumstances.

While price support is, under any circumstances, a grossly cost-ineffective method of supporting farm incomes, conditions when the CAP was initiated differed from those prevailing today in several vital respects. Not only was there full employment, but industrial sectors were generally growing so rapidly that they could use all and more labour than was 'released' by agriculture.

Thus while the founding principles of the CAP, worked out at a conference in Stresa in 1959, contain, in addition to goals for increased productivity and development of EEC trade, pronouncements about safeguarding the familial structure of European agriculture, nothing was said about maintaining employment or the number of farms. Indeed, precisely the opposite was intended, since industrialization and urban employment were seen as the means to solve the problem of the many farms which were considered too small ever to become 'competitive' (in the sense of viable *with* CAP price support) (BAE 1985:28–9). The intention of structural adjustment was thus to improve productivity on 'viable' farms, while encouraging marginal farmers to leave for industrial employment.

Another difference was that, at that time, such a policy was not particularly expensive in budgetary terms. Consumers of food paid through higher food prices, but consumers are always less well organized than producer groups, so this did not give rise to serious problems. In any case, real incomes were rising and food expenditures falling as a proportion of total expenditure. At the same time, costs of packaging and processing were rising rapidly enough to obscure the connection between agricultural protection and consumer prices of food.

Twenty-five years later, the situation looks very different. The world, including Europe, has been in a deep economic trough for some years and rates of unemployment are high. Given current trends in technology and productivity, there seems no reason to expect full employment to return in the foreseeable future. The much publicized 'recovery' of recent years, largely a result of US deficit spending and consumer credit expansion elsewhere, has not reduced unemployment significantly outside specific growth areas in the USA and Western Europe. Even if, as seems unlikely, it should be maintained and have more significant effects on world demand and trade than hitherto, the investment generated seems at least as likely to shed jobs as to increase employment. In short, the opportunities for finding non-agricultural employment for those squeezed out of agriculture are more and more limited, even though total employment in the agricultural sector is now significantly less than the aggregate number unemployed in the EEC.

In the long run, however, perhaps most important of all is the enormous structural change which has occurred in European agriculture as a result of technological change, increased control of agriculture from the outside, and policies, like the CAP, which have done all possible to hasten the change. It is not just that farm employment has fallen to about 40 per cent of its 1960 level (Comm., 1985: Table 52) or that farm size has increased very significantly over the same period. Use of purchased farm inputs has also increased enormously, including a switch from local grass to imported fodder for livestock production, and from human energy to fossil-fuel energy for both cultivation and fertilizer application.[5] There has been a similar growth in use of other agricultural and processing machinery on-farm, and all this together has led to a major increase in farm indebtedness, though here the variation between countries is enormous.[6]

The significance of the structural change is multiple. In the first place, European agriculture is far more dependent upon purchased inputs and equipment than it was 25 years ago, and thus more vulnerable to shifts in relative prices of products and inputs. In spite of the major increases in technical efficiency, there are still only a very few European farms which are sufficiently economically efficient to be able to survive without protection. Indeed, since most of this technical change has been premissed on heavy protection, much of it is unsuitable to non-protected agriculture. Moreover, since those farms which could survive without protection are almost without exception very large farms of at least several hundred hectares[7], the vast majority of all farms are just as dependent upon protection as before, if not more so.

Paradoxically, however, a sharp reduction in the level of protection would probably lead to a more than proportionate reduction in the number of large farms. This is because of the difference in decision criteria between 'family farms' and 'business farms'. However modernized, the family farm is business, workplace and livelihood for its operator, who normally lacks alternative opportunities. When prices fall he (or she) is more or less obliged to carry on, unless forced out by bankruptcy. The large business farm is quite another matter. Its operator, often a company and quite seldom involved in the day-to-day operation, has invested quite specifically as a business venture, after comparison of the rates of profit in farming and other sectors. A major reduction in the level of protection would probably reduce the rate of profit in farming below the returns to be gained elsewhere in the economy and thus lead to disinvestment, though falling land prices would make it difficult to dispose of the land except at a loss.[8] All this, however, is speculative since, rhetoric to the contrary, a major reduction in the level of CAP protection seems relatively unlikely.

Given the costs of the CAP, and its failure to assist those most in need – the EEC's smallest and poorest farmers – it is worth asking how it could have lasted so long in its present form, remembering that about three-quarters of total support went to the largest 25 per cent of farmers, not just the richest but those in the best position to survive with lowered protection.

The answer usually given is that producers are normally better organized that consumers or taxpayers. In the latter case, it is worth stressing that the main tax base for the CAP is a value-added tax, which is both regressive (that is, its incidence tends to decrease as income increases) and is paid in the form of an addition to consumption expenditure. Those who pay it are thus organized (or not organized) in a way similar to consumers. European agriculture has been protected, in many cases, since before the industrial revolution, and agricultural producers, their organizations and allies, have gained considerable experience in political lobbying. But even taking this into account, together with the over-representation of the rural vote in a number of countries, it would be surprising if a sector accounting for no more than 7.5 per cent of total civilian employment and only 3.8 per cent of total gross value-added at factor cost (Comm., 1985:195), could achieve such results on its own.

Of course, it does no such thing. Agricultural protection has the support of interests far larger than the agricultural sector as such – the large multinational firms and banks which make up agribusiness. CAP protection has stimulated an enormous growth in the use of inputs and equipment in agriculture, and thus of turnover and profits to these firms.[9] There has also been a concurrent concentration in both these industries and those involved on the other side in the processing, packaging and marketing of agricultural products. Fuel, fertilizers, agro-chemicals and, increasingly, seeds are dominated by the major oil companies. The major pharmaceutical companies are heavily involved in agro-chemicals. Farm equipment as a whole is somewhat less concentrated, though levels are still very high for particular items. The rapidly growing trade in animal feeding stuffs is largely controlled by the major international firms, five of which control some 80 per cent of world trade in grains and oilseeds, together with the oil-crushing and margarine-producing firms which separate oil from cake. In short, agricultural protection expands both markets and profits for some of the world's largest corporations.

From this, it can be seen that the supposed opposition of interests between 'agriculture' and 'industry', with which many economists operate, has limited relevance to the real world. Authors like Burniaux and Waelbroeck (1984) argue that European industry would benefit from the ending of EEC agricultural protection, since this would reduce food prices and thus wages, making European industry more competitive and allowing it to regain markets from the Third World. In this, they ignore a number of factors absent from economic theory but highly relevant in the real world. Capitalist firms are interested in profits (which is in all the economic textbooks), but they are quite as content to make them from protection as from optimization theory (which is not). Apart from the inter-relations mentioned above, it is often the same firms which produce manufactures in Europe and in the Third World, and their (highly complex) investment decisions will be based on group profits and growth strategies rather than on either geographical or sectoral interests as such. Of course, there are industries which have no connection to agriculture and no benefit from agricultural protection. But among the really large multinational corporations, most have at least a few fingers in this profitable pie.

How does the CAP work?

The basic instruments of the CAP are price support, protection against imports, and subsidies for the disposal of surpluses through exports or otherwise. Producer prices are set at levels agreed by the Council of Ministers, on the basis of proposals from the Commission and a subsequent process of bargaining among the various interests involved. For the most important crops (which include cereals, dairy products, beef and sugar), the prices arrived at are then supported by three mechanisms. Variable levies on imports prevent the latter from undercutting internal prices. Intervention arrangements involve purchasing surpluses, to prevent increases in local supply from reducing prices below the agreed level, and 'export restitutions' subsidize the prices of EEC agricultural products down to the levels prevailing on world markets (BAE, 1985:31–2). In addition, there are a number of other subsidies, which encourage the internal use of surplus products.

AN EXAMPLE FROM THE DAIRY INDUSTRY

Fifteen or twenty years ago, significant amounts of milk were either separated on-farm or sold to creameries which did so. The skim or butter milk was either retained by or returned to the producers for calf feeding. The same process now takes a less direct form. The milk is sold to dairies which separate cream and butter and turn the remainder into skimmed milk powder. Most of this is then 'sold into intervention', since there is no market for it at the high subsidized price. From here FEOGA provides subsidies for a certain amount to be sold cheaply to farmers, who replace the water and feed it to calves. The major beneficiaries from this technological progress are the dairies, since a previous waste product is now turned into a saleable item. Of course, the benefit derives from the heavy subsidy and not from the technical innovation itself.

Another major benefit to the dairy companies is the fact that skimmed milk powder (SMP) and butter oil (butter with the proteins boiled away to make it storable) are excellent export products (with the heavy subsidies). A previous limitation on international trade in dairy products was their weight (about 95 per cent water) and perishability, so that, even with the water content reduced (condensed milk), they had to be canned to prevent rapid deterioration. But SMP and butter oil have all the water removed and are both storable and relatively cheap to transport. They can thus be shipped to destination – increasingly in the Third World – where the water is replaced and the product sold to urban middle-class consumers. This has a further advantage for European producers of dairy equipment. It boosts exports of dairies for mixing and pasteurization of SMP and BO, again often with subsidies, this time normally from the state loans or export credit funds of member states.[10]

The variable levies which protect the most important CAP products from external imports are more effective for this purpose than tariffs. A tariff is normally set in percentage terms or as a flat charge per ton of produce, which risks that prices may fall so low in external markets that some imported produce can compete. A variable levy is set to fill the gap between the internal price and the cheapest imports on offer (often at weekly intervals) and can be increased as soon as there is any risk of local producers being threatened by imports. Being a more efficient means of stabilizing internal EEC prices, it also has a greater destabilizing effect on world prices (Koester and Bale, 1984:34). The precise details vary from product to product, involving different 'target', 'threshhold', 'intervention' and other prices, whose inclusion here would merely confuse matters.[11]

For other products, there is a further variety of protective instruments. Poultry, eggs and pigmeat are protected by different means, but with the same purpose, to compensate producers for the high cereal (and thus feed) prices which result from CAP protection. Oilseeds and sheepmeat are protected by the 'deficiency payments' system, similar to that operated by the UK prior to joining the EEC. Under this system, the market price is (broadly) allowed to be set by the price of imports, while producers receive a budgetary payment to make up the difference between this and the support price. This keeps prices to consumers relatively low, the producer subsidy coming direct from the EEC budget. It is particularly important for oilseeds, since oilseed cake is an important source of protein for animal feeding.

The CAP and agricultural surpluses

As already indicated, one of the major effects of CAP price support has been to increase agricultural production and generate surpluses. This was already evident almost from the start, but there was a brief respite from their inexorable growth in 1973, when the UK, Denmark and Ireland joined the EEC. For although the two latter are agricultural exporters, the former has historicallly been the world's major importer of agricultural produce, especially dairy products. This respite turned out to be extremely brief, for not only did production continue to outstrip demand in the other EEC countries, but EEC price support has generated a rapid growth in agricultural production in the UK itself. Table 7.2 shows degrees of self-sufficiency in certain products, in 1972–4 and 1981–3 for the UK and the EEC-10 as a whole.

Table 7.2: *Degree of self-sufficiency in certain agricultural products EEC-10 1972–4 and 1981–83 (percent)*

Item	EEC-10 1972–4	EEC-10 1981–3	UK 1972–4	UK 1981–3
Cereals (exc. rice)	91	109	68	105
Wheat	104	127	60	102
Sugar (EEC-9)	91	144	30	53
Whole milk powder	231	395	74	392
Skim milk powder	143	142	186	233
Cheese	103	107	61	71
Butter	98	122	19	65
Beef & veal	96	103	70	83
Sheepmeat	66	74	49	65
Vegetables oils	77	82	33	52

* The EEC-10 did not exist in 1972, so the figures refer to the countries which would form its future membership.

Note: Milk powder can be in surplus even when dairy products in general are not. That is, it is one way of transforming seasonal or local surpluses into a storable form. The UK was a net importer of milk powder until 1971 (FAO, *Trade Yearbook 1972*).

Source: Comm, 1985, Table 26, pp. 247–8.

For several of these products, the generation of surpluses has made the EEC into a major world exporter, in some cases, the most important exporting region. Table 7.3 shows EEC imports/exports by volume and as a proportion of total world trade for selected years.

Table 7.3: *EEC external trade in selected agricultural products 1973–83, by volume and percent of world trade*

YEAR	Volume of net exports					Proportion of world trade				
	Wheat	Other Grain	Sugar	Butter	Beef & Veal	Wheat	Other Grain	Sugar	Butter	Beef & Veal
	Mt	Mt	Mt	Kt	Kt	%	%	%	%	%
1973	0.0	−10.8	−1.1	204	−913	0.0	16.2	7.2	30.2	27.1
1974	−0.1	−12.0	−0.1	−40	−284	0.2	16.0	0.6	8.3	10.3
1975	2.1	−12.4	−1.3	−90	−118	2.8	18.8	0.6	22.2	4.3
1976	2.4	−11.4	−0.4	−6	−198	4.0	14.4	2.8	1.2	5.7
1977	0.9	−20.8	0.1	140	−210	1.3	27.6	0.5	23.0	5.9
1978	0.0	−8.6	2.0	119	−215	0.0	9.9	12.5	20.2	5.5
1979	3.9	−8.1	1.7	389	−63	5.1	8.6	10.2	47.9	1.5
1980	6.1	−5.9	2.5	481	333	6.6	6.1	14.0	50.1	8.7
1981	10.7	−3.4	3.3	373	389	10.8	3.3	17.4	43.4	10.0
1982	10.8	−2.3	3.7	292	160	11.2	2.4	18.6	36.5	4.1
1983	11.5	0.5	2.6	250	165	11.6	0.6	13.8	35.0	4.2

Source: BAE, 1985, Table 33, p. 142.

(Mt = million tons, Kt = thousand tons)

Notes:

1. EC-9 to 1980, EC-10 from 1981.
2. Net Imports denoted by – sign.
3. Proportion of world trade except for sugar, for which % of world free market trade.
4. Trade in beef and veal includes carcase equivalent of trade in live animals.

The changes shown are certainly quite spectacular. The EEC has gone from little more than self-sufficiency in wheat to being the world's third largest exporter and or other grains, from being one of the world's major importers to virtual self-sufficiency. For sugar it has moved from being a major importer to the most important exporter, as had also happened by 1984 for beef and veal. Butter exports peaked in 1980, but stocks have increased rapidly since then to 1.2 milion tons in 1985 (FAOb, 1986:71). The same is true for skimmed milk powder and cheese, for both of which the EEC was also by far the world's major exporter (ibid). (In net terms. Gross EEC exports come to between two-thirds and three-quarters of total world trade, because of the huge size of EEC intra-trade – see Chapter 5).

How does the CAP affect Third Countries?

Table 7.2 indicates clearly what the effects of the EEC are. Given the huge surpluses which it generates, it will evidently tend to lower world prices for the commodities in question.

The countries most obviously affected by CAP surplus generation are competing export producers of temperate agricultural products: the USA, Canada, Australia, Argentina and New Zealand. Worst affected are those other than the USA, for they bear the brunt of agricultural surplus disposal from both the EEC and the USA, without themselves having the economic or political power to retaliate. While US agricultural exports are not subsidized in quite the same way as those of the EEC, they are subsidized none the less, besides which the USA is an aggressive competitor in international agricultural markets.[12]

It is the Third World, however, and specifically tropical Africa, which is of interest in the present context and the interactions of different developed exporters are relevant only insofar as they affect conditions facing Third World countries. It is fairly obvious that subsidized competition between the USA and the EEC will have a greater effect on world prices and trading patterns than the policies of either alone. But what can be said about these effects? By how much do EEC surpluses reduce world market prices or its protection increase their instability, and how does this affect tropical Africa?

The plain answer is that the large number of different econometric studies which have tried to answer this question have come to such varying conclusions, that one cannot say with any certainty. This has partly to do with the different methods used to assess the effects of CAP protection and the different assumptions made. It also depends significantly on the particular set of years taken for study. If one chooses a period when the difference between EEC and world prices is at a maximum, for example 1975/6–1979/80, the effect of EEC protection will tend to be greater than during a period when, because of higher world prices or budgetary constraints within the EEC, the difference is less, for example 1980/81–1982/3.

Where price instability is concerned, there is also some disagreement. Koester and Bale (1984:34) cite their own previous work to the effect that EEC protection policies can be shown both theoretically and empirically to have had marked destabilizing effects on world grain prices. Josling and Barichello (1984), on the other hand, claim, on the basis of studying international grain markets between 1972/3 and 1982/3, that the major grain exporting countries have contributed to stability in cereal prices by building up stocks rather than exporting surpluses on to the world market, though their conclusions regarding the EEC are ambiguous. Harris (1984) does not explicitly disagree, but notes a number of important exceptions and points out that stock-holding policy is only part of the story. Certainly, one would wish to know what was stabilizing about US grain policy in the period 1972–4 (See Chapter 5).

Most of the econometric studies have as their purpose to measure the losses suffered by either third countries or Third World countries as a result of EEC protection. Reduced to its simplest terms, the methodology consists of the following steps. First, estimates are made of the change which could be expected in world prices of different agricultural commodities if EEC (or in some cases OECD country) agricultural protection were to be eliminated (or in some cases reduced by half). That is, the increase in EEC demand from lower consumer prices and the reduction in supply from lowered producer prices are estimated and combined to give an estimate of increased import demand (or decreased export surplus). Most studies then go on to estimate further adjustments as a result of this. For example, if the first-round effect of eliminating EEC protection on wheat was to eliminate the current export surplus, replacing it with a significant import demand (as might well be the case), then the overall balance of world supply and demand would be changed by some amount in excess of the current export (say, for example, by 15 million tons). Since this is also about 15 per cent of current world trade, it would obviously lead to a significant increase in prices. But this, in turn, would lead to increases in production and exports elsewhere in the world, not to mention that the EEC itself would probably readjust production upwards. After the new adjustments have been made, a new world price and pattern of trade are estimated, and it is from these that the gains or losses to different groups of countries are worked out. That is, their new levels of exports, in value terms, are compared with the 'baseline' situation.

Since it is clearly impossible to compute all these adjustments for every commodity and every country in the world, it is necessary to simplify. One way to do this is to restrict the study to one or a few commodities. This has the disadvantage that the interactions between different products is missed (changes in the price and demand for meat affect demand and prices of products used for feeding livestock, for example). The other way is to lump together several products ('dairy products', 'meat', 'other agricultural products'), or groups of countries ('OECD', 'LDCs', 'the rest of the world'). This can give rise to similar problems, if interacting products are included within a group, and also gives rise to categories which are rather broad if specific

Table 7.4: *Effects of trade liberalization on world prices: comparison of ten different studies*

Author	Period	% Reduction	Reference	Commodities in Study	Effect on Wheat	Barley/ma...	Sugar	Beef	Butter
Valdes & Zietz 1980	1975–77	50	OECD	99	4.9	7.6	7.7	6.8	
Koester 1982	1975–77	100	EEC	Grains	9.6	14.3			
Lattimore & Weedle 1981	1979	100	All	Dairy					100.0
Tangermann & Krostitz 1982		50	All	Beef				16.0	
"		50	EEC	Beef				5.0	
Roberts 1982	1981	100	EEC	Sugar			7.0		
Sarris & Freebairn 1983		100	All	Wheat	11.0				
"		100	EEC	Wheat	9.2				
Anderson & Tyers 1983	1980	100	EEC	Grain/meat	13.0	16.0		17.0	
von Massov 1984	1979–81	100	EEC	Beef				1.7	
Koester & Valdes 1984	1980	100	EEC	6	4.6		9.7	10.5	28.3
Matthews 1985	1978–82	100	EEC	13	0.7	0.5	6.0*	3.9	11.0

conclusions are to be drawn. There are a number of ways to reduce the effect of these problems, which are described at length in the different studies. When all is said and done, however, most depend on some fairly heroic assumptions about elasticity coefficients at a number of different stages.[13]

For purposes of comparison, Table 7.4 shows results from a selection of the studies which have been made. The table shows the estimated increase (or decrease) in world prices which would result from reduction or elimination of EEC (or in some cases OECD country or world) agricultural protection. That is, in many cases, it shows intermediate results, which the authors then use for further calculation of gains and losses.

One of the most influencial of these studies (Valdes and Zeitz 1980) reaches the conclusion that a reduction in the level of OECD agricultural protection by half would increase world trade by $8.5 billion, of which 36 per cent, or just over $3 billion, would accrue to 56 selected Third World countries. Of the total increase, 47 per cent is accounted for by raw and refined sugar and by meat and meat products, these being followed in importance by green coffee, wine and tobacco. The expected net income transfer associated with this (that is, roughly what the beneficiaries could expect to earn from the increased trade, when costs of production have been taken away) is estimated at just over $1 bn. (all the above are in 1977 prices and refer to gains in 1975-7).

Middle- and high-income LDCs reap the vast bulk of the expected gains, with $2.5 mn (83 per cent) of the total trade increase and $882 mn (84 per cent) of income gains. Geographically, Latin America is by far the major expected beneficiary (60 per cent of trade and 53 per cent of income). Sub-Saharan Africa gains only $253 mn (8.4 per cent) of the trade increase and $146 mn (14 per cent) of income transfer. Moreover, no account is taken of the impact on prices of items for which Third World countries are net importers. Valdes and Zietz' Table 2 indicates that the increase in export proceeds would cover only 39 per cent of 1975-7 imports of cereals (Third World) or 41 per cent four sub-Saharan Africa, but *without* adjusting for cereal price increases predicted within the model (ibid:30). Moreover, sub-Saharan African cereals imports have more than doubled since then. In terms of likely real effects the preponderance of sugar also gives rise to problems. Whatever happens to tariffs, it is hardly likely that the US change to HFCS will be reversed. Nor can one imagine that the EEC, having built at all its new plant for the past decade for refining beet sugar, would return to importing cane sugar. As soon as one starts to look at concrete cases, a number of such problems arise, but for the present we are concerned with econometric studies.

The next seven studies in the table are for relatively small groups of commodities and thus give no overall estimate of trade or income gains, though Anderson and Tyers estimate the losses to Australia from EEC protection. Koester and Valdes (1984) is a much shorter work, using a simpler methodology and clearly in part an adjustment of the results from the authors' earlier studies, Valdes and Zietz (1980) and Koester (1982). It provides no estimates of overall trade or income gains to Third World countries from liberalization, which is probably a good indication that they do not find them to be much, if at all, higher than did the 1980 article.

Matthews (1985) comes to markedly different conclusions. He arrives at somewhat lower figures for world price increases, in part as a result of the different time period chosen for analysis (above). But this is not the most important factor. Where Valdes and Zietz concentrate heavily on products for which developing countries are net exporters and largely ignore the effects of price increases on net importers, Matthews lays considerable weight on the latter. So important is this effect in his calculation that he finds the less developed countries as a bloc to *lose* $663 mn from the elimination of EEC protection in one of the scenarios he posits. In the other, where a higher LDC supply response to improved prices is assumed, there is a small gain to all LDCs of $256 mn.[14]

The reason for this, which is certainly relevant to Africa, is that LDCs would pay considerably more for their imports of cereals and dairy produce than at present, this more than offsetting the gains from increased exports. In the case of sugar, he finds (as do some other authors) that the gain from liberalization is more than offset by the loss of current quotas to certain ACP countries, to sell within the EEC market. Finally, LDCs lose export proceeds from oilseed exports, since elimination of EEC protection of its livestock sector would severely reduce the world market for oilseed cake.

Matthews does consider the possibility that low food import prices for developing countries may not be an unmixed blessing, but, at least where comparison with Valdes and Zietz is concerned, there is some logic in his conclusion that, if increased prices to exporters should be considered as income gain, then increased prices to importers should be considered a loss.

While there is room for (and has been) much discussion about the relative correctness of the different figures, this does not tell one much about how significant the estimated gains or losses are. It is thus worth giving a few comparative figures to indicate what an income transfer of between $1 bn and – $0.7 bn would mean for the Third World, or plus or minus $146 mn for sub-Saharan Africa.[15] To simplify matters, I have assumed that the figures of Valdes and Zietz can be applied to 1980-82, without adjustment. The basis for this is that figures in Koester and Valdes are sufficiently like those of the original article to make this a likely suppostiion, while most would agree that the impact of the EEC on world markets was reduced during the period 1980-83. In any case, the comparisons to be made do not require very precise figures.

In the period 1980-84, the annual cost of EEC agricultural protection to EEC tax-payers was about $25 billion (BAE, 1985:102).[16] The total long-term debt of all LDCs rose from about $450 bn in 1980 to $781 bn in 1985 (total debt $991 bn) and debt service from $84 bn to $133 bn (OECD, 1986:62-3). World military expenditure rose from about $564 bn in 1980 to $649 bn in 1982 at 1980 prices (SIPRI, 1986:270). Military spending of LDCs rose from $101 to $114 bn over the same period (ibid.). Total net financial flows from all developed to all developing countries were between $70 bn and $80 bn between 1980 and 1984, in 1982 prices (OECD, 1984:201), while the official development assistance (oda) (concessional) component of this was around $27 bn over the same period (ibid.). In short, even if the expected increase in net income transfers was twice as high as estimated by Valdes and Zietz, it would still be rather small beer by comparison.

Turning to sub-Saharan Africa, total indebtedness rose from $44 bn in 1980 to $82 bn in 1985, with debt service rising from $5.6 bn to $11.3 bn (OECD, 1986a:74-5). In 1984, the debts of Nigeria were $14 bn, those of Sudan and Ivory Coast in excess of $6 bn. Military spending declined in real terms from $14 bn in 1980 to $11 bn in 1984, from very incomplete data (SIPRI, 1986:274-5). Total net oda transfers from all OECD countries to sub-Saharan Africa were about $8 bn per annum between 1980 and 1983 (OECD, 1984:228-9) while total net financial transfers (including non-concessional) were about $12-14 bn. EEC Commission oda to African ACP countries was about $580 mn in 1980 (Raikes, 1984: Table 6). In short, within the context, the expected gains from liberalization are even less significant for Africa than for the Third World as a whole, if indeed they were positive at all.

Given this, it seems pointless to spend too much time looking at the finer points and assumptions of the different studies. One can conclude directly that, on this basis, EEC/OECD agricultural protection is not one of the most important factors affecting the poverty of Third World countries. This still leaves open the possibility that it may negatively affect resource allocation and income distribution within those countries.

This argument cannot, however, be made in terms of econometric studies with countries or groups of countries as actors. It is considered in the next chapter, since a similar case can be made regarding food aid.

It is, however, worth considering the unrealism of one of the assumptions which many of these studies (though not Valdes and Zietz) hold in common. In attempting to measure the effect of EEC protection alone, they compare 'the world as it is' with a hypothetical world in which EEC agricultural protection is removed (or reduced by half), while everything else remains the same. This is, to put it mildly, an unlikely eventuality. That the EEC would simply eliminate agricultural protection, having carefully built it up to 25 bn ECU in budgetary costs and nearly double that in consumer prices, is hard enough to imagine. That it should do so without corresponding reductions from the USA and other agricultural surplus producers is just about inconceivable. All of the authors concerned are well aware of this, but find it necessary to make the assumption in order to measure the effect of EEC protection.

It is also worth discussing briefly another model and series of estimates which are not included in Table 7.3, since the results are presented differently and are thus not comparable. Burniaux and Waelbroeck (1984) differ from all the other studies in using a general equilibrium model, which includes not only agricultural commodities, but industry as well. They compare expected results in 1995 as between a future in which EEC protection is eliminated and another 'base case forecast', which represents current forecasts made by the World Bank 'on a judgemental basis' (that is outside the model). This 'base case' assumes a further increase in EEC agricultural production and exports, combined with self-sufficiency in virtually all agricultural products, generating an estimate of an overall agricultural protection level for 1995 of 77 per cent compared with 54 per cent for 1978 (that is, roughly the level of agricultural protection currently existing in Japan).[17] For comparison, they offer several different 'runs' of a model assuming no EEC agricultural protection, but reflecting different assumptions about price trends, non-agricultural protection and balance-of-payments constraints. Their model offers estimates of income benefits or losses to groups of countries, to urban and rural sectors, and estimates of changes in per capita food consumption in urban and rural sectors.

Their first estimate of the effect of eliminating EEC protection draws the conclusion that per capita incomes in European agriculture would fall by 49 per cent, rural-urban migration increasing appreciably as a result (i.e. the rural areas would be more or less emptied of their remaining population). Agricultural production would be only 16.5 per cent lower than with protection, presumably implying elimination of all except very large farms and expansion of the latter. The European industrial sector would grow, as would industrial exports, cutting into the gains made in this area by NICs over the past decade. In very general terms, the model predicts a pattern of world production and trade which returns to the situation of the 1960s, with Europe more dominant in industry and the Third World back in its old role as agricultural exporter. The only available estimates of urban and rural incomes and food consumption in Africa refer to a different 'scenario' (lower world prices) and are not fully comparable. In this case, real incomes are 1.8 per cent higher in 1995 than they would be with the maintenance of protection, rural per capita food consumption is 2.5 per cent higher and urban per capita food consumption 1 per cent lower. Given that none of these are even known to within plus or minus 10 per cent, this seems an undramatic result, though if one took account of the fact that the fall in urban consumption would be concentrated on those thrown out of their jobs by the realignment, the effect could be very considerable. The main results occur in Europe, where rural per capita food consumption is estimated to fall by 17.5 per cent, while, in the towns, it rises by almost

30 per cent. Since this latter increase seems to be composed largely of meat and sugar (Table 2), an increase in the incidence of heart-disease is among the permissible conclusions which the authors do not draw. More seriously, such conclusions seem highly unlikely, given elasticities of demand for food products and trends in consumption over the past several decades.

The authors do not claim that this scenario is at all likely, but simply that it is a better predictor than the other models of what would happen *if* the EEC eliminated protection. That is, a general equilibrium model allows for the assessment of the full range of economic adjustments rather than the relatively narrow range allowed for in even the broader partial equilibrium models. Most of the other authors would probably assert, to the contrary, that what is gained in generality is lost in the use of categories so broad as to lose sight of reality. Dairy products, for example are not separated, but included as part of 'other foods'.

My own opinion is that the total elimination of agricultural protection would imply such drastic changes in so many relations other than the economic, that it would be about as realistic to try to model the economic consequences of nuclear war. It is also my opinion that the latter is, under present circumstances, about as likely to occur as a total elimination of EEC agricultural protection with no corresponding changes in other parts of the world.

What the authors do appear to claim is that the future without protection is in some way preferable to that with it, though to my mind it is difficult to draw this conclusion from the figures they present. Certain incomes rise. Others fall. All are conceived as aggregates or per capita averages, without regard to distribution. On such a basis, it is hard to see how the authors' claim that the liberal outcome is preferable can come from other than their prior assumption that free trade is preferable to protection. Thus they find the expansion of developed country industrial exports and shrinking of Third World industrialization to be a positive step, since this moves nearer to the pattern generated by market forces. Even allowing for the fact that a considerable proportion of protected Third World industrialization may have negative economic effects, this conclusion seems dubious, especially since it is the least heavily protected (export industries) in the Third World which would be most severely affected.

But the real problem with this model is that, while it may be a general *economic* equilibrium model, it excludes, as by its nature it must, any reference whatsoever to responses other than economic adjustments. And yet one knows full well (as do the authors) that there is hardly a single country in the world today which would follow such a (lack of) policy.

Indeed, one of the most useful contributions of this model, to my mind, is to show how totally unrealistic it is even to think of simply eliminating agricultural protection in Europe. To reduce agricultural per capita incomes by about 50 per cent, with this clearly concentrated among the smaller farmers, would generate civil war long before it had been implemented. Given current capital intensity in manufacturing, the increase in industrial production would probably not even reduce the present level of unemployment, to say nothing of employing the millions thrown out of agriculture. Perhaps even more explosive would be the results in the huge cities of the Third World, where industrial production and employment would fall, together with per capita incomes.

If these last remarks may appear unfairly dismissive, this is not from any wish to deride the work and mathematical sophistication which has gone into the various studies cited. But it is my strong opinion that they start from mistaken premises and attempt to measure the wrong things. That any large economic region will entirely eliminate agricultural protection, or reduce it by half, across the board, seems wildly unrealistic. That the response would occur simply in terms of economic adjustment to given price changes seems even more unrealistic. So does the assumption that one can divide the world into watertight geographical regions and sectors, when most of the largest companies span both. Far more useful than the statistical measurements are the discussions of what has happened in specific markets and what might actually be expected to happen in the future, as can be found in the discussions of, for example, Koester (1982), Matthews (1985) and Valdes and Koester (1984), despite my disagreement with the conclusions drawn by the latter (see below).

The other point relates to the measurement of benefits accruing to countries, and the implicit assumption that this necessarily implies benefits to a majority of their citizens. Here an example may be in order.

Zimbabwe is one of the beef-exporting countries in Africa, which would theoretically benefit from the freeing of agricultural trade. In reality, it probably would not, even accepting the terms of the argument, since its current quota to export beef to the EEC at well above world prices is probably worth more in export revenue.

But there is another side to this question. Most beef in Zimbabwe is produced on large-scale commercial farms, mostly owned by whites. Beef likely to meet EEC specification is almost exclusively so. Foreign exchange is an important problem for Zimbabwe, shortage of land for the African 99 per cent of the population is much more so. Not only is over 40 per cent of all agricultural land still under the control of large farmers, but so is three quarters of all land with rainfall above 750 mm per annum. A significant proportion of this land (and of land in the next best category) is used for grazing beef at very low stocking densities. Far worse land, with lower and less reliable rainfall, is grossly overcrowded and, thus, of necessity, overcultivated and overgrazed in the 'communal' (African peasant) areas.

Since Independence, a considerable amount of land has been transferred to African peasants, though very much less than originally planned. This has been strongly criticized by 'economic-minded' development experts, who claim that African peasant crop yields are so much lower that this will lose the country both food and foreign exchange. The evidence for this proposition is largely spurious and relies on two things. Firstly, large-scale and peasant yields are compared across the board, although the land occupied by large farms is dramatically better. Secondly, it ignores the large amount of large farm land which is not cultivated, but grazed at very low densities (Weiner *et al.*, 1985). It seems highly likely that Zimbabwe's recent allocation of an EEC beef quota will strengthen the hand of this faction and further slow down the transfer of land. It is hard to see this likelihood as anything other than a disaster for the majority of the population. In short, one cannot say what effect an increase in export proceeds or 'income transfer' to a country will have for its inhabitants, without considering what goes on inside it.

To summarize the conclusions of this section, it is not clear what effects eliminating or reducing EEC agricultural protection would have. Not only is the range of answers given far too wide for generalization, but the whole exercise makes such unlikely assumptions about the (lack of) political response as to be thoroughly unrealistic. If one compares the whole range of results with other relevant financial magnitudes, agricultural protection does not emerge as a major factor, but this conclusion is subject to just the same caution as the others. Finally 'gains' or 'losses' to a country tell one little about the effect on its inhabitants.

What are the options for reform of the CAP?

It seems opportune at this point to consider what are the realistic possibilities for reform of the CAP. One can start out with some limiting propositions. The first is that the likelihood of a total elimination of agricultural protection is so small as not to be worth considering. Another is that any partial reduction will not occur in the form of an across-the-board percentage reduction. On the contrary, the most likely form will be one which reduces the budgetary costs while retaining the maximum protective effect. To predict this requires no special genius. Not only is it the obvious policy, given current pressures, but it has already started to happen.

BUDGETARY PRESSURES

During the past two or three years, the cost of CAP protection has started to press seriously upon budgetary limits, and this is likely to continue, in spite of the raising of the proportion of VAT from January 1986, which increased the budget by about 22 per cent. Given increasing political strife within the EEC over the budget, it is clearly necessary to limit expenditures on the CAP, and to do this it is necessary to limit or reverse the growth of export surpluses, since it is these which load the budget most. They are also the aspect of EEC protection which provokes the most aggressive response from the USA, since the surpluses compete with US agricultural exports.

But price reductions sufficient to reduce production significantly would have to be so large as to be politically impossible. Recent estimates by the EEC Commission indicate that, in order to get rid of the cereal surplus, it would be necessary to reduce producer prices by about 20 per cent. Yet in 1985, in the face of serious budgetary constraints, with a record grain harvest, record stocks and the prospects of further increases, it proved impossible to get Germany – a country which is a net contributor to the EEC budget and which has complained about its size in the past – to agree to any price reduction for cereals at all. For dairy products, the Commission estimated in 1983 that a producer price reduction of 12 per cent would be needed to reduce production to the 1981 level, which had been defined as a tolerable 'threshold' (BAE, 1985:224).

The two most likely alternatives are the partial replacement of tariff/levy protection by quotas and attempts to 'round out' the pattern of production and protection, so that the EEC is self-sufficient in even more agricultural products than at present, hopefully by shifting production from areas currently in surplus. Before considering these two alternatives, it is worth noting briefly another policy change of rather less significance.

Co-responsibility levies have already been introduced for dairy products. They are a form of tax levied on products, and have the effect of a price reduction, with three differences:

- some producers are exempted from paying the levy (mostly small farmers);
- since the levy is paid into the EEC budget, it does not reduce consumer prices, as a direct price reduction would;
- one cannot be certain that a 2 per cent levy is the equivalent of a 2 per cent price reduction, since the Commission may take the existence of the levy into account when setting prices (and set them higher on that account).

It is claimed, on this basis, that co-responsibility levies are less effective than price reductions in reducing surpluses (BAE, 1985:311), but given the rather low demand elasticities for dairy products, the difference is probably not large. Co-responsibility levies are simply a means of introducing a (relatively small) hidden price reduction and paying the proceeds to the EEC budget. They do not constitute a very significant change in policy.

QUOTAS

The change to quota restriction in the dairy sector does, however, mark a significant change within the general framework of the CAP, although a quota system has been in operation for sugar protection from 1968 onwards.

The general idea with such a system is that the EEC sets quotas or 'threshholds' for total production and the protected price is paid only for produce within that quota. *If* the quota can be set and maintained at a level which reduces the export surplus, then the result should be a considerable reduction in budgetary costs, while keeping farm prices at a higher level than an equivalent reduction in production through price-cutting. Consumers continue to pay as high prices as before.

But sugar has been subject to a quota system for almost two decades, during which time the EEC has grown to be the world's major sugar exporter. One reason for this is the high level at which quotas have been set. Another is the multiple quota system in use. There are two separate quotas 'A' and 'B', for which different prices are paid ('A' being the full protected price). 'C' sugar, over and above the two quotas, has to be sold at world market prices, though it can be stored at EEC expense until the next year (so that if there is a shortfall then, it can be included as quota sugar). Both of these factors have helped to increase production. A third factor is that, given the high prices for quota ('A' and 'B') sugar, it could pay large and efficient producers to produce non-quota ('C') sugar, even for sale at world market prices, since fixed costs are covered by the former (Harris, 1980, cited in BAE, 1985:200).

In general terms, the setting of quotas is subject to the same processes of political bargaining as affect price-support levels. On the other hand, since agricultural producers lose less from quotas than from price reduction, their opposition could be expected to be less and the level of production and surpluses to be reduced somewhat. If quotas can be made to work, one might think them a reasonable compromise under the circumstances – so long as it is necessary to retain price support under the CAP at all.

The response from economists has generally been negative, however. In part this derives from the justifiable suspicion that since producers have shown their ability to bend quota systems in the past, they will probably continue to do so. But it also derives from the general opinion within the profession that free trade is preferable to protection, and that if there must be protection, tariffs are preferable to quantitative restrictions. Tariffs merely distort market responses, while quotas aim entirely to prevent them. Moreover, it is extremely difficult to work out the effects of quantitative restrictions with econometric models, though, as shown in the previous section, it is not that easy with tariffs, either.

Apart from this, it is said that quota restrictions will limit future structural change within European agriculture, freezing the pattern of production as of the time they are imposed. Personally, I cannot see this as a major problem, since structural change and the concentration with which it has been associated have been among the major problems in EEC agriculture, reducing agricultural employment and the number of farms. The experience with sugar also indicates that this fear may be exaggerated.

Another objection is that quantitative restrictions stabilize production, prevent it from responding to prices and so destabilize prices outside the EEC. But opinions are divided over whether EEC protection currently destabilizes world prices, and other exporting countries might welcome some destabilization, if accompanied by a higher price level. The setting of quotas might well follow a 'market' pattern, with pressures for expansion in the event of an improvement in world prices and the onset of the next round of shortages. Indeed, this sort of tactic will probably limit the success of quotas in reducing production, unless the Commission and member states are willing to take a much harder line than hitherto.

Another point made is that imposing quantitative restrictions on one or a few crops merely ensures that surpluses of others will increase, as squeezing one part of a balloon makes it swell in other parts. This seems quite likely to be true, though price reductions for surplus commodities would have the same effect. This relates to another criticism of quotas: that one ends up with a massive entanglement of bureaucratic regulations. This is again true, but would seem more of a problem if there was not already such a massive entanglement under the current price-support-oriented CAP. It is probably true, however, that quotas will gradually spread to cover more and more products, leading to anomalies and thus opportunities for some to speculate and profit from inside knowledge. In short, while it is hard to see why quota protection should be so much worse than the current system, neither does it offer hopes for any significant improvement.

DIVERSIFICATION

The idea of 'diversification' is to find alternative crops and end-uses for present ones, to give scope for increasing agricultural production without increasing over-production of the current surplus products. A number of suggestions are set out in a 'Green Paper' (Comm., 1985b:21–37), covering both new products and new outlets for existing ones. When one looks through the list, it becomes apparent that only oilseeds and possibly wood crops have much potential for substantial increases in production. The other products mentioned (nuts, medicinal plants, traditional small-fruit production, beekeeping, fish-farming) are clearly interesting ideas and would fit well within a re-oriented European agriculture, giving more weight to small-scale production, organic farming and lower input levels. They do not have much to do with 'mainstream' farming as it presently exists, and would require further back-up activities to be significant. On the other hand, proposals for the production of bio-ethanol as a petrol substitute, or of starch for non-food uses, are clearly directed at large-scale producers and processors, but would require very heavy subsidies. These two latter proposals seem more likely to be taken seriously than those for small-scale agriculture, by those in a position to make decisions. This has everything to do with backing from large commercial interests and nothing to do with common sense or economics. They would almost certainly cost far more and generate far less income than the small-scale proposals. But the latter are easily derided by the hard-nosed as 'fringe' activities and are unlikely to be taken seriously by mainstream decision-makers. The fact that the large-scale proposals are far less economic and would generate little employment may be no great objection in such circles. After all, the same could be said of the CAP itself, and it has grown steadily for nearly thirty years.

It would seem that concrete proposals for significant change are limited to the following:

- quota restrictions to be continued for dairy products and probably expanded to other crops (beef, possibly cereals);

- restrictions on imports of oilseeds and cereal substitutes, allowing increased EEC production of the former and increased use of EEC cereal surpluses instead of the latter;
- a number of smaller changes and introductions, depending on changes in the structure of European agriculture.

In quantitative terms the most important, and in political terms the most explosive, of these proposals relate to protection against imported livestock feeds.

PROTECTING LIVESTOCK FEEDS

The EEC is already self-sufficient in most major crops, the one remaining significant area being livestock feeds. Large quantities of cereal substitutes and oilseed cake are currently imported. High EEC protection of livestock products increases feed demand, and high protection of EEC cereals makes them too expensive to feed to stock. Oilseeds and oilseed cake are imported because livestock rations require a high-protein component (especially if the basis is protein-poor cereal substitutes) and because the EEC currently does not produce enough oilseeds for its own consumption. The problems with going for self-sufficiency in oilseeds are multiple. The EEC is an expensive producer, and this would increase both the direct cost of agricultural protection and the cost of raising livestock (so increasing protection costs there too). Another problem is that the major supplier of oilseeds to the EEC is the USA, which has already made it clear that any protection which affects its soya bean exports will be met with retaliation. Finally, one cannot produce oilseeds for the cake without producing vegetable oils. The EEC is not self-sufficient in vegetable oils as a whole, but its current requirements are more than covered by crushings from imported oilseeds. It also produces a huge butter surplus which competes directly for markets with vegetable oils. The case of cereal substitutes has already been touched upon above. Here again, retaliation from the USA is a strong likelihood, and the effect on livestock production costs as important as for oilseed cake.

Most of the signs are that there will be adjustments to the CAP in the future, but that they will not be particularly significant. The form of protection will change slightly, to limit surpluses (though how effectively remains uncertain). There will be efforts to achieve self-sufficiency in most areas where it has not been achieved already. But EEC prices will remain far above world levels. Structural change may slow down but will almost certainly continue, especially with bio-technology just around the corner. Thus farm concentration seems likely to proceed, and with it the exodus from farming. One slightly more hopeful sign comes from an aspect of Koester and Valdes' analysis of the impact of quotas. They conclude that, with prices and output levels fixed, improving returns or profits will depend increasingly on cost-minimization. This could lead to more careful use of inputs and some diminution in pollution. But this depends firstly on the effectiveness of the quotas (which they doubt) and on savings in materials not being over-ridden by the costs (in extra machinery time, for example) of making them.

What happens to world prices of agricultural commodities depends at least as much on US policy as upon the EEC. But there seems no likelihood of dramatic change emerging from within agricultural policy. Changes in EEC policy may limit the further growth of surpluses, or even lead to a slow diminution. It will almost certainly not eliminate them, so that their depressing effect on world prices will continue. The USA is a far larger cereal exporter than the EEC, while the effectiveness of agricultural policy in the USSR will also have significant effects. The question has still to be discussed whether lowered world grain prices are a benefit to importing Third World countries (notably in tropical Africa) or not.

If one thing can be asserted with a fair degree of certainty, it is that the food, agricultural or economic problems of Africa will not be among the major factors affecting decisions over the future of the CAP. If the EEC Commission (or the member states, for that matter) could see a solution to their surplus and budgetary problems through replacing yet more of Africa's remaining export crops, they would scarcely hesitate a moment before plunging in. As indicated by Junne (1986 – Chapter 5), future developments in bio-technology make this highly likely.

An alternative approach to agricultural support

Previous sections have shown some of the ill-effects and plain absurdities of the Common Agricultural Policy. There seems no prospect of significant change within current mainstream policy thinking. Yet there is no reason to suppose that an agricultural free market is either a realistic possibility or likely to be beneficial. Most likely, it would drain the countryside of its remaining population, leaving a few huge monocrop and factory farms and further increased environmental pollution.

A completely different solution was, however, briefly put forward in the Commission's 1985 Green Paper, though it was rejected almost before it had been proposed, as politically impossible. This was to change the basis of EEC agricultural protection (in part) from price support to income support. That is, to relate payments to farmers to their income requirements rather than by increasing the price of what they sell.

Price support is an extremely cost-ineffective method of supporting farm incomes, since by far the largest proportion of the support goes to the largest and richest farmers. It would cost far less to support farm incomes directly, targeting the support specifically to those who need it. The budgetary costs would be reduced and it would no longer be necessary to maintain high food prices. Consumers within the EEC would benefit – and at current levels of unemployment and poverty in Europe, there are plenty of consumers who would benefit significantly. It would probably lead to considerable 'de-modernization' of agriculture. At world market prices, it would not pay to use current levels of fertilizer or farm equipment. But there is nothing optimal about high levels of input use based on heavy subsidies. It would also stimulate research into technologies more relevant to areas of the world where producer prices are not subsidized (tropical Africa, for example).

The Green Paper and most other authorities, however, write off a change from price support to income support as totally unrealistic, and as things stand at present they are probably right. The argument is that farmers would never accept it, and this is certainly true of large farmers, who would lose heavily. It is probably true of the majority of small farmers. There are two reasons for this.

The first is primarily ideological. Farmers, of whatever size, normally hold strongly to the belief that, in contrast to the urban unemployed and other spongers on the state, they are independent, self-sufficient and make their living by producing something of value. Where the EEC is concerned, this is largely a fiction. What they produce would not have anything like its present value without protection. They are, that is to say, in receipt of transfers both from the state (EEC) budget and from consumers who are forced to pay higher prices. However, this belief among farmers would give rise to very strong opposition to having their support from the EEC transformed into a form in which it was more easily likened to social security or unemployment benefit.

There is another, entirely realistic, component to this opposition. As unemployment has risen in Europe, so governments have introduced measures to limit payments to the unemployed and make them more difficult to collect. In at least some EEC countries this has been accompanied by government and media campaigns to shift the blame for unemployment on to the shoulders of the unemployed themselves. Farmers no doubt fear, and with very good reason, that if they were to receive payments having any relation to social security, they would be subject to similar cuts in the real value of the payments and similar denigration to go with it.

However, as mentioned above, the farm sector as a whole accounts for a smaller proportion of the European population than the unemployed. Even allowing for the organization of farming constituencies, farmers would probably not prevail without their powerful industrial allies. And there is no doubt where the latter stand on this issue. Reducing levels of inputs and equipment may be socially beneficial, but it would certainly not benefit agro-industry. Reducing price support and replacing it with income support would send land prices tumbling – and leave the banks which lend to agriculture on the verge of bankruptcy (if not over it).

Given the strength of this opposition, there is no chance whatever of a change from price to income support, so long as the CAP and agricultural protection continue to be discussed within the existing framework and among the existing decision-makers. The two go together, moreover, since the CAP and the EEC in general are topics so complex that most EEC citizens neither can, nor wish to, make head or tail of them. The CAP increases food prices, but most Europeans have been paying high prices for food for so long that they have become used to it. Nor would it make such difference for individual countries if they left the EEC. National governments would be forced by the same pressures as at present to give price support to agriculture and, in a number of cases, it would cost taxpayers and consumers more than now, though in others, like the UK, it would fall.

So long as the CAP is presented as being a question of supporting farmers and the cost thereof, the 93 per cent of the EEC population who are not farmers will continue to switch channels, turn over the page, or turn away in boredom at its mention. The issue of EEC membership arouses more interest, but as much on nationalist and rhetorical grounds as in terms of real policies. The EEC is only a part of the problem and not necessarily the most important one. The real issue is the combination of capitalist agriculture and protection through price support and the bizarre absurdities to which it leads. Prices are kept high to keep rich farmers rich and to generate production which cannot be sold. Increasing productivity does the same and throws still more people out of work. Productivity is increased through means which involve major social costs of pollution, inhumane factory farming and the uprooting of hedgerows to turn the countryside from an amenity into a featureless prairie. While this costs billions of dollars in taxation and high food prices, environmentalist criticism (and a much cheaper solution to farm support) are sneeringly turned aside as uneconomic. Only when far more than at present can see through this intellectual three-card-trick will it be possible to plan for a more realistic and human alternative, based on income support.

NOTES

1. The issue arises because agricultural support is set in terms of a common EEC currency, previously known as 'units of account', now as the European Currency Unit (ECU). So long as most international exchange rates were fixed, as was the case under the Bretton Woods Agreement which lasted from not long after the Second World War until the late 1960s, this

presented no problem, since the exchange rate between, say, Francs and Marks and 'units of account' did not change. But with the breakdown of the Bretton Woods system, the different EEC national currencies began to vary in relation to one another. The problem was compounded by the fact that two different sets of 'units of account' were in use, calculated on a different basis, so that changes in relative exchange rates between national currencies also led to variations between the 'normal' and 'green' (i.e. agricultural) EEC 'units of account'. This in turn gave rise to trade between the member states (intra-trade) in agricultural commodities, whose sole purpose was to profit from these variations. Typically, the solution was not to simplify but to add another layer of complication in the form of 'monetary compensatory amounts' (MCAs), whose purpose was to eliminate variations between 'green' and 'normal' currencies. It has not been entirely successful, as witness the fact that currently some two-thirds of the *world market* in skimmed milk powder (an almost entirely homogeneous commodity) is EEC intra-trade, much of it involving the movement of milk powder back and forth between Holland and Germany. The EEC is in process of phasing the system out, since it had also become a means for including further hidden price increases, not to mention a considerable bone of contention between the member states. A (relatively) simple explanation of the system up to about 1979 can be found in Tarrant (1980:101ff.). A more complete, complex and up-to-date explanation is provided in BAE (1985:39–46).

2. EEC regulations limit what seeds may be traded and used by commercial producers, this having the effect of eliminating certain previously common varieties and of reducing choice and variety. The ill effects of this and other factors relating to property rights in planting material, are discussed in Mooney (1979,1983).

3. This is not the same as normal VAT. It is a similar, but entirely separate, tax, levied on an agreed 'harmonized list' of goods and services, on the same basis in all member countries (as ordinary VAT is not). I am grateful to John Medland for this point.

4. This could be a bit on the high side. I made a calculation for Denmark (where I was then living) in 1986, adjusting in respect of per capita income and VAT collections (from Comm. 1985: Table 25, p. 245). Since then, I have not had access to more recent data.

5. Nitrogenous fertilizers are made through extraction of atmospheric nitrogen, by application of heat in the form of fossil fuels, and account for about 85% of all energy consumed through fertilizer use. The other two major nutrients applied – phosphates and potassium – are also exhaustible minerals.

6. In this case, Denmark 'leads the field', with debts accounting for 89% of operating capital and interest payments taking 58% of income (net farm value-added) in 1983. This was almost three times as high, in percentage terms, as the next highest country, Holland (Comm. 1985a:45).

7. Intensive factory livestock production units may be far smaller in area terms, but the capital invested per hectare is correspondingly far higher.

8. In reality, such investment decisions are enormously complicated by tax minimization.

9. The effect of price subsidies has probably been to increase profit margins and profits more than in proportion to turnover, even if one excludes the direct profits arising from speculation on factors like discrepancies in green currency rates of exchange.

10. The uninitiated (myself included) might wonder why SMP, butter oil and water need pasteurizing at all, since the first two have been through a sterilization process, while the latter can simply be filtered and boiled (which requires far less expensive equipment than pasteurization). The only answer that I have been able to get from asking this question is that some, though often not much, local raw milk is included in the process. In cases like Dar es Salaam's dairy plant where 98 per cent of throughput has been SMP and butter oil, this seems a poor economy.

11. Reasonably simple explanations for the most important commodities are contained in BAE, 1985, one of the most useful general sources which I have found on the CAP. Coming from Australia, which, it estimates, loses about $US 750 million per annum (at 1984 exchange rates, so more now) from EEC protection, it cannot be considered positive towards the CAP. But it is a thoroughly professional piece of work, with much useful documentation, citation and analysis.

12. Apart from the Egyptian wheat sale of 1983 and the BICEP 'initiative' mentioned in Chapter 5

the USA has in recent years both complained vociferously about European (especially French) 'mixed credits', and offered its own heavily subsidized loans for investment projects thought previously to be certainties for French firms. Mixed credits are considered further in Chapter 9. Euphemistically referred to as a combination of aid and trade, they involve using aid funds to sweeten trade deals by subsidizing interest rates on loans.

13. Elasticity measures the percentage response of demand (or supply) to a given percentage change in a price (or income level). If the price elasticity of demand for a particular commodity is 1, this means that a 10% reduction in price would give rise to a 10% increase in demand (and vice versa). It would be 0.5 if a 10% fall in prices increased demand by 5%. Direct demand elasticities are normally negative (an increase in price reduces demand). *Cross-elasticity* measures the change in demand (or supply) of one commodity in response to a change in the price of another. If the two are competitors for markets (say poultry and pigmeat), the elasticity will be positive (increasing the price of pigmeat will increase the demand for poultry – other things being equal). If they are complementary (say cereal substitutes and oilseed cake for animal feeding) the cross elasticity is likely to be negative. Supply elasticities (the percentage increase in supply from a given percentage increase in price) are normally positive (increasing the price increases the supply). Cross elasticities of supply are generally negative for substitute products (increasing the price of wheat will reduce the supply of barley), while for complementary products they tend to be positive (increasing the price of vegetable oil increases the supply of oilseed cake). *Income elasticities of demand* for food products are generally less than one, that is, as people get richer, they spend a smaller proportion of their income on food. Price elasticities for foodstuffs tend to fall as incomes rise, for similar reasons – that the price of food is a less important determinant of what people buy. Apart from problems of measuring elasticities, which are considerable, they have important implications for policy. If the price elasticity of demand is less than one (as it usually is for food products), it pays (the producer) to limit supply and hold prices up, or to hold prices up (as under the CAP), even if this reduces the amount sold. In this respect, the political analogue of Engels' Law (income elasticities of demand for food fall as incomes rise) is also relevant, since it means that food prices are a less burning political issue and policies like the CAP, which raise them, can be implemented. Where urban consumers are poorer, even failure to subsidize food prices can give rise to riots.

14. These figures come from a preliminary version of Matthews' work which, I understand, has since been revised to show a more positive effect from reducing protection. None the less, it remains the work which estimates the lowest effect.

15. Matthews gives no separate figures for sub-Saharan Africa. He finds 'low-income LDCs' to be net losers from the elimination of EEC protection, by about $214 mn, of which losses from current gains under the EEC Sugar Protocol account for $190 mn. Since the latter refers largely to Africa, which is also a major net cereal importer, the calculated losses might rise as high as $300 mn.

16. This rose from 22.3 to 25.6 million ECU during the period, but declined in dollar terms from $31 mn to $23 mn, as a result of the falling dollar value of the ECU. As cited in Chapter 5, US agricultural protection was of a similar order in the early 1980s.

17. Japan gives very heavy subsidies to small and relatively inefficient rice farmers who provide the political backbone of the ruling Liberal Democratic party.

8

Food Aid & Food Security

Introduction

This chapter considers food aid and takes up questions about the impact on local food production in Africa of cheap and grant-aided food imports. None of the studies looked at in the last chapter had anything to say about this. They either simply ignored food imports to the Third World or assumed that cheap food imports were a benefit, as, in a direct sense, they are. This leaves unanswered the more important but more complicated question of the long-term effect of food aid. Does it discourage local food production? Does it accustom people to new foods which cannot easily be grown in their own countries and which will continue to be imported? Does it increase or decrease their long-run food security?

These questions have been debated in relation to food aid for most of its thirty years of existence, and without reaching agreement. This is not for lack of evidence, though in part because it points in both directions. Critics of food aid (especially that of the USA) have built up a huge file of cases where the effects of food aid have been to induce food dependency and where arbitrary or political variations in its incidence have been harmful. This has not always been the case, and there have been claims of success, but it becomes increasingly clear that dependency *can* easily become a serious problem – especially since it can leave the dependent country vulnerable to unpredictable variations in disbursements.

The defenders of food aid have had to rely on some rather weak arguments. One of these emphasizes the 'potential' of food aid – what it could achieve in ideal circumstances.[1] This obviously generates positive conclusions. Any resource transfer *can* have positive effects. But if looking at the 'potential benefits' means ignoring all the things which can (and often do) go wrong, then it gives a somewhat rosy-tinted view. A certain defensiveness also shows in an aptitude for countering by attribution rather than argument. Thus Clay and Singer (1982:9) contrast the 'more complex and constructive views of food aid that predominate in the professional community' with 'the increasingly critical attitude to food aid among development lobbyists'. In point of fact the critics include people with every bit as much professionalism and experience as the 'constructive' and include views every bit as complex and sophisticated. Significantly, they omit to mention in this context work like that of Mitchell B. Wallerstein (1980) which is highly professional but which draws critical conclusions. They also omit to mention lobbyists *for* food aid, who are at least as numerous and far better paid than the opposition.

Of a slightly different order is the 'practical' argument: that food aid is going to continue, whether one wishes it to or not and that it makes more sense to see how it can best be made to work than to criticize it. Of course, it is worthwhile to see what can be done to improve the administration and effectiveness of food aid. But it is not clear to me how one can hope to improve without, criticizing. Nor does this, to my mind, preclude one's questioning whether it does more harm than good.

It is worth noting that even if one finds that food aid does more harm than good, it does not necessarily follow that it should immediately be stopped. If dependence is the problem, 'cold turkey' withdrawal is not necessarily the answer. Precisely the problem with dependence is that it places the recipient country's population (or part of it) at risk of abrupt withdrawal. Another important practical consideration is that there are countries in Africa and elsewhere (Lesotho or Cape Verde, for example) which are totally dependent on food imports under virtually any circumstances one can imagine. Clearly it makes no sense to eliminate food aid in such cases. On the other hand, the practical danger of abrupt withdrawal lies *not* in the overall cessation of food aid, but rather in political decisions by donors to cease food aid to one particular recipient.

What is Food Aid?

Food aid is food (mostly from surplus stocks) which is given or lent as aid to developing (and some developed) countries. Since the early 1970s, between 60 and 70 per cent of the total has been in the form of grants. During the previous decade, the loan component was much higher. Most donors other than the USA operate purely grant programmes, and their importance has increased relatively, since 1970. The main loan food aid programme is the USA's PL 480 Title I which is composed of loans on soft terms but in hard currency. In most cases, recipient countries pay for transport costs, though some donors (the EEC, for example) deliver free to the recipient port for the poorest and landlocked countries. In all cases, however, food aid is significantly cheaper than commercial imports even when, as is widely the case, the latter are subsidized. Food aid takes three main forms.

Programme food aid, or direct transfers – The major proportion of all food aid – between 60 and 70 per cent – is transferred to Third World governments, or their food purchasing and processing agencies, for them to distribute. It thus contirbutes primarily to normal consumption and to reducing the foreign-exchange cost of importing food. Whether the food aid is grant or loan, it is common for donor agencies to make some charge to the purchasing agency, the funds from which (the counterpart funds) are placed in a government account to be spent on approved purposes. In a few cases donors place conditions on how the food is to be distributed. More often they place conditions on the use of counterpart funds, though often not very effectively. Almost all of this sort of food aid is sold to consumers, though it may be sold at subsidized prices, in ration-shops or through other rationing systems. While the recipient country may (but may not) have to pay transport costs, it is usually much cheaper than normal commercial imports, though it usually takes longer to arrive.

The main criticism of direct transfer food aid is that it generates dependence on imported foodstuffs – often of sorts which cannot be produced domestically. The main defence is that it is a form of foreign-exchange transfer, thus increasing the capacity of the local government to perform its functions and implement development policies (or spend the money in other ways). Food aid constitutes a net foreign-exchange transfer to the extent that it substitutes for commercial imports (or for the foreign-exchange costs of local production, like fuel and fertilizer). In the majority of cases, it seems that food aid does substitute at least in part for commercial purchases in the short run, although this is (or at any rate was) forbidden under provisions that food aid transfers should be over and above 'usual marketing requirements'.[2]

There are two main points to be made about the transfer of foreign exchange and funds. Firstly, the foreign-exchange transfer is not the 'face-value' of the food aid, which is often denominated (by the EEC, for example) in terms of high internal prices. Nor is it even the significantly lower world market value. For food which would have been imported in any case the net foreign-exchange saving is the cost of the commercial imports for which it substitutes, less any costs involved in receiving food aid. For food aid additioanl to 'normal' imports, the saving is more difficult to assess, unless it substitutes for some other import. It it substitutes for local cereals, the effect will be to reduce local producer prices and thus be negative. On market criteria, it would be less than the commercial value, since one can assume that the country would have preferred (given the foreign exchange as such) to use it for some other purpose. But there may be more complex cases where, for example, a country's government has decided not to import food, in spite of a serious need among the population. In this case, one cannot estimate the value of the food aid without some idea as to its distribution. In all cases though, the foreign-exchange value of food aid will be substantially lower than the 'face value'.

The second point is that the 'free' funds transferred – the counterpart funds from the sale of food aid – are not foreign exchange but local currency. While donors increasingly try to place conditions on how such funds should be spent, a more serious problem in many cases, is to get them spent at all, since foreign-exchange shortage is the crucial factor.[3] There is also the danger that spending the large sums of local currency generated will increase the rate of inflation.

Project food aid is the next largest use, typically accounting for some 20 per cent of the total value of food aid. It takes two main forms. One of these is 'food-for-work', in which public (and sometimes private) works are underaken by a labour force paid in food aid. By far the major supporter of this sort of programme is the FAO World Food Programme (WFP). The other form of project food aid is 'supplementary feeding', whose aim is to improve the diet of special groups within the population: nursing mothers with children, small children, school children etc. Here again the WFP is the major donor, though a number of large NGOs, often religious, are also involved.

Jackson and Eade (1984: Ch. 3) criticize food-for-work programmes in terms of the (lack of) development impact of the work performed. They find that 'public works' projects only too often benefit the rich alone, while there are cases where food-for-work labour has been done directly for large private interests. In either case, there is all too often no benefit for the needy sections of the local population, once the work (and food-payment) stops. They also find that large programmes result in poor motivation, low productivity and 'make-work' schemes including cases where small but useful projects are swamped and ruined by food aid. They conclude that 'truly successful FFW projects . . . in which the benefits show *after* the work has finished, are rare' (ibid., p. 40). One defence of FFW is that, whether or not the programmes achieve anything, they do provide food to people who need it.[4] Defences have also been made of the long-term results of specific projects.[5]

Where supplementary feeding programmes are concerned, Jackson and Eade cite a number of examples and quotations from aid agency personnel to show that many of the programmes achieve no nutritional improvement at all. One problem is that the family of the child or mother-and-child in question tend to reduce food purchases to off-set (or even more than offset) the nutritional supplement provided. Another frequent problem has been that foodstuffs provided cannot be obtained (or afforded) without food aid, leaving people where they were before, as soon as the project finishes.[6]

Emergency, disaster and refugee feeding is the third category, and typically takes between 10 and 15 per cent of the total, though it has been over 20 per cent in some recent years in Africa. There is no dispute about the necessity of this sort of aid, though considerable criticism of its actual working, as for example Gill (1986) and other sources cited in Chapter 4. Jackson and Eade (1984: Ch. 2) cite a number of cases where food aid for a specific emergency has arrived just in time to reduce producer prices for the following harvest. They also note that food aid is commonly sent in the case of disasters which do not involve major food shortage and to which it is thus not relevant. Almost all concerned are agreed that, where possible, disaster relief needs to be followed up by help to get the survivors back on their feet economically and/or improve conditions to reduce the risk of a future occurrence. But since the latter comes under 'normal' rather than exceptional aid, it is often not forthcoming.

The origins and development of food aid.[7]

Food aid was started by the USA and, for its first decade or so, remained an almost solely North American phenomenon. There was nothing especially philanthropic about it, though food aid was sold to the American public as such. As Wallerstein (1980:3) remarks, 'US (food aid) policy can be understood adequately only through close examination of the intersecting relationship between the requirements of domestic agriculture and the imperatives of US foreign policy'.

The Agricultural Trade Development and Assistance Act[8], also known as Public Law (PL) 480, was passed in 1954. Its primary purpose, as indicated by the name, was to find and develop markets for surplus US agricultural produce. US agricultural production had been stimulated by expanded export opportunities with lend-lease during the War and the Marshall Plan after it. High support levels and an explosion of technical advance in the early 1950s had led to the emergence of huge surplus stocks, which became increasingly expensive to store and difficult to justify. But food aid also had an important secondary purpose, which sometimes overrode the first – to support anti-communist (or anti-Soviet) regimes and movements, especially those bordering on the Soviet bloc.

PL 480 institutionalized a procedure for shipping surplus agricultural commodities abroad, which had previously been embodied in various short-term measures (Marshall Plan, Mutual Security Acts). It was initially hoped that three years would be

Table 8.1: *Top ten recipients of US food aid 1954–61, ($ million)*

Rank	Country	
1	India	1,408
2	Yugoslavia	585
3	Pakistan	416
4	Italy	398
5	Spain	391
6	Rep. of Korea	314
7	Brazil	241
8	Egypt	215
9	Israel	196
10	Turkey	167

Source: USAID, 'US Overseas Loans and Grants in the period 1.1.1949 . . . 30.7.1979, Washington.

enough, as President Eisenhower said at signing, to 'lay the basis for a permanent expansion of our exports of agricultural products' (Wallerstein, 1980:35). But this idea soon faded as the enormous contribution of food aid to cereal exports became clear. Table 8.1 on the previous page shows the distribution of US food aid during the first seven years of PL 480.

These ten countries received almost 70 per cent of total PL 480 aid during the period and show both the Cold War orientation and the market development aspects of PL 480. European countries (including Greece and Portugal) accounted for 25 per cent of the total in continuation of the Marshall Plan. Yugoslavia became a major recipient after the break with Stalin. Korea and Turkey (a wheat exporter at the time) also reflect Cold War objectives, while Israel and Egypt reflect Middle Eastern policy considerations. At the same time, Spain, Korea and others became major markets for American cereals. India and Pakistan were and are desperately poor countries, but their inclusion as recipients of food aid reflects primariy their 'exposed' position bordering China (then as hated and feared as the USSR), and the perceived need to keep them Western-oriented, this being perceived as a major problem in the USA at the time (Wallerstein, 1980:131).[9]

But by competing agricultural exporting countries, PL 480 was seen primarily, and with some justice, as dumping, provoking a stream of complaints over 'the unfair competition and practices of the USA' (Cathie, 1985:21), which took most of the time of the (Consultative Sub-) Committee on Surplus Disposal (CSD) of the FAO.[10] Nor is there any doubt about the success of PL 480 in this respect. Between 1953 and 1957, US agricultural exports rose from $449 mn to $1,900 mn, and by the latter date, PL 480 was directly responsible for 40 per cent of the total, having peaked at 54 per cent the previous year (Cathie, 1985: Figure 5). For the period 1955-7, PL 480 shipments accounted for some 18 per cent of total world exports of wheat, rising to 25 per cent in 1963-5.[11] Over the same two periods, food aid accounted for respectively 89 and 51 per cent of US exports of skimmed milk powder and 63 per cent (1955-7) and 34 per cent (1963-5) of total world exports (FAOb, 1969:39). Moreover, an increasing proportion of this increased trade was in new markets, notably those of the Third World. South Korea was one country where wheat consumption rose very rapidly during this period. Colombia was another.

There were problems, however. Food aid was costly to the US budget, as was the cost of storing the increasing surpluses brought about by high producer prices. The most important component of PL 480 was (and is) Title I, which was then supplied on loans which were to be repaid in local currency. This in turn was to be made available to US businessmen for investment ventures. In spite of various schemes to offer the counterpart funds on concessionary terms to US businessmen with interests in recipient countries, and although US multinationals availed themselves of this offer, large unusable sums of money built up. This was changed to sale at concessionary terms but in hard currency, by a later administration in 1966.

By the beginning of the 1960s, a number of factors had combined to generate some broadening of horizon for the aims of food aid. Appreciation of the foreign-policy potential of food aid had already led to its redefinition as 'Food for Peace'.[12] The incoming Kennedy presidency was more interested in world affairs and less interested in agriculture than its predecessor.[13] Moreover, the complaints of competing agricultural exporters about dumping could be more easily dealt with, if food aid was primarily transferred to developing countries. Not only did this stress humanitarian and developmental concerns, but it stressed the development of new markets rather than competition for existing ones.

Multilateral food aid

In this changed environment, it began to make sense to invite other exporters and even developed country importers to join in giving food aid. This removed more agricultural produce from the world market, developed new markets for the benefit of all (exporters), and spread the cost. Thus the USA pressed hard and successfully in the 1960s for the initiation of multilateral food aid programmes.

The formation of a multilateral agency to distribute food to needy countries and groups had been discussed within the UN framework after 1945, but had lost impetus as a result of US dissatisfaction with UNRRA, the primary agency involved (thought to be infiltrated by, or soft on, communism). But by the end of the 1950s, a proposal for a 'world food programme' had been made by the Republican Administration and this was taken up by that of Kennedy.[14]

The World Food Programme, which resulted, came under the aegis, but not initially the control, of FAO. It was funded by pledges in money terms from OECD members and subsequently other donors, and had three main functions. It was to provide food for emergencies, and to operate projects for nutritional improvement and food-for-work (FFW) projects. It thus covered humanitarian and developmental aims, while keeping the allocated food off world markets, since none of the above groups of recipients were likely commercial consumers. The USA initially provided over half the total funding, subsequently reduced to 50 per cent and then 40 per cent. By 1981–2, it accounted for 26 per cent of total pledges.[15]

The USA, having pushed for the setting-up of WFP as a means to get other donors to contribute food aid, seems rather rapidly to have lost its own enthusiasm for the body. In part this arose, from the inefficiency of WFP – which faces 'double-bureaucracy' problems, in that administration of all the food it disposes of has to pass through both its own bureaucracy and that of its national sponsors. This was a particularly serious problem with the USA itself, which 'has consistently tended to view the WFP as an extension of its own bilateral programme', demanding the right to prior scrutiny of all WFP projects (Wallerstein, 1980:178). The WFP has also had efficiency problems of its own, often stemming from clashes between its own directors and those of FAO.[16] A problem in relation to the USA has been that WFP has no local support base and is not directly represented in the system of lobbying which is so crucial to any body seeking US government funds.[17] Another problem has been that (in common with PL 480) its resources are denominated in money rather than commodity volume terms. When international food prices rise (as, for example, during the 1973–5 food crisis) the amounts of food available to WFP fall. This latter has prevented it from becoming 'global emergency relief co-ordinator',[18] leaving it with the main job of co-ordinating project food aid.

The Food Aid Conventions, the other initiative towards multilateralization of food aid, came a few years later and took a very different form. The Food Aid Conventions are fora for the pledging of food aid (in volume terms), but do not handle the food aid themselves. Apart from specifying that it be grant food aid, it is left to individual donors to decide how to allocate their pledges. The vast majority is distributed bilaterally. Pledges to the different Food Aid Conventions are biennial, and for the biennia since 1969, the amounts pledged remained at slightly over 4 million tons until 1980/81, when they were increased to 7.6 million tons. Prior to 1980, the US provided some 45 per cent of this. Since then, the US proportion has risen to just under 60 per cent. In the region of 4–5 per cent of FAC resources have been earmarked for distribution through WFP, this accounting for between 6 and 15 per cent of WFP's resources, though a higher proportion of the cereals transferred (FAO, 1984: Tables 24 and 25).

The history of the FAC's formation is also very different. Its day-to-day management is under the International Wheat Council in London, a body normally more concerned with regulating wheat trade than with aid in any form. The FAC was first started during the 'Kennedy Round' of GATT negotiations on tariffs. Here the USA forced a number of very reluctant partners to pledge food under the first FAC, by making tariff concessions conditional upon their agreement to do so. The USA itself was by far the largest donor under the FAC, but this is misleading in that its pledges under the FAC were not additional to food aid already being given. Since the USA was already providing more grant food aid than its commitment of 1.89 million tons, the pledge involved it in no extra food aid (as was also the case for the increase in 1981). For the other members, the pledges did, for the most part, involve new allocations. Major complainants were food importing countries (Japan and the UK) together with the EEC, which was at that time a net importer of grains. The beauty of this, from the point of view of the USA, was that not only did it involve other countries in opening new markets, but that net importers would have to increase their own cereal imports in order to do so.

US food aid from the 1960s

While the USA was active in involving other donors in food aid during the 1960s, the trend in its own shipments was downward after a peak in 1965/6. (Figure 8.1). The Johnson Administration followed more or less the same line as that of Kennedy. But once US wheat stocks had begun to fall, in the mid-1960s, the US was increasingly concerned to shift food aid, and especially that directed via WFP, on to other donors. Another change was the introduction of conditionality (other than the political conditionality which has marked US food aid from start to finish). This concerned 'self-help' (i.e. a greater emphasis by recipients on increasing their own food production) and was largely a result of US resentment that India was using US food aid cereals to permit a continuing emphasis in its plans on industry. In fact, as Wallerstein (1980:185–93) shows, this involved an almost obsessive policy of keeping India on a 'short tether', by allocating food aid on a very short-term basis (often month-to-month). Another aspect of conditionality, also applied to India, was to delay food aid as a means of opening up the fertilizer market to US investors.

If the Johnson presidency somewhat narrowed the visions of 'food-for-peace', that of Nixon altered the whole policy much more radically. Both Nixon and his Secretary for Agriculture were concerned to cut state expenditure and increase commercial agricultural exports. The discontinuation of the Bretton Woods agreement provided the basis for this, in devaluation of the dollar. The USA was thus concerned to reduce US grain stocks and food aid. Taking four-year periods, average annual shipments of cereals under PL 480 declined as shown in Table 8.2:

Table 8.2: US cereal food aid shipments 1961–64 to 1981–84 (million tonnes, average for four-year periods)

Period	Million Tonnes
1961–4	16.2
1965–8	15.3
1969–72	9.2
1973–6	5.0
1977–80	5.9
1981–4	5.4

Source: Wallerstein (1980: Table 3.6), FAOe (1984, Table 39).

The proportion of world cereal food aid supplied by the USA fell from 96 per cent in the early 1960s to about 77 per cent at the end of the decade, remaining between 60 and 70 per cent during the 1970s. Concessional sales had accounted for some 30 per cent of US agricultural exports from 1960 to 1965, but this fell to just over 15 per cent by 1970, and to only about 6 percent by 1980 (Sexauer, 1980:994). More recently it has risen to about 10 per cent, with an increase in food aid shipments and a larger reduction in US commercial exports (FAOe, October 1986). These trends can be seen in Figure 8.1, from which it can be seen that the really massive reduction in food aid transfers dates from 1972. In 1973/4 US grain transfers fell to 3.2 million tons at the worst of the Sahelian, Ethiopian and general Third World food crisis of 1973–5.

The change in US agricultural policy of that period is considered in Chapter 5, and in more detail by Tarrant (1980) and Garcia (1981). Both find this to be the most important factor behind the world grain shortage of 1972/3–74/5. An effective devaluation of the dollar, reduction in farm support and efforts to reduce stocks (assisted by a large sale to the USSR) achieved a rapid threefold rise in the price of wheat just as the famines of the Sahel and other parts of Africa were reaching peak seriousness. All PL 480 shipments fell, reflecting the fact that they were and are denominated in money terms, while cereal prices had risen sharply. But Title II shipments, which cover grants for emergencies, fell by 50 per cent between 1973 and 1974. Moreover, two-thirds of all US food aid was directed to Vietnam and Cambodia, which took, between them, more than the whole of tropical Africa. This had less to do with food than with evading Congressional restrictions on military spending, through the use of counterpart funds for this purpose (Wallerstein, 1980:45–50, 193–7).

US food aid shipments picked up again, but have never approached the levels of the 1960s, rising above 7 million tons only in 1984/5 and 1985/6.[19] The US proportion of total grain food aid has varied between 55 and 70 per cent since 1973. In 1983/4, the top ten recipients of US food aid in cereals received 75 per cent of the total, as shown in Table 8.3. Ethiopia received 13.6 tons of cereals from the USA in that year (0.2 per cent of total US food aid), coming 37th out of 69 countries receiving cereal food aid from the USA. While Mozambique received somewhat more, of the African countries affected by famine only the Sudan was among the major recipients of cereal food aid from the USA.

Table 8.3: *Top ten recipients of US cereal food aid 1983/4, (kg per capita and percent)*

Country	% of total	Kg per capita
Egypt	26	57
Morocco	8	20
Bangladesh	8	4
India	7	0.5
Sudan	6	16
Sri Lanka	5	17
El Salvador	5	47
Indonesia	5	1.5
Bolivia	4	35
Peru	3	6

Sources: FAO, Food Aid in Figures 1985, Table 18, World Bank 1986, Table 25, for 1984 population.

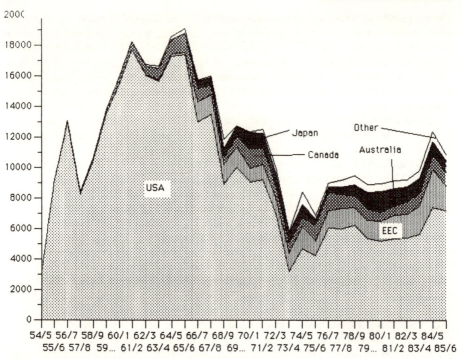

Figure 8.1: *Cereal food aid by major donor 1984/5–1985/6 ('000 tons)*

Other major cereal food aid donors

As Table 8.4 shows, no other donor begins to approach the importance of the USA where cereal food aid is concerned. During the 1960s, Canada was the second largest donor, but has since been overtaken by the EEC (Community and member states combined). Of others, only Australia and Japan are at all significant in world terms.

The EEC may have been reluctant to start giving food aid, but with the growth of surpluses it has become an integral part of EEC agricultural and development policy. Both the Commission and the member states provide food aid, though, as shown in Table 8.2, the growth of 'Community actions' has been far more rapid, reflecting that it is at EEC-level that surpluses pile up and have to be disposed of. Thus while Community actions have increased by over ten times during the fifteen years recorded, transfers by member states only increased by about 50 per cent.

Table 8.5 shows the distribution of EEC cereal food aid between major recipients and reveals a gradual shift over time towards Africa. The distribution for 1969–79 gives a strong impression of following the (US) leader. But 1977–9 shows some divergence in emphasizing post-war Vietnam, Lebanon and Jordan (for refugees) and three tropical African countries instead of only Somalia. By 1981/2, there were five African countries in the top ten, counting Madagascar. The same was true of 1983/4, while their combined proportion of the whole had risen from 17 to 25 per cent. Figures for 1984/5 would show a further significant increase in the proportion going to Africa (as, it would for the USA). The absence of Egypt from the 'top ten' in 1983/4 may reflect the US grain deal mentioned in Chapter 5, rather than EEC policy.

Table 8.4: *EEC cereal food aid deliveries, national and community 1970/1–1984/5 (million tonnes and percent)*

Year	Community	National	Total	Community as percent of total
1970/71	105	783	888	12
1971/72	196	783	978	20
1972/73	321	666	986	32
1973/74	603	614	1,218	50
1974/75	871	542	1,413	62
1975/76	297	631	928	32
1976/77	555	576	1,131	49
1977/78	709	665	1,374	52
1978/79	612	547	1,159	53
1979/80	597	609	1,206	50
1980/81	714	564	1,278	56
1981/82	847	733	1,580	54
1982/83	842	729	1,571	54
1983/84	1,024	866	1,890	54
1984/85	1,259	1,207	2,466	51
1985/86 (preliminary)		..	1,580	..

Sources: FAO, *Food Aid in Figures*, 1983, 1985, Table 6. FAO, *Food Outlook*, October 1986, Table A.14.

Table 8.5: *Top ten recipients of EEC food aid, rank and percent of total 1969–79, 1977–79, 1981/2, 1983/4*

Rank	1969–79	1977–9	1981/2	1983/4
1.	Bangladesh	Bangladesh	Egypt (16)	Mozambique (10)
2.	India	Egypt	Bangladesh (15)	Egypt (10)
3.	Egypt	Vietnam	Pakistan (5)	Bangladesh (10)
4.	Pakistan	Pakistan	Somalia (4)	Pakistan (6)
5.	Vietnam	Sri Lanka	Sri Lanka (4)	Somalia (5)
6.	Indonesia	Lebanon	Tanzania (4)	Sri Lanka (4)
7.	Sri Lanka	Somalia	Jordan (4)	Ethiopia (4)
8.	Turkey	Jordan	Mozambique (3)	Tunisia (4)
9.	Somalia	Mauritania	Madagascar (3)	Sudan (3)
10.	Tunisia	Zaire	Ethiopia (3)	Angola (3)
Top ten as percent of total			59	59

Sources: 1969–79, 1977–9, Stevens (1983), Table 2.4 and Commission papers : Com (81) 41 final and Com (81) 804 final. 1981/2 and 1983/4, FAO, *Food Aid in Figures 1983*, Table 18 and 1985 Table 18.

By far the major recipient of *Canadian* cereal food aid is Bangladesh which took 36 and 40 per cent respectively of the total in 1981/2 and 1983/4. Egypt and Pakistan are also major recipients. But in 1983/4, 20 per cent of the total went to China which, one must assume, relates more to market development than anything else. Another variation from the US pattern is the inclusion of Nicaragua as one of the top ten recipients. Among African countries, Ethiopia, Mozambique, Sudan and Somalia have been important recipients. The top ten recipients received 72 and 84 per cent of the total in 1981/2 and 1983/4 (FAO, 1986a: Table 18). Bangladesh, Pakistan and Egypt are also the major recipients from *Australia*, the first receiving 25 per cent of the total in 1981/2 and 26 per cent in 1983/4. Among African countries included were Tanzania, Ethiopia, Kenya, Sudan, Mozambique and Zambia. The *Japanese* programme is about the same size as that of Australia and as concentrated, with 80–90 per cent going to the top ten recipients. In 1981/2, Korea received 48 per cent of the total, followed by Tanzania (11 per cent), Pakistan (7 per cent) and Sierra Leone (4 per cent). In 1983/4, Korea received none; Indonesia was the major recipient with 31 per cent followed by Bangladesh (21 per cent), Pakistan (12 per cent) and Kenya (5 per cent).

There is evidently considerable following of the leader or of fashion. Bangladesh, Egypt and Pakistan are top recipients from all the major donors except Japan, which excluded Egypt. But apart from this, it is clear that all donors make specific agreements as part of foreign or commercial policy, whether that involves food sales or (in the case of Japan) more general trading factors.

The composition of food aid

By weight, cereals are by far the most important products shipped as food aid, accounting for between 90 and 95 per cent of the total. They are also by far the most important in nutritional terms. Other products shipped include vegetable oils,

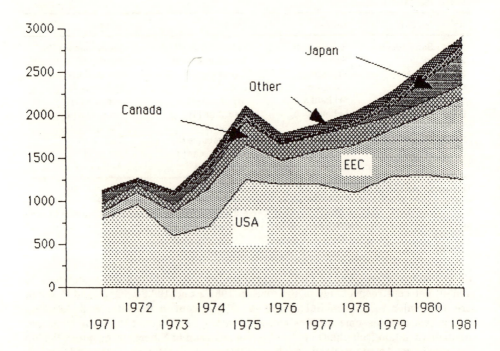

Figure 8.2: *Food aid disbursements by major donor 1971–81 ($ million)*

skimmed milk powder, butter oil and other dairy products, together with very small amounts of canned meat and fish. The EEC provides most of the dairy produce (60–70 per cent), reflecting its enormous surpluses. The USA provides about 90 per cent of all vegetable oil for similar reasons. The remaining products are shipped in much smaller quantities, but also reflect what countries have in surplus. Denmark, for example, is a major donor of canned meat and cheese.[20] The USA is the largest donor of 'other dairy products', much of this being in the form of 'blended foods', which are usually cereals 'fortified' with milk powder.

As Figure 8.2 shows, the prominence of the USA is somewhat reduced when other commodities are included and the total valued in money terms. Skimmed milk powder costs about ten times as much by weight as cereals, meat rather more and vegetable oil three to four times as much. Part of this higher value is illusory. In the first place, the much higher *market* value enormously exaggerates their extra nutritional value. Apart from this, food aid commodities are normally valued by donors at their domestic market prices so that, the more heavily these are subsidized, the more will the nominal value exaggerate their real 'market value'. Skimmed milk powder and butter oil compose over half of EEC food aid. From 1976 to 1980, over half the total value of this was represented by refunds under the CAP. In 1983–4, the difference was even greater and the EEC intervention price (at which food aid was valued) was three times the world market price. That is, recipients could have bought similar amounts commercially for less one third of the nominal value of the dairy food aid they received.[21]

Dairy food aid has been subject to more criticism than any other form, and not just because of the disparity between its price and its market value. Skim milk powder goes mostly into the recombination dairies of recipient countries, to provide fresh milk for the middle classes. In so doing, it generates demand for a commodity which usually cannot be produced locally on a year-round basis without use of feeding-stuffs (imported or diverted from human use). But the problems are still worse where it is not provided for such dairies, for it is then likely to be used (whatever the intentions of the donors) as baby food. The problems with powdered baby foods in tropical Africa are well-known. With clean water scarce, firewood for cooking expensive and refrigeration completely unavailable to the poorer sections of the population, any baby food made up from powder and water is bound to be unhygienic. Moreover, poor mothers are inclined to put in too much water to save money, with the result that children are malnourished. But if this is true of balanced baby foods or whole milk powder, the situation is still worse for skim milk powder, which is unsuitable as a baby food even if not over-watered. While dairy food aid is the most criticized, there is little to be said for canned meat, jam and other similar items, which are sent as food aid, because donors have spare stocks to off-load. However, the quantities of the latter shipped are usually smaller.

The EEC food aid programme

EEC food aid started, as recorded above, unwillingly and at the insistence of the USA that it join the Food Aid Convention. But those days are long gone. With its own huge surpluses, the EEC is the second most important food aid donor and uses it, to some extent, as a means to compete for markets with the USA. Food aid composes the highest proportion of total aid for any donor.

Over half of all EEC food aid, by value, is composed of skimmed milk powder (SMP) and butter oil, making the EEC by a long way the world's major dairy food aid donor. It accounts for 50–70 per cent of milk powder shipments and 95–100 per cent of butter oil shipments. Of these over 90 per cent are shipped by the Community, since it is here that they accumulate as surpluses. EEC cereal food aid deliveries have more than doubled between 1970–71 and 1985–6 and the Community proportion has generally been over 50 per cent since the early 1970s. Among the member states, France and Germany are the major bilateral donors of cereal food aid.

About 15 per cent of EEC cereal food aid goes to the regular programme of the World Food programme, with the USA and EEC each contributing about 25 per cent of WFP's funds. In addition, there is a contribution to the International Emergency Relief Fund run by the WFP and transfers through NGOs. But the large majority of EEC food aid is composed of direct deliveries (programme food aid).

There have been two major criticisms of EEC food aid. The first relates to the high proportion of the total composed of dairy products. The other relates to the inordinately long time which EEC food aid has taken to arrive. Some years ago, the Court of Auditors estimated the average time from ordering to arrival for different products and found times of up to one and a half years. As in the case of the WFP, 'double-bureaucracy' has been one of the problems. The EEC Commission makes proposals on the allocation of food aid, but final decisions rest with the Council of Ministers. In addition, the Directorate for Agriculture, which has been formally responsible for food aid until very recently, is more concerned with surplus disposal than with the problems of developing countries.

This has been considered as purely a problem of administration by some of the consultants who have commented on EEC food aid (ABC/IDS, 1982, for example). But as Talbot (1979) makes clear, it is part of an ongoing power-struggle between the Commission and the Council. The ABC/IDS report makes a number of recommendations for improved and streamlined administrative control, but these all require the concentration of more control in the hands of the Commission, and are likely to be resisted by at least some members of the Council. To complicate matters further, the European Parliament has become concerned over food aid in recent years, and now wishes to extend its control over the programme. This may introduce a much-needed element of public discussion, but could well provide further administrative steps. It is claimed however, that procedures have been much speeded up in the last years, so that the EEC no longer compares unfavourably with other food aid donors.

Against this, it must be said that, where political pressure is applied through EEC food aid, it is done much more subtly than by the USA. The pattern of EEC deliveries may contain a certain arbitrary element, and favour certain clients or good customers, but is far more stable than that of US food aid. There have, to my knowledge, been no cases of outright withdrawal or massive reductions in deliveries on political grounds. There have been, as for other donors, variations over time which reflect the standing of African governments and donor attitudes to their policies, rather than direct food needs. But this, when one comes down to it, is what 'conditionality' is about, and this is widely regarded by aid donors, if not recipients, as not just acceptable but necessary.

There is most certainly an element of export promotion in EEC food aid, not only for the products themselves but for processing plant and equipment. As it was put in an answer to a question in the European Parliament about Operation Flood in 1979, 'with the support of the Community, the project will help establish factories and plants for reconstituting milk and thus open up markets for us in the long run' (Alvares, 1985: 20). This is particularly the case in the dairy sector, where about 60 per cent of EEC food aid goes to Third World dairy processing industries to subsidize their development, the most prominent example being India's Operation Flood (see below). There is also an important element of export promotion in EEC cereal food aid, especially where larger recipients like Egypt are concerned and in competing with the USA for markets.

But this cannot be blamed on the EEC *as opposed to* the member states. One major aspect of the export-promotion strategy is the provision as 'aid' of dairy factories and other food-processing plants at heavily subsidized interest rates, and this comes primarily from national aid or export-credit agencies, operating on behalf of national rather than EEC interests. EEC or WFP dairy food aid helps to sweeten the deal. In the early years of EEC food aid, it was specifically stated to be primarily a programme for surplus disposal and export promotion, and as such operated by the Directorate General (DG) for Agriculture, rather than for Development. In recent years, it has been officially stated by the EEC that export promotion and surplus disposal should not be among the purposes of food aid, and operation has been transferred to the DG for Development. But so long as there are huge surpluses to dispose of and most food aid is transferred directly as programme aid, it is hard to see how this can be avoided.

Food aid and policy

Before looking at recipients and concrete food aid policies, it is worth summarizing four basic policy aims for which food aid can be used.

Export promotion through the development of new markets or bidding markets away from others. There are formal safeguards against both of these, but they seem to be ineffective and there is ample evidence of the effectiveness of food aid for this purpose.

'*Foreign policy*', which can cover whatever foreign policy aims the donor has in respect of the recipient. The USA has used food aid to defend 'friendly' governments against expected left-wing pressure, to secure contracts for US firms and to secure agreement on matters of international policy. By the same token, it has used cutting off the supply of food aid as a sanction both against regimes considered 'unfriendly' and as a warning to otherwise 'friendly' regimes whose leaders speak against the USA in UN fora or nationalize US firms.

'*Developmental goals*', can cover a variety of things, but most discussion of food aid in development refers either to the foreign exchange which it saves, or food-for-work programmes. Some would also include items like assistance to the building of wheat mills or dairy plants to process imported foodstuffs. Other proponents of food aid as a development instrument find such claims an embarrassment and would exclude them. There is considerable lack of clarity in the whole discussion of this topic, which has been generated to a large extent to counter criticism of the anti-developmental effects of food aid and its manifest failure in most cases to improve nutrition for those most in need.

Nutritional goals relate primarily to the use of food aid in special feeding campaigns for specific groups, whether they be considered especially mal- or undernourished or for some other reason. There are, for example, numbers of projects for improving the diets of school children, although in many Third World countries only the better-off are able to send their children to school. While there is no doubt that there are some projects which achieve some nutritional improvement for those in need during the life of the project, there is no evidence that this occurs either on a significant scale or that it lasts beyond the project period. *Emergency Feeding* could be considered under nutrition, though its purpose is rather to keep people alive and reasonably healthy until the food situation improves. No-one could possibly object to the idea of this, though in practice, as mentioned above, there are cases when food arrives so late as to be worse than useless (by reducing producer prices for the next harvest).

It is sometimes said that there is no necessary conflict between these different aims and that food aid can help even when the motives for giving it are less than pure. But there are some rather obvious points of conflict. Building export markets can hardly mean other than encouraging countries to become dependent on food imports. Potential commercial markets are not usually the poorest or most needy countries and certainly not the poorest and most needy people in them. Political pressure cannot be exerted effectively without the threat, and if necessary reality, of withdrawal of food aid. This seems clearly inimical to either developmental or nutritional goals. Yet long-term contracts which avoid this danger seem still more likely to generate dependence on imports. All the same, this seems the lesser of two evils, since, in making a contract, the donor abjures certain possibilities for direct political pressure.

Recipients of food aid

Figure 8.3 shows the main recipients of cereal food aid for the period 1970/71 until 1983/4, by volume. The most dramatic feature is the enormous contraction in food aid shipments in 1973–4, a period of serious food crisis in Africa and other parts of the Third World. This was a direct result of the denomination of US food aid in money terms and the sharp rise in cereal prices of that year, largely engineered by US policy.

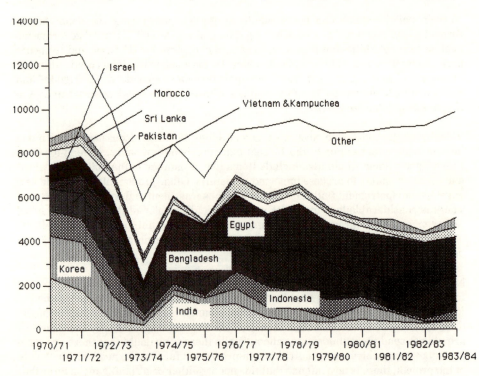

Figure 8.3: *Major recipients of cereal food aid 1970/–1983/4 ('000 tons)*

It also provided a turning-point in other respects, as the period up to 1973/4 saw the running-down of the patterns of the 1960s with declining food aid. By 1970, the European recipients of the late 1950s were already gone and large Asian countries were the major recipients. In 1970, India, Korea, Indonesia and Israel were major recipients. But by the mid-1970s, India was approaching self-sufficiency in cereals, while Korea and Israel had used food aid to develop substantial intensive livestock industries and received smaller amounts henceforth. By the early 1980s, Egypt and Bangladesh were by far the largest recipients of both cereals and dairy products, and African countries, Morocco, Sudan, Ethiopia and Mozambique, together with El Salvador, were joining the major recipients. While current recipients of food aid are generally poorer countries than those of the 1960s, let alone the 1950s, it is less clear that the alleviation of hunger is among the major factors determining its distribution. This still seems based largely on political clientelage (Egypt, El Salvador, Sudan, Bangladesh) and market factors (Egypt, the Maghreb).

In the case of *Bangladesh*, no-one can doubt that the country is poverty-stricken and that many of its people are hungry. Whether food aid has helped the situation is another matter. During the 1970s, it was notorious that a large proportion of all food aid was used to feed the army, police and favoured civil servants. In addition, since food aid counterpart funds made up a very significant proportion of total government revenue, there was reason to suppose that the Bangladesh Government could not afford to achieve self-sufficiency, since this would reduce its revenue. There was also an occasion when, with a serious food shortage in progress, the USA held up food aid shipments until Bangladesh cancelled a contract to sell jute to Cuba. Clay (1985) indicates that the administration of food aid in Bangladesh has improved to the extent of its being a reasonably effective safeguard against shortage, though in an earlier article (1979), he is less sanguine about the long-term impact. Cutler (1985) presents a much less reassuring picture, though the shortage he considers (1979) is earlier than the 1984 floods considered by Clay. Cutler points to the frequency of unrecognized famines and to the function of famines in still further increasing the already enormous number of 'functionally landless' rural households – over 50 per cent according to figures he cites. While there has been much discussion of Bangladesh's 'lagging food production', he finds that 'vulnerability to food crises lie(s) more in its grossly inequitable agrarian structure, coupled with its dependence on richer countries for fiscal and material economic aid, than in its record of food production', (1985:208).

India was by far the major recipient of cereal food aid during the 1950s and 1960s, this combining the interests of the USA in disposing of stocks and of the Indian Government of the time in investing in industry. For this reason, many of the studies which aim to assess the impact of food aid upon local production have been done in India. Since the late 1960s, changes of policy in both the USA and India, combined with production increases related to the Green Revolution, have led to a major reduction in both commercial and food aid cereal imports.

However, India became in the early 1970s and remains the world's largest recipient of dairy food aid, taking on average some 20 per cent of the total (FAOa, 1986: Table 16), with most coming from the EEC. A high proportion of this goes to Operation Flood, a massive programme to develop a modern dairy processing and distribution system, aimed at the cities and especially their middle classes. The 'White Revolution', as it has been extravagantly termed, has been hailed by food aid enthusiasts for many years as one shining example of the successful use of dairy food aid. More recently, however, there have been a number of far more critical assessments, including a number which have attacked the programme and its leadership for concealing unpalatable facts and disseminating misleading information (Alvares, 1985; Terhal and Doornbos, 1983).

The project has been much more successful in starting dairies and processing plants and in expanding urban middle-class demand than in increasing production, while a significant proportion of increased local milk deliveries are said to have come from reduced family consumption rather than increased production. Though the programme was specifically intended to help poorer milk producers, recent information seems to indicate a severe bias towards larger producers, and particular areas. Moreover, as pointed out by Narayan Nair and Jackson (1985), the plans for increased production focus heavily on imported exotic breeds, high technology and thus a high import content, while taking insufficient account of the limited availability of fodder. At the same time, not only are dairies under Operation Flood among the most important producers of baby foods in India, but the project and its leader have been among the most prominent in rejecting WHO guidelines on advertising and sales methods for baby foods. This, it may be noted, is despite a large body of documented evidence on the disastrous consequences for child nutrition and health of replacing breast milk with powdered formulations.

But while there are many aspects of Operation Flood which give rise to concern (only a few of them mentioned here), there seems little doubt that it has been run exceptionally efficiently in terms of pure administration. Since this can hardly be expected to be the case in most of tropical Africa, it follows that the results of similar (but much smaller) projects there are likely to be less favourable. As some examples below indicate, this is the case.

Egypt was among the major food aid recipients during the late 1950s. This continued during the early 1960s, though shipments were cut or delayed on a number of occasions, reflecting US disapproval of anti-Israeli statements and the intervention in South Yemen by the Egyptian Government. In 1967, with the war against Israel, US shipments of food aid ceased, having previously covered 10–20 per cent of total cereal imports. This continued until the mid-1970s and the improvement of relations which culminated in the Camp David agreement. With the shift in Egyptian policies towards Israel, US food aid re-commenced and Egypt rapidly became the major single recipient.

In recent years Egypt has received an average of slightly under 2 million tons of food aid cereals per annum, accounting for between 25 and 55 per cent of total cereal imports and up to 25 per cent of total domestic grain consumption (Thomson, 1983a, Table 4). This has been used by the government to subsidize wheat, flour and bread prices to the mass of the population, significantly increasing bread consumption and acclimatizing the population to a diet dependent on subsidized imported wheat. This both helps to reduce the level of popular discontent and assists the country's industrialization policy by holding down the level of wages. It can be seen as beneficial for the Egyptian population and government so long as it continues. But its continuation is dependent on the government's pursuing policies satisfactory to the USA, notably with regard to Israel. Given the current level of food aid, it seems highly unlikely that any Egyptian government could withstand the reaction to its withdrawal.

The case of Chile underlines this point with brutal clarity, though food aid was not the only issue at stake. In the period 1969–70, Chile received some $290 million in US aid and credits, of which about $45 million was food aid. In late 1970, a Marxist president, Salvador Allende, was popularly elected to power, despite the fact that 'perhaps a billion dollars in public funds had been committed by the US during this period . . . largely . . . to prevent the left coming to power'.[22] Following the announcement by the

Allende Government of its intention to nationalize extensively, the USA unleashed a strategy for economic warfare designed, as President Nixon said privately, 'to make the Chilean economy scream'.[23] This involved reducing US official aid, including food aid, to a fraction of its former level, and persuading the World Bank, the Inter-American Development Bank and foreign private investors to do likewise. Over the next few years the screws were further tightened as goods and food became scarcer, and in the final weeks before the coup, the White House imposed pressure to prevent a cash sale of wheat to Chile. Within weeks of the Pinochet coup, not only were financial aids unlocked, but an agreement was signed for an immediate PL 480 credit for eight times the total for the entire Allende period. Cereal transfers for 1974/5 and 1975/6 were over ten times the levels of the previous four years (FAO, 1983: Table 12). But there was no question of food aid improving the diets of the poor. A study of income and nutrition concludes, 'In Chile, between 1968 and 1976, the urban poor attained a significant improvement in their nutritional status until the end of 1972, only to have it decline seriously and reach even lower levels than those of 1968' (Ruiz, 1980:143).

That EEC food aid is less subject to direct political pressures of the sort which have lain behind some changes in PL 480 transfers, has partly to do with the fact that the EEC is a less political body than a single state like the USA. It also reflects differences in the way economic policies are made and implemented in the USA and European countries. In the USA both President and Congress intervene in the making of day-to-day economic decisions, to a far greater extent than is generally chracteristic in Europe. For example, while US food aid is still referred to as PL 480, that law has in fact been revised, renewed and added to at least ten times. The Congress has voted several times on matters like the maximum income of countries receiving US food aid, or cuts in aid because of nationalization of US-owned property. By contrast, something so dull and technical as food aid seldom surfaces in the national parliaments of Europe, and almost never in terms of specific decisions. While there has been rather more discussion of food aid in the European Parliament, national interest is far less obviously to the fore. One obvious reason for this is the divergence of national interests among the member states.

AFRICAN FOOD AID RECIPIENTS

As can be seen from Figure 8.4, until 1980 Egypt received more cereal food aid than the whole of sub-Saharan Africa combined. Food aid to Egypt rose rapidly with the resumption of US deliveries, to a level of about 2 million tons per annum, at which level it has remained for the past eight years. For tropical Africa, there was a first expansion around 1973, followed by decline and then a really major increase from the late 1970s. 1984/5 saw a further increase by over 40 per cent, almost all of which went to tropical Africa. Deliveries for 1985/6 are estimated to be down considerably.

Table 8.6 and Figure 8.5 show cereal food aid deliveries to major tropical African recipients from the beginning of the 1970s. This shows the major increase in 1973/4, followed by three years at below the 1973/4 level, and then a gradual build-up to 1984/5, when food aid transfers were ten times the level of before 1973/4, and accounted for 46 per cent of total cereal imports into tropical Africa (FAO, 1986; Table 1). As Table 8.4 shows, a large proportion of the whole goes to a few major recipients. In 1973/5, about 60 per cent went to the top five recipients; by 1980–83, this had fallen to 50 per cent, but the huge transfers to Sudan and Ethiopia in 1984/5 have pushed it up to 78 per cent.

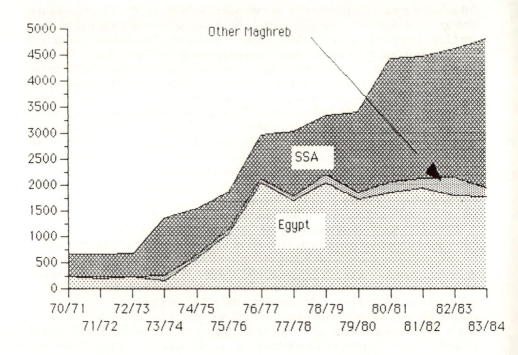

Figure 8.4: *Cereal food aid to Africa 1970/1–83/4 ('000 tons)*

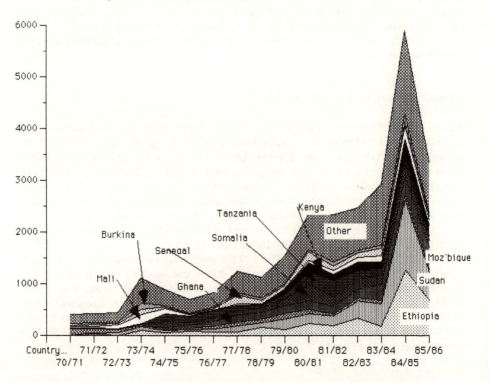

Figure 8.5: *Major tropical African recipients of cereal food aid 1970–85 ('000 tons)*

Table 8.6: *Cereal food aid to tropical Africa 1970/1–72/3 to 1985/6 ('000 tons)*

Country/ year	70/1– 72/3	73/74	74/5– 76/7	77/8– 79/80	80/81	81/82	82/83	83/84	84/85	85/86
Angola	0	0	5	13	25	75	60	69	72	91
Burkina Faso	38	108	22	46	51	81	45	57	143	54
Cape Verde	0	0	15	38	31	54	35	63	49	53
Chad	3	67	19	29	14	29	36	69	203	45
Ethiopia	10	96	72	117	228	190	344	172	1265	675
Ghana	85	36	40	89	94	43	58	74	110	61
Guinea	28	9	31	30	34	39	25	43	34	27
G. Bissau	0	0	12	19	26	30	35	19	23	6
Kenya	2	1	6	35	173	127	165	122	346	193
Lesotho	20	25	14	30	44	34	28	50	77	60
Mali	41	180	49	26	50	66	88	111	239	113
Mauritania	21	95	35	36	106	86	71	129	236	94
Mauritius	23	33	17	13	21	43	13	22	17	15
Mozambique	0	0	62	136	155	149	167	297	350	428
Niger	21	195	68	17	11	71	12	13	284	80
Senegal	30	101	25	96	153	83	91	151	97	69
Somalia	5	16	82	99	330	186	189	177	262	128
Sudan	32	79	42	138	195	194	330	450	1350	548
Tanzania	8	11	131	82	236	308	171	136	118	93
Zaire	21	3	12	58	77	98	110	53	100	103
Zambia	0	0	13	78	85	100	83	76	117	119
Other	34	66	41	87	196	252	316	573	400	294
Total SSA	425	1121	812	1312	2335	2338	2472	2926	5892	3349

In the early 1970s, by far the major recipients, apart from Ghana, were Sahelian countries, together with the Sudan and, from 1973/4, Ethiopia. In 1974/5 Tanzania became a major recipient, as did Somalia. Mozambique joined the group about two years later. With the addition of Kenya and Zambia in the late 1970s, this more or less completes the current list of major recipients. In 1984/5, Sudan and Ethiopia took over half the total. What shows from Figure 8.5, apart from the enormous increase of 1984/5, is an apparent 'ratchet effect'. That is, food aid deliveries increase during emergencies and although they may fall afterwards, remain well above the pre-emergency level. By 1984/5, food aid accounted for over 45 percent of total cereal imports into tropical Africa, so that the ratchet effect may well apply not just to food aid but to total imports. As pointed out in Chapter 2, this latter effect is likely to be lagged by one or two years, because delayed food aid shipments cover part of 'normal' import requirements in the years after an emergency.

The impact of food aid in tropical Africa

Before presenting some examples of how food aid functions in particular countries, a study is discussed briefly, which purports to show the effects of food aid, with regression analysis.

Vengroff (1982) looks at US (PL 480) food aid to Africa and attempts to relate it statistically to need, to US trade interests, to US foreign policy interests, to the level of dependence on the USA and to the subsequent rate of growth of the economy. This is done on the basis of a sample of 32 black African countries which became independent in or before 1966 and with a variety of measures and proxy-variables. One could spend some time criticizing the methodology, but an example may suffice. The author reaches the surprising conclusion that the *less* dependent on the USA a country is, the more likely it is to receive food aid. Measures of non-dependence include anti-US statements in the UN and purchases of arms from the USSR, Czechoslovakia or China. Whatever one thinks of such indices, food aid transfers are measured over the period 1962–78 with (as has been seen in Table 8.4) a very large proportion of the total concentrated in the mid-1970s. The 'disagreement scores' and arms trade were computed for 1966–71 and 1964–73 respectively. By excluding it from black Africa, the author avoids attributing Egypt's massive post-Camp David food aid receipts to Nasser's anti-US speeches and arms purchases during the period when Egypt received no food aid. But the anti-US posture of Nkrumah is credited with food aid transfers to Ghana under those who ousted him. Somalia's large food aid transfers of the pro-US period are attributed to the socialist statements of the early Barre years etc. The author himself does not take this finding very seriously. But it does indicate a serious problem with the whole procedure. In comparing different series of largely arbitrary and non-contemporaneous quantitative indicators, all sorts of odd conclusions may emerge. That one can accept those which agree with common sense and reject those which do not, leaves one no wiser than by using common sense alone.

SOME EXAMPLES OF FOOD AID IN TROPICAL AFRICA

The following examples do not aim to cover the whole 'food aid story' for any of the countries mentioned. They select some cases which demonstrate relevant issues.

Ethiopia. While much of the writing on food aid in Ethiopia focuses on 1984/5, the country has been a major food aid recipient since the late 1970s. Unlike most recipients, rather little of this has been in the form of programme deliveries, the major proportion going to emergencies and a huge food-for-work (FFW) programme, operated by the World Food Programme.

Holt (1983) describes the FFW project, whose purpose was reafforestation and erosion control in four of the most drought and food-shortage affected regions in the country (Eritrea, Wollo, Tigrai and Harerghe). The project was started shortly after the 1973 famine, as a rehabilitation project, and grew into the largest FFW project in tropical Africa. The author's aim is to show that the project had, at least in part, the character of emergency relief. Whatever its productive outcome, it provided essential elements of subsistence to a population which had been through many droughts and which was moving towards another while the article was being written. He finds that this overcomes at least some of the normal criticisms of food aid, though he is careful not to present this as a 'proof' of the worth of FFW.

He shows that the FFW programme provided a useful income supplement to those involved without attracting labour away from own-farm production (a maximum of 90 days per annum was available and labour was organized by Peasant Associations which suspended work during peak labour seasons). He makes clear that organization by Peasant Associations was the crucial factor allowing a huge programme to be organized with relative efficiency over a short period. From interviews, it appeared that food aid cereals did not depress the prices of local grains, though this was partly because the project was situated in drought-prone areas and because the most recent harvest had been relatively poor. The FFW programme was specifically concerned with soil conservation, terracing, bunding, and reafforestation on an almost entirely labour-intensive basis. One might suppose therefore that it was achieving something positive.

Unfortunately it was too early to tell in 1983. There is reference to excessively hasty planning and to the fact that, while the very ambitious targets were not kept to, the amount achieved was still impressive. This is attributed in part to food shortage making wages in food more attractive than otherwise. There is mention that locals seemed to have accepted soil conservation methods, though with certain reservations, but no mention of how much was changed or how it affected subsequent production. According to agricultural authorities with whom the author spoke, it will take at least ten to fifteen years before one can notice any major difference – though one would at least expect some reduction in erosion before then, I would have thought. Holt does indicate that strong pressures to get the project moving as a response to food shortage were among the reasons for lack of thorough planning or monitoring of its long-term productive aspects. So while terracing, bunding and reafforestation under the control of Peasant Associations sounds, on the face of things, a more than usually favourable combination from which to expect positive results, there was little information on that in 1983. Maxwell (1986a:8) refers to studies in 1982 which asserted that 'rates of return were . . . adequate, though much higher rates could have been obtained with better project planning . . .' How a 'rate of return' could have been assessed on this type of project at all, let alone so early, is not explained.

Another point on which there is no information is military activity, although Eritrea and Tigrai were among the areas in which the project was carried out. Apart from its obvious general implications, the issue of war and military action is highly relevant to any project depending on Peasant Associations. Official figures for 1985 show the number and membership of PAs to have been very much lower in Eritrea and Tigrai than in any other parts of the country (Teke, 1986:Annex 1). Again, while this article looks at land reclamation in the northern highlands, large amounts of food aid have been used in schemes to settle highlanders from these areas (especially those affected by the war) in the south. There has been considerable criticism of these enforced settlement programmes, both on technical grounds and because of their political implications. Some experts have supported the programme on the grounds that the lands from which people are moved are so worn out and eroded that they cannot recover without reduction in population. But since there are cases of people walking the 700-odd miles back to their old homes, they are clearly not very popular. Another source, as of late 1986, was of the opinion that, to the extent that these projects had been successful at all, this would apply only to Hararghe.[24]

Maxwell (1986c) looks more generally at the issue of the disincentive effects of food aid in Ethiopia. A previous survey in 1982 had not discovered any negative effects on local production or state policy. But by 1986, while not specifically finding ill-effects from food aid, he writes that 'warning lights are flashing' in respect of two aspects of state pricing policy. The first, as mentioned in Chapter 4, is the low procurement price paid by the monopoly Agricultural Marketing Corporation, together with a disturbing trend towards 'administrative measures' (i.e. force) to get grain from peasants. The second is that the consumer price of wheat is subsidized, thus increasing demand for a cereal, some (but by no means all) of which has to be imported. He could also have spread the policy net more widely and mentioned the fact, which he documents, that at least 80 per cent of government recurrent expenditure on agriculture went to state farms and co-operatives during the period between 1976 and 1983.[25] But while such policies clearly have negative effects on the availability of marketed food, it is not easy to link them to food aid. It does seem likely that food aid mitigates some of their impact, by providing cheap imported cereals. But these sorts of policies are also standard for African 'Marxist' regimes,[26] and it may well be that they would have been pushed through even without food aid. In particular, this may well be the case because food aid has never come to more than 5 per cent of total cereal supply, except in 1984/5, and because most of it goes into food-for-work or emergency supplies (Ibid., Tables 14 and 16).

Somalia. Where Holt discusses the use of food-for-work as emergency relief in Ethiopia, Anne Thomson (1983b) looks at some of the effects of the development in Somalia of a 'long-term emergency', where, for various reasons, food-for-work was not an option, and at the considerable quantities of food aid involved. The factors leading to emergency were genuine enough. Somalia suffered the same severe drought in 1973–4 as Ethiopia and a band of countries across the Sahel, turning many pastoralists into refugees. In 1977, there was a war with Ethiopia (the Ogaden War), which led to more refugees, and in between there were rain shortfalls and floods.

Whatever the reasons, both imports of cereals and food aid have increased rapidly in Somalia, which has received on average, over the past decade, about the same amount of food aid as Ethiopia, although it has only one eighth the population.[27] The evidence on dependency and producer incentives is ambiguous. On the one hand, Thomson finds that there is little evidence of major effects on producer prices of local cereals. But there is clearly significant leakage of cereals from the refugee camps to the surrounding populations. In 1980, when a new monitoring system was brought in, this led to the immediate halving of the estimated number of refugees (who had previously been registered in but not out). Even so there were those who thought the number still over-estimated by over 50 per cent. There is a clear problem of dependence on food aid within the refugee camps, though it can hardly be ascribed to food aid, being rather a reason for it. The government will not provide land for the refugees to cultivate, partly because of shortage of land, but also because this would weaken Somalia's claim on the areas in Ethiopia from which they come.

At the national level, there can be no doubt whatever that Somalia is dependent on food aid. Thomson shows that cereal imports increased as a proportion of aggregate supply from about 29 per cent in 1970–72 to over 50 per cent in 1978–80, with about half being accounted for by food aid (p. 211). During the first half of the 1980s, imports again accounted for half of total cereal supplies, with food aid accounting for half to two-thirds of the total. By 1984/5, food aid accounted for nearly 90 per cent of total cereal imports (FAO, 1986b: Table 1).

Thomson also describes the crop marketing system in terms familiar from other tropical African countries. In spite of problems since the early 1970s, official deliveries of sorghum have continued to increase, but not fast enough to keep up with the growth of the urban population and of demand. A number of increases in producer prices have failed to solve the problem. With severe foreign-exchange shortage, the government has relied increasingly on food aid for sale.

There is thus no way of disentangling the effects of food aid from the mass of other things going on within the Somali economy and society. The agricultural base of the country is relatively weak and would, under any circumstances, be pressed to cater for increased urban demand. Food aid had probably hastened that process, but it is not clear what the result would have been without it. Certainly refugee food aid seems necessary, although excessive amounts and leakages may have had negative effects in the rest of the economy.

Tanzania. Recorded food aid shipments to Tanzania go back to the colonial period, according to Tibaijuka (1981), both for famine relief in Dodoma Region in the dry centre of the country and for various feeding programmes operated by the Catholic Church and supplied by US Catholic charities. She cites Maletnlema (1980) on the development of the latter into a widespread 'school-feeding programme', providing lunch for (Mission) school children, with PL 480 food. This was built around bulgur wheat and skim milk powder, until US stocks of the latter ran down and it was replaced with a soya product.[28] Malctnloma, Head of the Tanzania Food and Nutrition Council, was in no doubt as to the negative effect, since this accustomed future members of the urban elite to diets based on imported items.

Tanzania became a major food aid recipient in 1974/5, after the peak of its most serious food crisis. This declined in the late 1970s with increased local deliveries, and the country imported no maize at all in 1977/8. Then, with a sharp deterioration in the balance of payments, war with Uganda and other problems, official deliveries fell seriously and food aid began to increase again, along with commercial purchases. Food aid peaked in 1980/81, at about 300,000 tons, since when transfers have fallen to about one-third that level. The most recent series of famines hit other countries harder, though Tanzania probably suffered more in 1984 than Kenya, which received far more food aid, because of unofficial exports from Tanzania to Kenya.[29] But it may also be due to the general 'donor fatigue' which affected all sectors in Tanzania until the signing of an agreement with the IMF in 1986.

Although maize is by far the most widely consumed cereal in Tanzania, most of its cereal food aid has been wheat and rice. In most years since 1976, food aid has covered 100 per cent of the country's wheat imports, which can be seen as a substantial foreign-exchange saving. There has been no marked upward trend since the early 1970s, which presumably means that the government has decided that only food aid wheat should be imported. Rather over half of rice imports have been covered by food aid since the late 1970s, rising in some years to 100 per cent. Here there does seem to be a rising trend in both food aid and commercial imports. For maize the proportion of food aid is far lower. In 1977/8, food aid accounted for 100 per cent of maize imports, but for a special reason: the country achieved self-sufficiency and did not need to import maize. The food aid shipment was probably a late arrival. During the most recent years, with large deficits, Tanzania has had to buy the majority of its maize requirements commercially.

During the 1970s, Tanzania received considerable amounts of dairy food aid, in relation to the building of a number of urban dairy plants. The first of these, commissioned in 1969 in Dar es Salaam, was intended to operate with at least 50 per cent local production (which had been developed for a previous non-pasteurizing dairy). But the cost of pasteurization and packing in tetra-paks, with government unwillingness to raise consumer prices, led to producer prices being reduced by half, so that most producers went out of business. Much the same happened with the construction of a pasteurization plant in northern Tanzania, except that in this case local producers continued production for the unofficial market, at substantially higher prices than paid by the dairy. This has generally been the case with urban dairies, almost all of which operate almost entirely on imported SMP and butter oil, much of it food aid. This was the case in spite of a series of World Bank-funded dairy development projects (state farms) which increased foreign indebtedness but not the milk supply. That this has not led to major increases in commercial imports during the period up to 1986, mainly reflects Tanzania's critical shortage of foreign exchange to purchase either raw materials or spare parts for the dairies.

As in the Ethiopian and Somali cases, it is virtually impossible to work out whether food aid has had any direct impact on state policy and if so what. It is well-known (and discussed above) that Tanzania has had problems with agricultural policy. Official crop prices have been low in relation to prices of consumer goods and inputs, discouraging sales through official markets. But since export crop prices have fallen more than those of food crops, there is no special reason to attribute this to food aid. It is certainly true that food aid has been a major boon to the National Milling Corporation, allowing it to continue operating as inefficiently as it has. But it is far from certain that the absence of food aid would have forced greater efficiency. The result might as easily have been still lower levels of urban consumption.

Senegal is a major food aid recipient, but it is also a major commercial food importer. Most of the commercial imports are of rice, and the major growth took place before food aid became important. This reflects a deliberate policy to maximize export earnings (largely groundnuts) and import food. Josserand (1984) recounts that consumer prices of rice are low, without subsidy, because of low international prices and low transport costs from the USA and Europe. This in turn is an important underpinning for the export-crop economy since, without the possibility of buying cheap rice, many local producers would not risk high proportions of export crops, but would rather grow more of their own food. A small sample survey of three villages seems to bear this out. The richest and poorest villages consumed more rice than those of medium income, the richest because they could afford it, the poorest because they had not produced enough millet to cover their needs.

Maxwell (1986b) comes to similar conclusions, especially as urban rice consumer prices were increased significantly in 1985, to a level well above import parity. He does note though, that producer prices for millet, the most widely grown local cereal, are low and that a government floor-price system seems not to work well. But since the markets for millet and for rice/wheat are segregated to a considerable degree, he does not find food aid to have been a significant factor – arguing indeed for increased programme food aid for balance of payments supplementation.

Mauritania is among the worst hit of the Sahelian countries, by droughts and food shortages during the past decade. Only a very small strip in the south of the country is cultivable at all, most being arid country suitable for grazing, or outright desert. The droughts and famine of 1970–73 meant enormous loss of livestock, leaving thousands of pastoralists destitute and driving them to 'sprawling squatter camps on the edge of the towns'. Local cereal production covers no more than 6–25 per cent of total requirements, with food aid covering from 10–90 per cent of imports. Mauritania's main source of foreign exchange is mineral exports and food aid from the EEC (and other donors) 'relate only haphazardly to the country's needs'.[30]

Mitchell and Stevens (1983) have as purpose not to describe the food situation in general but to consider the 'cost-effectiveness' of EEC food aid transfers in different forms. They compare wheat, SMP and butter oil. The comparison starts out by assuming that, had food aid not been available, the items or substitutes would have been imported. It then compares the cost to the EEC of giving them as food aid with the saving to Mauritania of not having to import the cheapest close substitute. This gives the result that SMP is 'good value' (under assumptions which they admit are favourable), while wheat and butter oil are not. This may be of interest to the EEC Commission, but it tells the general reader little about the impact of food aid. Whatever the 'cost-effectiveness' of SMP in these rather technical terms, cereals are almost certainly more important to the country and most of the people. That SMP is relatively cheap in no way detracts from its general problems as food aid and in any case relates closely to the price conditions prevailing when the study was made. The authors make it clear that the article aims to say nothing about the general impact of food aid or its effect on food dependency. The figures they provide make it plain that Mauritania is heavily dependent on cereal imports though, because of mineral exports, less dependent on food aid than most of its neighbours. While urbanization is clearly one of the factors increasing demand, this has to be seen in the light of the decimation of pastoralist herds during several droughts. Food aid clearly substitutes for some commercial imports but, accounting for under a third of total imports in most years, it is probably not a major factor in their increase.

Such brief country examples miss a number of important points referring to *details* of the distribution of food aid, which are seldom available to those not closely involved themselves. There are large numbers of examples where food aid shipments have been held up for lack of transport or hijacked by local merchants or politicians. When direct transfers are distributed through rationing systems, these tend to favour groups like the civil service, police and army. The only FFW project I have visited myself was a tree-planting project in Mozambique. In a two-hour visit, it was not possible to find out much more than that more trees had been cut than planted, that those being planted were not the sort which the local population could use for building or firewood, and that relationships among the local and foreign experts were strained to breaking point. Food was a thoroughly acceptable wage-good with the existing shortages in the country, but it was far from clear that all the food was getting to those who did the work. A number of veiled accusations of corruption were made during an embarrassingly sticky meeting. As so often, pressures of time in consultancy work prevented any follow-up.

An article in the British *Observer* from 1979 cites a retired WFP administrator on widespread misappropriation of food from projects in western and central Africa. Reading the list of different accusations reminds one that, where food is in short supply, it makes an effective basis for patronage and corruption. This is especially the case since WFP has no control over food, once it reaches its country of destination, although it is also supposed to take part in the organization of FFW projects.[31] But while many, if not most, newspaper articles on food aid focus on this sort of factor, they seldom appear in academic or professional papers except as oblique references. While such factors may not fit well into theoretical anlayses, they certainly do affect the impact of food aid.

FOOD AID, DEVELOPMENT AND FOOD STRATEGIES

The more responsible proponents of food aid have for some years been concerned to improve its developmental effect, at least to the extent of limiting or offsetting the growth of dependence on food imports. This has normally implied trying to use the food or counterpart funds for investment in increased production and setting conditions on transfers with the avowed intention of improving their use in this respect.

During the past three or four years, the EEC has taken the lead in this, with its efforts to integrate food aid into more general 'food strategies' for certain countries. This development had its genesis during the period when M. Edgard Pisani was Commissioner for Development. Pisani was highly critical of certain aspects of EEC and Lomé aid, most notably food aid and expenditures on large-scale irrigation projects. Moreover, his period as Commissioner coincided with the serious food shortages of the early 1980s. The idea of food strategies was to integrate food and project aid with local policy for the purpose of improving the food situation and to do so within the context of an overall plan for the food sector.

This in itself constituted something of a departure from standard donor policy at the time, the latter being heavily influenced by the World Bank's Berg Report and by IMF conditionality, both stressing the importance of increasing export-crop production.

But in other respects, the thinking behind food strategies does not depart so much from the norm. Where food production is concerned, the emphasis is on increasing its aggregate level and specifically the marketed portion. In relation to marketing, the EEC seems both to share the general negative donor opinion of state grain marketing monopolies and to rely on these selfsame monopolies as the basis for its policies. Its strategy is quite specifically a programme aimed at the *national* food situation, which is hardly surprising, since it is states to which the EEC relates.

Responses to this initiative have been varied. Donor circles have generally welcomed it, since it makes obvious sense to co-ordinate policies. ACP governments have been much less enthusiastic, seeing the 'policy dialogue' involved in their setting up as the means for inserting yet more conditionality into aid transfers.

As to the results, although studies for the first food strategies were started in 1982, it is still difficult to assess their impact. When the policy was initiated, it was decided to concentrate in the first instances on pilot projects in four countries, Mali, Kenya, Rwanda and Zambia. It was not the intention that this should involve increases in the amounts of funds transferred. Rather it was to make more effective use of them, through concentration of efforts, through integrating the use of different transfers (project and food aid), the programmes of different donors and, via 'policy dialogue', the policies of the recipient countries.

For *Mali*, the central aspect of the programme has been privatization of grain marketing, combined with efforts to improve the efficiency of the state grain monopoly, thus allowing it to pay higher prices to producers and attract more food on to the official market. In the meantime, until these savings are achieved, producer prices have been improved by providing food aid, whose lower purchase price would allow subsidies to local production without increasing consumer prices. Successes have been claimed, at least in respect of privatization, though, as cited in Chapter 3, this seems to have increased problems of distributing food aid to those who need it. It is also claimed that there has been some (unspecified) improvement in the efficiency of cereal marketing through OPAM, the state agency (*Courier* No. 100:20–23, December 1986). On the other hand, this source admits that there has so far been no increase in food production related to the strategy – though aggregate food production has increased significantly in the past two years as a result of improved climate. A more recent report (Speirs, 1987) also indicates that progress in the improvement of production has been very slow. He cites the use of food aid counterpart funds for purchase of agricultural inputs though since these are in local currency, this cannot refer to increased levels of imports. From this and other reports, it would appear that increasing production is obstructed by another set of state agencies, the Operations de Développement Rurales. These are described as a 'dramatic failure', leading to declining production, even in the Niger Valley, regarded by some as a potential 'breadbasket of the Sahel'. As indicated in Chapter 3, much of the Malian food strategy can be seen as a part of the donor effort to persuade Mali's conservative but corporatist military government to privatize the parastatal corporations which dominate and apparently strangle its economy – so far, it seems, with limited success.

In *Rwanda* the problems are very different, among the most obvious being high population density and land shortage. The focus of the food strategy has thus been on agricultural intensification of small farming combined with 'the encouragement of small- and medium-scale rural enterprises to provide alternative sources of revenue and employment for the rural population'. These sound, on the face of it, laudable aims, but it is not possible to say much more than that on the basis of information available (*Courier* No. 100 December 1986).

In both Mali and Rwanda, the EEC is a major donor, and thus capable of exerting significant influence on the government and other donors. This is less so for Kenya and Zambia, the other two 'pilot' countries, for which other donors, notably the World Bank and USAID, are far more influential. Thus in Kenya, 'the EEC food strategy initiative has *coincided with* important Kenyan food policy reform measures' (ibid. emphasis added). EEC support has been concentrated on channelling food aid and other resources to the National Cereals and Produce Board, via a government and mutli-donor committee. While this has doubtless helped Kenya during the shortage period of 1983–4, there is not much evidence of a major independent impact from the EEC food strategy, especially as Kenya is currently (though probably temporarily) a cereal exporter. The 'policy reform measures' referred to would probably have been deregulation of the marketing of cereals. This was done partially during 1986/7, but many of the controls (notably the monopoly of the NC&PB) were subsequently re-imposed. The current situation seems to be that, while a few private traders are licensed in some areas, most producers can only sell legally to the NC&PB, except for small amounts in local markets. Given that payments from the latter are up to six months late in some areas, there are many who sell illegally. Among the reasons why Kenya has not yielded to donor pressure and decontrolled cereal marketing, is that 'deduction at source' from a monopoly agency is among the means sought to achieve repayment on credit programmes, many of them started by the same donors. By early 1988, in return for substantial EEC aid payments the Kenyan government had formally agreed to privatise *part* of the domestic cereals market.

For *Zambia*, the review cited refers largely to dialogue and 'paving the way, for the many food policy reform measures taken by the government', while tacitly admitting that World Bank and IMF conditionality have been far more important factors. Here again 'marketing problems are considered, to form the most important constraints on agricultural growth', and many donors have written reports and exerted pressure in this respect – apparently to limited effect. An article describing the almost unbelievable chaos in Zambian maize marketing in 1985, makes no mention of the EEC at all (Good, 1986). However, where the EEC review follows most donor opinion in approving recent changes in Zambian agricultural policy, Good, having considered a wider range of policies (and notably emphases on export crops and large-scale capitalist farms), sees them as more likely to bring 'deepening poverty for the peasantry and uncertainties for domestic food production'. How the 'dialogue' has been affected by Zambia's more recent rejection of IMF terms, I do not know, though I would hazard a guess that no fully fledged food strategy has yet been agreed.

It would appear from available evidence that the EEC has not been able to achieve the position of food strategy co-ordinator in Kenya or Zambia, though it may well be that its own co-ordination with government and other donors has improved. Neither in these two cases, nor for Mali and Rwanda, does it appear that the food strategies have had any marked effect during their first three years, though they may have contributed to positive developments. But a further difficulty in assessing this is that the food strategies seem mainly to concentrate on improving the efficiency of food marketing, which is by no means the same as improving the food security of the population. There is thus the risk of co-ordinating the same old set of policies which have contributed to the problem.

TRIANGULAR OPERATIONS AND LOCAL PURCHASES

If the verdict on food strategies must so far be 'not proven', another recent innovation in food aid policy seems more hopeful. This is the practice of providing funds for the purchase of cereals in African countries, for distribution either in other deficit areas of the same country (local purchases) or to supply other African countries (triangular transactions). In some cases donors supply cash, in others local agencies make cereals like maize or sorghum available in return for food aid wheat or rice. In either case, the impact is preferable to that of direct food aid, since this supports rather than reduces the prices of locally purchased cereals.

The major problem with either of these initiatives is size and likely dependence on wheat food aid. The volume of such transactions has increased significantly in recent years, but amounted in 1985/6 to some 0.6 mn tons, almost equally divided between triangular and local operations. While 0.31 mn tons of triangular operations is significant, when compared with some 0.7 mn tons of commercial cereal exports from African surplus producers, it is rather less significant when compared with 3.3 mn tons of cereal food aid in that year. But if such transactions are to grow further in the future, they will probably be tied rather closely to transfer of wheat food aid destined for the towns of African producers of surpluses of maize and sorghum. Nevertheless, this is a positive development and much to be preferred over straight food aid transfers.

The EEC has been prominent in both such types of operation. In 1985/6, the Commission provided 36 per cent of combined triangular and local purchases and the Commission plus member states some 61 per cent, compared with only 13 per cent each for the USA and the WFP, the latter in any case receiving some of the cereals from the EEC. The real test of the genuineness of recent EEC statements will come if and when the quantities available for triangular transaction are sufficient seriously to interfere with the quantities of food aid transferred.

DEPENDENCY AND DISINCENTIVES

If it is hard to draw conclusions on the claimed positive effects of food aid, it is no easier when it comes to negative effects. Certainly there are countries in Africa which are heavily dependent on food aid. And while some, like Lesotho or Cap Verde, have been in that situation for decades, others, like Somalia, have become heavily dependent during the past decade or so. Yet it is difficult to find cases where food aid can clearly be identified as the causal factor. In most cases, it is at least arguable that absence of food aid would have meant the same policies but increased hunger. In reality, it is impossible to tell how the availability of food aid has affected policies. It appears to have made life easier for cereal marketing parastatal monopolies, providing physical grain supplies for them to distribute and cheap grain from which they can offset some of their losses on other operations. Whether the fact that there is food aid to fall back on has prevented governments from taking hard decisions – like increasing producer prices at the expense of some other use for the funds – is impossible to tell. Even if it was the case, none of those concerned would admit the fact to outsiders, possibly not even internally.

The various studies looked at are unable to answer this point, while most find no significant local 'disincentive effect' (lowering of local prices through flooding the market with food aid). I do not doubt that cases could be found where this did happen, but I can find no evidence that it is widespread, let alone inevitable. To some extent this may result from absence of detailed study, though even then it is always difficult to disentangle food aid as such from the environment within which it is delivered.

Conclusions

From the above, it might seem that one can conclude very little about the general impact of food aid, and there is some truth in this. In a sense it is mistaken even to expect food aid to have general effects. It would be hard to generalize about the effects of transfers of money. One would have to know about the size and conditions of particular transfers, the socio-economic systems into which they were inserted and a large number of other factors. This is also true to some extent of food aid. But not entirely so. Food aid substitutes directly for domestically produced food, as transfers of money do not. Moreover, the impact of developing a taste for imported food has rather different effects from developing a taste for hard currency (which generally needs no development in any case). Another difference is that food (and other commodity) aid allows donors artificially to boost the apparent value of their aid transfers by valuing it at subsidized domestic prices.

Direct-transfer food aid tends to be popular with recipient governments. It saves foreign exchange which would otherwise be spent on importing food, provides items like milk and wheat which are popular with urban consumers and, most importantly, puts resources at the disposal of the government. These can be realized as government revenue, though in the restricted form of 'counterpart funds', or in the form of political patronage through cheap distribution either to the urban population as a whole or to specific groups within it. It also takes some of the pressure off African domestic marketing agencies and systems, allowing them to reduce losses or even making profits without improving efficiency. It can be concluded that it may have a tendency to discourage them from making the 'hard choice' to improve the agricultural marketing system. Sure enough, there has been a distinct unwillingness to make such decisions, but this would be expected with or without food aid. The choices facing many African governments are hard – and some of them involve risks of rioting or of coups d'état.

Under the circumstances, food aid keeps or helps to keep a number of African governments in power. 'Political stability' is seen by some as an unalloyed good, though this must surely depend on the government whose stability is being assured. On the other hand, it seems better for external forces like food aid donors to provide the aid routinely rather than, as has happened with US food aid, withdraw it suddenly as a sign of displeasure over one or other aspect of policy.

When one turns to the formation of taste for necessarily imported items, the ground is a bit firmer. Although the major staple cereals in Africa are maize and sorghum, food aid is largely composed of wheat (two-thirds to three-quarters of total cereal food aid). Even during the recent emergency, food aid pledges to the 24 most affected African countries appeared to be two-thirds composed of wheat. One might suppose that famine victims will never become part of the commercial market for wheat, and that is probably true. But it is a widespread current practice for state grain agencies and emergency aid donors to exchange food aid wheat for local grains. The emergency victims get a cereal they know and the urban population get the wheat. This is also done by the WFP in a number of cases for similar reasons – and perfectly sensible ones in their own context. But they do transfer wheat to where it can build up people's tastes. All in all, it seems clear that food aid plays a part in developing tastes for wheat, milk and other 'import-intensive' products. But to the best of available knowledge, a taste for wheat products well in excess of local supply has already developed in many of the towns of Africa. Almost all sub-Saharan African countries which have the foreign exchange to spare (and some which do not) use some of it to import wheat for the urban population, whether or not they receive food aid. At present the number of countries which have any spare foreign exchange is small and declining. If and when the food aid stops, the urban populations of several countries will be forced to eat less wheat products, regardless of their tastes or their governments' predilections.

When one turns to food-for-work, the problems seem to be of a rather different order and relate primarily to the effects (or lack of them) from the works undertaken. To the extent that food is not misappropriated before it reaches the project, FFW would seem one of the fairer ways of distributing food aid. Having to work for it reduces the interest of the rich, though if the implicit wage in food is well above standard local levels, it may be that wealthier peasants get more than their share of work and food. There are major problems in the organization of large labour-intensive projects with or without food aid. But there is a lot to be said for labour-intensive capital formation *if* it can be made to work. Again, while accepting Jackson's criticism that FFW may lower the morale and motivation of genuinely grass-roots voluntary initiatives, this is not a salient characteristic of most 'self-help' in tropical Africa. It is at least as common for this to be one or other form of forced labour, in which case FFW seems a preferable alternative. Nor is the swamping of local initiative peculiar to food aid, it is probably at least as commonly swamped with outside money or other materials.

Where recent initiatives are concerned, the current trend towards 'triangular trade' and local purchase seems strongly to be encouraged, since this both moves food towards where it is needed and provides markets to stimulate local production. It seems over optimistic to suppose that this will ever substitute for large quantities of food aid so long as the donors have huge surpluses to dispose of, but the initiatives to date are certainly well worth support.

The major problem with EEC food aid is the high proportion of dairy products, which are most unsuitable as food aid. If they are fed into commercial dairies, the end-product (recombined milk) is directed primarily at the upper and middle classes, where it generates new demand for a product which costs foreign exchange either to import or to produce locally. If distributed as powder, it tends to be used as infant foods, with appalling consequences for child health. This results *both* from hygiene problems *and* from the fact that SMP is not suitable as baby food, even if not watered down too much (as it often is). The proportion of dairy products in EEC food aid should therefore be reduced, though, given the huge surplus, it is unrealistic to expect this to happen. Where cereal food aid is concerned, the major complaint against the EEC has been slowness of delivery. There have been major improvements in this respect in the last year, however, and if this can be maintained, it would appear to be no worse than other food aid. This does not mean perfection – delays have been characteristics of all food aid programmes.

The strongest argument in favour of EEC food aid is that it is less liable to arbitrary or politically determined cessation than is the case with US food aid. Despite a certain tendency to follow the USA in its pattern of food aid donation, the EEC has a far better record on this score. That said, EEC food aid, like that of any other donor, derives from the existence of surpluses and thus contains the same dangerous combination of aid and export promotion. Moreover, as EEC internal agricultural prices are far higher than those of most other major food aid donors, the element of over-valuation is greater.

The EEC's food strategies have been a sensible initiative, though they are only one among many efforts by donors to use their aid as a lever to co-ordinate the efforts of others and of the recipient state. Given that IMF/World Bank conditionality has been by far the most powerful external lever applied during recent years, it is hardly surprising that they have made rather slow headway, so far as can be told. They are, in any case, at least as much a matter of general aid strategy as of food aid specifically and, as such, considered further in the following chapter.

Longer-term food aid commitments and contracts are generally seen as one of the more important means to improve the stability and usefulness of programme food aid.

But there is a major problem here. To the extent that food aid induces dependence on imported food, the more regular and long-term the deliveries, the more effectively they achieve this purpose. This is perhaps the basic contradiction of food aid – that efforts to improve its reliability and potential value as aid have the tendency of increasing its dependency-inducing characteristics.

Finally, although I strongly disagree with some of the more Panglossian statements of the supporters of food aid and feel that they grossly underplay the contradictions between export promotion or political pressure and development or getting rid of hunger, I am not able to follow some of the more extreme critics into total negativity. There do seems to be *some* cases where food aid does as much good as harm. There also seems a tendency to attribute specifically to food aid problems which are likely to arise with any sort of aid. Some of these are considered in the next chapter.

NOTES

1. See, for example, Maxwell and Singer (1979).
2. Requirements on this score were brought in, in response to complaints from other cereal exporters over the dumping (in the form of food aid) by the USA, in the 1950s. Although the formal requirements still exist, they do not seem to be taken very seriously now, when most exporters have food aid programmes and many African countries have no foreign exchange for 'normal' imports.
3. A number of food aid recipients have large balances of counterpart funds which cannot be used, either because they need foreign exchange to make them usable, or because donors have placed conditions on their use which cannot be (or are not) met.
4. Holt (1984) makes this point rather guardedly for one specific project in Ethiopia (see below).
5. Such claims are made in some (not all) of five background studies of US PL 480 food aid (in Sri Lanka, Bangladesh, Egypt, Peru and Jamaica). But Rogers and Wallerstein (1985), writing as co-ordinators of the studies, find the only agreed 'positive' effect to have been a contribution to 'political stability'. Apart from questions about what this means (Sri Lanka, Peru) and whether political stability is beneficial regardless of the nature of the regime in power, this raises questions about the effects of cessation of food aid. Since even in the sample in question, it has happened a number of times (Egypt several times in the 1960s and in 1967–74, Sri Lanka 1963), it is not simply hypothetical. One could well argue that stability achieved on the basis of food aid was more than usually vulnerable to its withdrawal.
6. Since Jackson is among those who would be considered a 'critical development lobbyist' by Clay and Singer (1982), it is worth pointing out that he has worked for many years for OXFAM, an NGO which started out in the post-1945 period to distribute food relief, though it has placed more focus on development work than famine relief for many years now. Jackson's own disenchantment with food aid derives from close contact and experience.
7. Main sources for this section are Wallerstein (1980) and Bard (1972).
8. *Not* the Agricultural, Trade and Development Assistance Act, as misquoted by some authors.
9. This implied not only to giving food aid but to cutting it off (Yugoslavia 1950s, Egypt and India 1960s) should the recipient country's political leader make an anti-US speech or undertake some other disapproved act.
10. This FAO committee, which always met in Washington rather than Rome, came up with various formulae relating to the 'normal marketing requirement' in attempts to solve this conflict. It had largely failed to do so when the expansion of the 1960s and the success of PL 480 in opening new markets took some of the heat from the debate. Bard (1972) provides a detailed treatment of this, Wallerstein (1980) a shorter one.
11. Computed from Wallerstein (1980:Tables 3.4 and 3.5) and from FAOb (1969:Table 4).
12. The term was coined by Democratic Senator Hubert H. Humphrey, but taken up as a slogan by the Republican Eisenhower-Nixon Administration.
13. In addition to which, Head of the 'new' Food for Peace was thought to be a good job to prepare a political crony, George McGovern, for a later senatorship. (Wallerstein, 1980:41).
14. The timing of this relates largely to efforts to demonstrate internationalism during the 1960

presidential election. Wallerstein cites Kennedy's Secretary of Agriculture on the 'difficulty and embarrassment' that the USA, 'having initiated all this activity has no policy whatsoever' (pp. 168–9).

15. FAO 1983:Table 26. On the other hand, US contributions to the International Emergency Fund for Relief (IEFR), another part of WFP, have increased recently to 46% of cereals and 75% of other foods.

16. During its first years, WFP was largely autonomous from its 'parent body', FAO. It appeared that in recent years this had been somewhat reduced, but Gill (1986) cites conflict between the leadership of FAO and WFP as among the (many) factors making for a delayed and inadequate response to the 1984 famine in Ethiopia.

17. It is (or was) represented by the 'Inter-Religious Task Force', a grouping of US religious charities concerned with food aid. Not only do these have their own interests, but there are indications of suspicion of a non-religious body. For example, Wallerstein cites complaints from them that the WFP has an 'unfair advantage' in non-Christian areas.

18. Wallerstein (1980:97). Another reason is that the task is formally allocated to another UN body, the United Nations Disaster Relief Organization (UNDRO). Given the proliferation of UN bodies in this area and the considerable degree of rivalry and tension between them, it is rather difficult to sort out cause and effect. (See Gill, 1986).

19. FAOe (1985:Table A.14, expected figures for 1985/6). Food aid transfers in money terms grew steadily during the 1970s, with only a minor downturn in 1973 and a larger one in 1975/6 (FAOe 1983: Table 1). Comparison with volume figures, as in Figure 8.1, shows how misleading this is.

20. A special new product, 'meat loaf', has been developed for this purpose and, if successful, may be marketed commercially. As of early 1986, some was already reported as being marketed informally in Cyprus, having presumably been misappropriated from the countries or refugee camps to which it had been supplied. But the significance of the example is that food aid provides the chance for a company to 'test-market' a product at state expense and over a much wider market than it could afford with its own funds.

21. Up to 1979 see Court of Auditors (1980:37), since then FAOb (1985:79,136).

22. Wallerstein (1980:158), citing Fagan (1975).

23. This remark has a particularly hideous ring in view of the widespread use of torture by the Pinochet regime and the fierce reduction in living standards which it ushered in.

24. Woldemeskel, seminar at CDR Copenhagen November 1986. The seminar was primarily concerned with the ecologically disastrous consequences of one resettlement scheme.

25. Maxwell (1986c:Table 10). Since the figures are divided into 'state farms', 'co-operatives' and 'government budget', the proportion may well be higher. For comparison, by 1981/2, co-operatives were responsible for 1.3% of production and state farms for 6.2% (Ibid. Table 4). Moreover, since these are the 'modern' and more state-controlled sector, the figures probably exaggerate their importance.

26. Mozambique also invested a very high proportion of all government funds into producer co-operatives and state farms, and with very little indeed to show for it.

27. Until 1984/5 when Ethiopia's receipts increased sevenfold. Somalia, like Kenya, is an important strategic link in the Indian Ocean strategy of the USA.

28. Bulgur is a form of parboiled wheat, deriving orginally from the Middle East. In its modern form, it is largely a development by US corporations as a 'food aid product'. While a student at Stanford in the mid-1960s, I attended a seminar where a senior manager from one such corporation explained how it had been developed to overcome the reluctance of rice-eating people in southern India to accept PL 480 wheat. Bulgur can be boiled and eaten like rice, though it tastes very different.

29. See Chapter 3. While Kenya was seriously hit by drought in 1984, its large food aid shipments from 1980 onwards seem to reflect more on its strategic importance to the West than on the severity of its food situation.

30. Mitchell and Stevens (1983) assert this only for 1978, the year in which they studied food aid. A glance at Table 8.3 indicates that the problem is more general.

31. Geoffrey Lean, 'Scandal of UN's Food Aid in Africa', The Observer, 17 June 1979.

9
EEC & Other Project Aid

Introduction

This chapter looks at project aid, both that of the EEC and of other donors. Its main purpose is to try to give some rough idea of how aid programmes and projects operate, what they can be expected to achieve and why they so often go wrong. A secondary purpose is to compare the EEC with other aid programmes, though I find the similarities more striking than the differences. The chapter starts with a brief review of the overall aid situation followed by a discussion of EEC and EDF aid.

What is Aid? Aid is defined by the OECD to be money or materials transferred to developing countries on concessional terms – that is, in the form of grants or loans at subsidized interest rates. It is also part of the definition that it should have 'the promotion of the economic development and welfare of developing countries as its main objective', whatever that means. Aid is divided by the OECD into two parts. By far the larger comes from state or international donor agencies and is referred to as 'official development assistance' or ODA. The much smaller part consists of transfers by non-governmental agencies (NGOs), also referred as 'private voluntary agencies'. The large majority of both of these takes the form of outright grants. The 'grant element' of an aid programme is computed by taking the proportion of outright grants and adding to this a further percentage to take account of interest subsidies on loans.[1] For OECD Development Assistance Committee (DAC) members (roughly the 17 richest developed countries) the overall grant element of their ODA comes to about 90 per cent, ranging from 100 per cent each for Australia and New Zealand down to 77 per cent for Japan and 69 per cent for Austria. Outright grants are said to amount to about 80 per cent of the total (OECD, 1985a:106).[2]

How large are aid flows? Figure 9.1 (from OECD, 1985a: Table VI–2) shows the evolution of total resource flows from the developed to the developing countries from the 1950s to 1984 in constant (1983) prices. As can be seen, total flows, in real terms, increased up until 1981, by which time they amounted to some $125 bn and three and a half times the level of 1960–61 (the 1950–55 figures are not comparable). But since then, total resource flows have fallen by some 30 per cent and non-concessional flows (especially private) by almost 45 per cent. The major growth (and decline) has been in non-concessional transfers (and especially private transfers which compose some two-thirds of total non-concessional flows). That is, there was a massive increase in transfers of funds from the early 1970s, which started to taper off at the end of the decade and has declined since then. Where ODA is concerned, there has been little

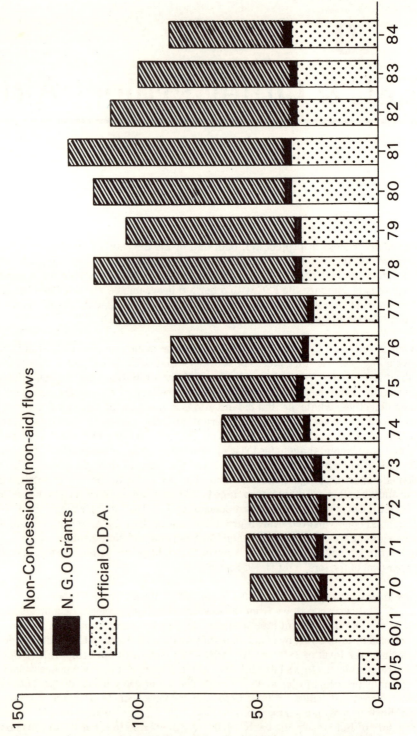

Figure 9.1: *ODA and other financial flows to the Third World 1950–55 to 1983 ($ billion)*

appreciable increase since the mid-1970s and a slight decline since 1981. Nor has there been much change in the very much smaller NGO transfers.[3]

But while transfers to the Third World have declined since about 1980, transfers in the other direction – and the indebtedness to which they are related – have continued to grow. The total long-term indebtedness of developing countries has increased from $391 bn in 1979 to $691 bn in 1984 (total debt in the latter year being $902 bn), while their debt service (the payments they have to make, covering interest and principal repayment) rose from $71 bn to $133 bn over the same period (OECD, 1986. Table V.1). While the fact that the latter figures are in current prices means that the growth of debt service cannot be compared directly to the growth in ODA and non-concessional transfers to the developing countries, it is clear that the direction of the overall net flow has turned from positive (to the Third World) to negative. By 1984, debt service alone exceeded total transfers to developing countries by 43 percent, and came to some *four times* the level of ODA transfers. Thus, even ignoring quite substantial other transfers, the total value of money and materials transferred from the developing countries exceeds that to the developing countries by a large margin.[4]

If one looks specifically at sub-Saharan Africa, gross aid disbursements for 1982/3 amounted to $8.6 bn (OECD, 1985a: Table III.13), which compares with total debt of $61 bn and debt service of $8 bn (OECD, 1986:74–5).[5] By 1984, this balance had apparently become negative, with gross disbursements of $7.1 bn and debt service of $9.8 bn.[6] By 1987, a different source estimated Africa's total debt at between $150 bn and $200 bn, with debt service for 1986 at $15 bn, compared with total development assistance and NGO grants combined of some $16 bn.[7] One can conclude from this that, even if debt service has not already grown to exceed inflows of aid, it is already taking almost every penny which comes in, and will take more in future. It seems almost certain by now that Africa is paying out more than it receives – or that it would be if there were not so many countries which are simply unable to maintain their debt-service payments.

EEC aid

Including the individual member countries, the EEC is by quite some way the largest aid-donor group in the world, accounting for some 42 per cent in 1984, compared with 28 per cent for the USA, the next largest donor.[8] But most of this is accounted for by member bilateral aid. The EEC as such does not account for more than 3–4 per cent. It is only the world's eighth largest programme, smaller than four of its members, though if double counting is eliminated by considering only bilateral programmes, the EEC becomes fifth largest, after the USA, Germany, France, and Japan. On the other hand, if the EEC is considered as a multilateral agency, then it is the third largest provider of concessional loans and grants after the IDA (the concessional loan wing of the World Bank) and the combined UN agencies.[9]

It is far from clear what significance the size of the EEC aid programme has, for, at least in principle, it is *not additive*. Any extra payments by the member states to the EEC are subtracted by them from some other part of their aid programmes, either bilateral aid or (more commonly) funds previously transferred through multilateral agencies, like FAO, UNDP or the World Bank. In reality, it is impossible to tell whether the existence of the EEC aid programme has any effect on the total ODA transferred. It seems highly likely, for example, that Italy's programme, which has nearly quadrupled since 1970–71, with a high proportion going through multilateral agencies, has probably been affected by the EEC (ibid. Annex Table 12). But, in general, it does seem to be true that funds transferred via the EEC are taken from some other aspect of the member's aid budget. Thus the significance of the EEC programme lies less in its size than in the way it is used. Is it allocated to different regions, sectors and aims from those of the members or of other donors?

Over the past ten years, exactly one third of EEC aid spending has gone to food aid, rising to about 40 per cent at present. This is the highest proportion for any major donor, being almost twice the proportion of US aid which takes the form of food aid[10] and also far more than is the case for the individual EEC members. The reason, in the latter case, is that produce is bought into surplus stores through the CAP, so that the large surpluses 'belong' to the EEC rather than to the members. Rather less formally, the most active members in food aid affairs are the major surplus producers, who gain from having the EEC give their surpluses as food aid rather than doing it themselves. It thus seems highly likely that one effect of having an EEC aid programme is to increase the proportion of food aid in the total transferred.

The larger portion of EEC aid is composed of grants for items other than food. This usually accounts for some 40–50 per cent of the total, the rest being made up of loans and capital spending and administrative costs. Since food aid is given entirely on a grant basis, this brings the grant element of EEC aid up over 90 per cent, though this computation is based on the EEC valuation of its food aid at internal prices, which is well over double its market value. Moreover, as for other donors, it is not clear what proportion of the loan subsidy element should really be considered as grant aid.[11]

EEC aid is divided into two programmes. The smaller is financed through the EEC budget, and is directed mainly to 'non-associated states', these being defined negatively as Third World countries which are not members of the ACP. This part of the EEC aid programme is mainly concentrated on a number of large recipient countries, with a focus on the Mediterranean basin and south and east Asia. Africa receives relatively little project aid from this source. But almost all food aid is funded through the budget, and sub-Saharan Africa received 52 per cent of cereal food aid from the EEC in 1983/4, but only 18 per cent of the more 'valuable' skimmed milk powder (FAO, 1986a: Tables 18 and 22).

The other and larger part of EEC aid goes through the Lomé Convention, which is the contractual framework for 'development co-operation' between the EEC and its associated African, Caribbean and Pacific (ACP) countries.[12] This group is based around the ex-colonial possessions of the member states but has spread to include other countries in the areas and now includes all independent sub-Saharan African states. The Convention is agreed at five-year intervals, and covers both aid and trade. The European Development Fund (EDF) is the account out of which the project aid (and other transfers like Stabex) agreed under the Lomé Convention is funded. Unlike food aid and aid to non-associated countries, the EDF and other funds transferred through the Lomé Convention are not part of the general EEC budget. Specific agreements for funding by the member countries are made during the negotiations for each successive Convention and subsequently paid directly into the Fund (albeit often with delays).

When the first Lomé Convention was signed in 1975, the combination of aid and aid-like transfers with agreements on protection and preference levels was the topic of much inflated rhetoric about its prefiguring a 'new international economic order'.[13] There was precious little basis for this at the time, though the easy money and low interest rates of the period gave rise to some excessive optimism.

Already, during the negotiations for the Second Lomé Convention in 1980–81, it became clear that the EEC member countries regarded existing arrangements as the maximum they were willing to offer, rather than as a basis for further development. Despite increasingly bitter negotiations, no significant improvements in trading arrangements were forthcoming, while it proved not even possible to maintain the previous level of real transfers from the EDF. The figures in Table 9.1 may appear to show a significant increase, but taking inflation and population increase into account, plus the expansion of the ACP group, Hewitt and Stevens (1981:53) find a reduction in total funding of 19 per cent and in the EDF of 21 per cent. Even without considering population growth, the reduction in real transfers was about 10 per cent.

The same has been true of the most recent Third Lomé Convention, which was signed in December 1985, after a long-drawn-out and sometimes bitter series of negotiations, marked by two features. Firstly, the ACP countries got nothing like the amount of aid which they had hoped for. Secondly, a major sticking point was 'policy dialogue'. This was presented by M. Pisani, the then Development Commissioner for the EEC, as a means to ensure co-ordination and the best use of aid funds. It tended to be seen by the ACP leaders as yet another form of conditionality. In the final Convention, a number of adjustments have been made to specific aspects, with the ostensible aim of improving co-ordination and the flexibility and speed with which funds can be disbursed. But the level of funding has barely been maintained in real terms, and, with several new members, has fallen in real per capita terms. There have been no significant changes in the trade provisions.

The European Development Fund(s)

There have been five EDFs to date, and the Third Lomé Convention involves a sixth. The first EDF (1959–64) was agreed under the Treaty of Rome, and the next two in the two Yaoundé Conventions (EDF II – 1964–70, EDF III – 1971–6). These concerned primarily francophone African countries (ex-French and Belgian colonies). The Lomé Conventions (EDFs IV–VI) include anglophone (and some lusophone) countries in Africa, the Caribbean and Pacific. Their number has grown somewhat during the life of the Conventions, mainly through status transfers of OCT (Overseas Countries and Territories) to independent states eligible for ACP membership. With the accession of Mozambique and Angola under Lomé III, all independent countries in tropical Africa are now ACP members. Table 9.1 shows amounts allocated under the three Lomé EDFs, and their sectoral breakdown.

Table 9.1: *Funds allocated under EDFs IV–VI, ECU million and percent*

	Lomé I		Lomé II		Lomé III	
	ECUm.	%	ECUm.	%	ECUm.	%
Grants	2,058	60	2,998	54	4,860	57
Special loans	446	13	524	9	600	7
Risk capital	97	3	284	5	600	7
Stabex	380	11	557	10	925	11
Sysmin	380	11	282	5	415	5
Total EDF	2,890	86	4,645	84	7,400	87
EIB (European Investment Bank)						
Subsidized loans	390	11	685	12	1,100	13
Ordinary loans	–	–	200	4	–	–
Total Lomé	3,370		5,312		8,500	

Notes: Of the total allocated for grants, 6% is reserved for emergency aid and refugee assistance under Lomé III, as compared with 7% under Lomé II. A further 5% is for interest-rate subsidies (for special loans), compared with 6% under Lomé II.
Sources: Lomé I, Stevens (1981). Lomé II & III,
Courier, No. 89 (Jan–Feb 1985), p. 24.

About two-thirds of EDF funds are used for grants, though some of this goes to subsidizing loans. Stabex and Sysmin (see Chapter 6) involve grants (to the poorest countries) and heavily subsidized loans (to others), giving them a high grant element. Funds from the European Investment Bank are negotiated and agreed under the Conventions, although they are not ODA. The EIB is not an aid-giving body, but a bank lending at commercial rate, both in the EEC and other parts of the world, with promotion of EEC exports as its major aim. The subsidized loans under the Lomé Conventions use some of the EDF grant money to reduce the interest rate and turn loans into 'aid' or 'unfair competition', according to viewpoint. They are mainly used for large industrial or agricultural projects, and primarily directed towards the richer ACP countries. As of 1984, Kenya and Ivory Coast were the major African recipients of EIB loans, receiving, between them, some 20 per cent of the total under the Lomé Convention (*Courier* No. 83, Jan–Feb 1984:9). Over 90 per cent of the funds under the Lomé Conventions are directed to Africa, and together with the high proportion of food aid to Africa, this means that the overall proportion of EEC aid directed to Africa is above that of its member countries, or most other donors.[14]

As a whole then, EEC aid is focused upon Africa, its grant element is higher than most, and the proportion of food aid is higher than for any other major programme (over half of this in value terms being dairy food aid). The preponderance of food aid implies that its value per dollar of ECU officially stated, is exaggerated. The high proportion of grant aid in the EDF would seem to point in the other direction, though there are serious questions about, the extent to which subsidized loans should be considered 'aid' rather than subsidies to export promotion. But when all is said and done, figures of the proportions and amounts spent on different categories of aid do not tell one very much, because they say nothing about *how* the funds are used.

EDF COMMITMENTS IN THE PERIOD 1981–4

In a first attempt to approach this question, this section of the chapter looks at EDF project aid in the period 1981–4, based on as complete a listing as possible of all projects on which decisions were made by the EDF Committee in that period.[15] This committee, composed of members from (the aid agencies of) member countries, decides upon project and other expenditure proposals put forward by EEC Commission staff. While this is the point at which formal decisions are made to commit funds to particular purposes, most of the real decisions are made some time before in negotiations between the EEC and recipient countries. The EDF Committee has rather limited scope for changing projects once presented to it, and no scope for initiating decisions itself. Comparison of lists of projects 'to be discussed' with those 'on which decisions have been taken' shows so little change that one gains the impression of a rubber stamp body.[16] Nevertheless, this provides the most complete list of EDF projects.

During the period January 1981 to December 1984, 566 different funding decisions were taken by EDF Committees of which 442 refer to African countries, wholly or in part.[17] In going through them I have categorized them as 'rural' and 'non-rural', finding 225 (185 in Africa) to be 'rural-oriented' in one way or another. This distinction is inevitably partly arbitrary and far from watertight. For example, I have taken feeder roads and 'rural roads' to be 'rural', but not trunk roads joining major towns or countries, although some of the latter probably have a significant rural impact. Agricultural processing industries have been included as 'rural' on the 'normal' assumption that by increasing the added value of agricultural products, they contribute to the development of agricultural production or incomes. But there are enough cases in tropical Africa of agricultural processing and input-producing industries which seem to subtract rather than add value to make one wonder about their 'contribution' to rural development. The same can sometimes be said about other projects, however 'rural-oriented' they may be. There are a number of other problems, like which training programmes should be considered 'rural', or whether water-supply projects are rural or aimed at regional towns. In short, one could make reasonable arguments for increasing or decreasing the proportion of 'rural-oriented projects' by up to, say, 10 per cent.

'Funding decisions' are not always the same as projects. Some large projects are funded and implemented in several phases. Some of the decisions made by the committee refer not to specific projects, but are general 'authorizations to commit'. Thus annual commitments are made for trade promotion activities, allowing the office in charge to distribute the funds without referring back to the EDF Committee. Similarly, the funding for micro-projects is made in one annual block. Table 9.2 shows, in summary form, the number and value of funding decisions for 1981–4.

Table 9.2: *EDF committee funding decisions 1981–4 (number & million ECU)*

| Year | Total – All Sectors | | | | Rural – Oriented Only | | | |
| | All ACP | | Africa | | All ACP | | Africa | |
	No.	ECUm.	No.	ECUm.	No.	ECUm.	No.	ECUm.
1981	105	472.8	89	458.1	40	197.4	35	192.1
1982	152	674.5	116	623.4	66	360.0	57	340.2
1983	154	677.2	109	607.4	65	176.1	39	151.3
1984	155	536.9	127	470.2	69	183.7	58	160.2
Total	566	2,361.4	441	2,159.1	240	917.2	185	843.8

Source: Own calculation from *Courier* and *Telex Africa*

Africa received 91 per cent of total project funds and 92 per cent of rural project funds, so the Caribbean and Pacific countries (mostly small islands, and with 1.5–2 per cent of ACP population) are substantially over-represented in per capita terms, receiving 8–9 per cent of total funds. 'Rural-oriented' projects take about 40 per cent of total funds and compose some 42 per cent of all decisions. One might expect rural projects to have been smaller than industrial and infrastructural projects, but the difference turns out not to be large. 'Rural' projects cost on average 3.8 million ECU as compared with 4.2 ECUm. for all projects. For Africa, the overall average cost was 4.9 ECUm. and for rural projects 4.6 ECUm.

The size of EDF projects. For purposes of comparison, I have divided EDF decisions into four size categories: 'Very Large' – 10 million ECU and upwards; 'Large' – from 3 to 9.99 ECUm.' 'Medium' – from 1 to 3.99 ECUm. and 'Small' – under 1 ECUm. 'Authorizations to commit' have been put in a separate category, except for micro-projects, which are included under 'small' projects. Table 9.3 shows the size distribution of all projects.

46 per cent of all the funds committed to EDF project aid during this period went to 53 'very large' projects costing on average 16.7 ECUm. This underestimates the real size of the projects, since several were phased and covered more than one 'decision', while projects of this sort are often partly financed through EIB loans, which are not included here. Even so, projects costing more than 3 ECUm. took 71 per cent of total

Table 9.3: *Size of EEC funding decisions 1981–84 (number & million ECU)*

Year	All	V.Large	Large	Medium	Small*	Authorizations
			(A. Number of decisions)			
1981	90	11	25	27	13 + 2	12
1982	116	17	38	27	17 + 1	16
1983	109	19	26	25	23 + 1	15
1984	127	6	40	43	27 + 1	10
Total	442	53	129	122	80 + 5	53
As percent of total projects:						
	100	14	33	31	22	--
			B: ECUm. committed			
1981	458.1	207.4	121.3	46.0	22.0	61.4
1982	623.3	283.9	234.9	51.7	16.1	36.8
1983	607.4	313.0	124.1	44.6	33.4	92.3
1984	474.0	83.3	227.1	83.8	32.0	47.7
Total	2,162.8	887.7	707.4	226.1	103.4	238.3
As percent of total projects:						
Total	100	46	37	12	5	

* Project decisions plus micro-project allocations.
Source: as for Table 9.2.

funds, and even with micro-projects, small projects, costing under 1 ECUm., accounted for only 5 per cent. About 65 per cent of the largest projects are non-rural, mostly roads (52 per cent of funds), other transport (12 per cent) and other infrastructural projects like water supplies.[18] Transport also takes about 50 per cent of funds for 'large' and 'medium' project. Table 9.4 shows similar data, but for 'rural' projects.

Table 9.4: *Size of EDF 'rural' projects 1981–84 (number & million ECU)*

Year	Very Total	Large	Large	Medium	Small
A. Number of projects					
1981–4	185	18	75	61	26 + 5
B: ECUm. allocated					
1981	192.3	76.8	77.8	15.7	22.0
1982	364.5	168.7	152.1	29.3	14.5
1983	151.3	51.9	53.3	27.9	18.2
1984	216.6	15.2	123.5	46.2	31.7
Total	924.7	312.6	406.7	119.1	86.4
Percent of total		34	44	13	9

The proportion of very large projects is smaller, but those over 3 ECUm. still take 78 per cent of all funds. Of eighteen 'very large' rural projects, eight concerned export-crop production, either by peasants or on plantations. There were four integrated projects, most having an 'anchor' export crop. Three were irrigation projects, including this period's contribution to the series of huge dams on the Senegal River, leaving two rural water-supply projects and one veterinary project. Table 9.5 breaks down rural expenditures into broad categories.

Table 9.5: *EDF 'rural' projects by type 1981–84 (percent)*

Category/Project	Medium	Large	V. Large	Total
Unspecified	1	6	0	8
Food Crops	1	4	0	5
Export Crops	1	7	9	18
Livestock	1	4	2	7
Fisheries/Forestry	1	2	0	4
Irrigation	2	3	10	14
Integrated	1	6	10	16
Training/Research	1	4	0	5
Inputs	0	2	2	4
Credit/Marketing	1	4	0	5
Processing	..	1	1	3
Rural Infrastructure	3	4	3	11
Total	15	49	37	100

Notes: Totals do not add, due to rounding.
    ".." = less than 0.5%
Source: As for Tables 9.2–9.4.

While some interesting things show from this table, it should be interpreted with great caution. Firstly, there is considerable overlap between categories. For example, many of the food-crop projects concern irrigated rice and could have been put under 'irrigation'. Many 'integrated projects' (among which I have included settlement schemes) are built around an 'anchor crop', which is usually an export crop. The 'non-specified' category consists mostly of titles like 'rural development in Mali' or 'development of agricultural production', which can mean just about anything.

But it does seem significant that export crops took three and a half times the funds allocated to food crops. It seems especially significant since the EEC has been careful to play up the presence of food crops in its project titles and descriptions wherever possible since the early 1980s. This is a response to criticism that its rural spending during the 1970s was heavily concentrated on export crops (as it was). The table shows a heavy emphasis on irrigation, and especially in projects costing over 10 ECUm. Integrated projects are also clearly a major focus of EDF project aid. During the late 1970s, plantation projects were more numerous than now.[19] It seems that integrated projects have taken over that particular aid 'niche'. While the pattern of production is very different, being based on peasants rather than large company estates and plantations, both involve considerable centralized management and are thus 'suitable' for operation by consultancy and management firms.

Beyond this, however, it is doubtful how much one can deduce from such a table. It is not just that the categories overlap; they contain a wide variety of different items. 'Livestock', for example, includes large ranches, tsetse-clearing, other veterinary programmes and work on the multiplication of the N'Dama tsetse-resistant cattle breed, in addition to projects for sheep and goats and much more. 'Training and research' ranges from practically-oriented research and training to courses and meetings whose impact on local agriculture is probably confined to local foodstuffs in the delegates' meals (if any). 'Inputs' covers both supply of imported inputs and developing local production. 'Rural infrastructure' refers primarily to rural health and water-supply projects, but could have contained more roads. But even if these problems could be sorted out – and this would require almost as many categories as projects – it would still say nothing about either how the projects fitted into the recipients' agricultural strategies or how they worked and what their effects were.

The next section works towards this by looking at the EDF programmes of specific countries and at some individual projects. There is no reason to expect the EDF project 'package' to form an integrated whole. It is only one small part of the recipient's expenditure on rural, development and could serve as much purpose by 'filling in holes' as by striving for internal coherence.

EDF project aid to specific African countries

Tanzania – From 1976, when EDF funding started, up to the end of 1984, almost 180 ECUm had been committed to Tanzania, of which 103.6ECUm. was during the period 1981–4.

By some way the major user of EDF aid was the transport sector, accounting for 38 per cent of total funds committed. The major part of this went on roads, including a four-lane asphalt road from Mwanza to the Kenya border (then closed) via the President's home village. The remainder was split between railway renovation and repair and assembly of trucks. Both these latter are much needed by Tanzania. The railway renovation project (a multi-donor affair) was held up for some time, because certain donors (though not, so far as I am informed, the EEC) refused to provide funds until Tanzania accepted IMF conditionality regarding devaluation (which it did in June 1986).

The agricultural part of the programme, some 34 per cent of the total, is entirely composed of two large programmes: a coffee improvement and rehabilitation programme, focussed on the north of the country, mostly Kilimanjaro Region, and an Integrated Rural Development Programme for Iringa Region in the Southern Highlands.

The coffee programme is credited by the EEC with a record coffee harvest in 1982, though since coffee production did not continue to rise, this remains unclear. Generally price policy seems to have been a more important determinant of coffee production than technical factors. The programme has been criticized for concentrating heavily on agro-chemicals and thus a pattern of production very heavy in import costs. However, given that one of the major problems facing coffee in Kilimanjaro when the project started was coffee berry borer, this may be understandable. But the project has operated a form of 'input-supply' quite common among modernization projects, in which they are distributed 'free' with payment in the form of a deduction from the price of all coffee, regardless of whether the producer got or used the inputs. The idea of this is to encourage innovation by allowing the 'laggards' to subsidize the 'innovative'.[20] In reality it more often works out that the poor subsidize those with political influence to get hold of limited input supplies. This also probably encourages those who do have access to inputs to overuse them, which could be problematic with poisonous chemicals in a densely populated area like Kilimanjaro, where the same streams which pass through the coffee fields are used for drinking water. In Kagera, Tanzania's second largest coffee-producing region, the EEC programme has been far less active, since the local producers show less interest in fertilizers and pesticides for coffee (response to fertilizer seems uneconomic).[21] There is a really serious insect pest, the banana weevil, which is undermining the whole food economy of the area, shifting the relative prices of coffee and bananas and causing peasants to lose interest in coffee (Tibaijuka, 1984).[22] One encouraging sign is that, since the period under consideration (in early 1985), the EDF has started planning a programme for banana improvement.

The Iringa programme is considered below. Reports on its progress and quality are varied in the extreme. Some observers find it to be among the better integrated projects (scarcely extreme praise). However, an evaluation in 1984 found the programme to be poorly managed, weakly co-ordinated at the regional level, and to have low 'beneficiary involvement' and 'institution-building' and a bias toward infrastructure. It was found to have failed to meet its objectives and was to be totally restructured with a focus on 11 selected villages. There is no basis for an estimate of the overall effect of the programme on Iringa Region, since the price and other policies of the Tanzanian Government are certainly a more important influence. Apart from that, the region is a focus for a large number of different aid organizations and projects, being among the cooler highland Regions in the country.

Most of the remainder of the Tanzania programme has gone to the building of a canvas factory and to urban water supplies to three medium-sized towns, Mwanza, Mtwara and Mbeya. The remaining 6 per cent of the funds are divided between regional projects (mostly situated in Zambia), training and promotion of trade and tourism. The canvas factory in Morogoro was based on the reasonable idea of developing a use for second-grade (stained) cotton which fetches poorer export prices. However, the factory has been a long time in the construction phase. Funds were first committed in 1977 and it was only recently said to be opening. In the intervening period, cotton production, and especially ginnery capacity, have fallen significantly so that the economics of the factory seem unsure.

What can one say about this programme as a whole? These projects seem to fall within the general range of standard projects which have been implemented by different donors. But there is no sign of a serious commitment to food production, except for the banana project, which was started later, though the Iringa Project has focussed, to some extent, on maize.

Mali is a case of some interest, since it was the first ACP country to become involved in the EEC's 'food strategies'. One might thus expect this to be reflected to some degree in aid commitments. At an overall level, this would seem to be the case. Some 57 per cent of funds allocated nationally are directed to one or other rural orientation. This drops to 52 per cent if one excludes an (urban) abattoir and urban grain storage. Rural health centres and village water supplies account for about 30 per cent of rural expenditure, the remainder going to production.

Within agricultural production, most of the funds are allocated to a major rice-production scheme (Segou), which is in its fourth phase, having also had EDF funding for earlier phases. The scheme was started in the 1960s, had used some 40 million ECU by 1984 and covers some 15,000 rice-growers in the internal delta of the Niger River. Total irrigated area has grown from 18,000 ha in 1972 to 35,435 ha in 1983, but production and yields are rather low – 28,500 tons from 35,435 ha, or about 0.8 tons per ha. It would seem highly likely that not all produce passes through channels where it is available for measurement, since irrigated rice should get at least several tons per ha. The Office du Niger, which runs the scheme, is among Mali's many large parastatal corporations and has often been criticized for bureaucratic methods and high costs. The project 'has by no means live(d) up to expectations' (*Courier* No. 84, Mar.–Apr. 1984:17–18), or, in an alternative formulation, 'been a dramatic failure' (Speirs, 1987). Funding during the period 1981–4 was 14 ECUm. (for phase 4). Another dam for rice production (Selingue) was also partly financed by EDF at an earlier period, with about 23 ECUm. out of a total estimated cost of 127 ECUm. Which of these two schemes it concerns (or perhaps yet another) I do not know, but *The Courier* of Sept.-Oct. 1984 (p. 59) shows a picture of an almost dry dam, entitled 'an irrigated rice project in Mali', bearing the caption, 'It makes no sense to spend more aid money on irrigated rice project(s) until someone has managed to make the existing schemes work'. Precisely.[23]

Most of the remainder of Mali's agricultural aid from the EDF goes to an experimental ranch based on the small and tsetse-resistant N'dama type of cattle. In late 1984, under the 'hunger in the world' allocation of 58 ECUm. Mali received 3 ECUm. for food storage and 0.3 ECUm. for reafforestation.

The national programme is dwarfed, however, by a regional project, financed by the EDF, in which Mali is involved. This is the huge Diama and Manantali dams project on the Senegal River, planned to improve the navigability of the river and to irrigate 375,000 ha of land. Total cost of the project is estimated to be 583 million ECU, though it will clearly have far exceeded that before completion. The allocation from EDF V during 1981-4 was 55 ECUm. It was this project which elicited remarks from M. Pisani, when Development Commissioner, about 'cathedrals in the desert' and the high cost and dubious usefulness of large-scale dam projects. Because of desertification in the watershed of the Senegal River, flow is significantly and (apparently permanently) reduced, so it appears unlikely that the dams will ever be used to their full technical capacity. With financial support under the 1st, 2nd, 4th and 5th EDFs, the present level of irrigation is apparently 1,500 ha and there were recently plans to 'rehabilitate' 6,000 ha and irrigate 32,000 more (*Telex Africa*, 1984,228:3). A friend who recently visited Senegal reported conversation with an engineer involved, who indicated problems with backflow of saline water despite the building of an expensive barrage to stop it.[24]

Whatever is planned for the future, the present form of the EDF's aid programme to Mali seems to focus largely on irrigation and large-scale dams, whose prospects of producing very much seem dubious. As indicated in Chapter 8, the EEC's food strategy seems not to have had a major impact to date. Claims are made that privatization and improving the efficiency of OPAM, the state grain monopoly, have improved marketing and prices, though, given the impact of drought, it is hard to see any impact in deliveries. From the production viewpoint, the programme would appear to be stuck pouring funds into local bureaucratic structures which do as much to reduce production as to increase it.

Ethiopia is a rather striking example of a country whose EDF programme is mainly composed of large projects. Of the almost 200 ECUm., over which decisions were made in 1981-4, the *average* amount was 16.6 ECUm. The average size of the projects concerned was even larger since two concerned the same project, while three others were supplements or continuations to already existing projects. Table 9.6 below lists the decisions in order of magnitude.

While the EEC is only one of the donors dealing with Ethiopia, and it is possible that Ethiopia chose to get infrastructure from the EDF leaving agriculture to other donors, it is still striking that only 15 per cent of the programme should be aimed at agriculture, and that almost solely at cash-crop production (coffee). It is still more striking when one considers that Ethiopia was one of the countries worst hit by the 1973-4 drought, that there have been a number of minor droughts and shortages in the interval, and that almost everyone with connections to the country refers to the enormous problems of highland peasant agriculture and its reflection in frequent food shortages and not infrequent famines.

Table 9.6: *Projects funded in Ethiopia under EDF IV 1981–84 (ECU million)*

Project	ECUm.
Addis Ababa water supply	53.45
Amarti diversion (expand hydro-electric dam)	37.00
Coffee improvement programme (phase 2)	27.20
Power station	24.50
Djibouti-Ethiopia Railway (2 fundings)	24.80
Hospitals	15.40
Multi-annual training programme	5.00
Supplement to geothermal exploration	4.39
Finalization of (EDF IV financed) road	4.40
Continuation of Arnibara Irrig. Scheme	2.90
Equipment for Alemaya Agric. College	0.21
Total	199.25

This would appear to have been recognized by all parties concerned, since the new Indicative Programme under Lomé III proposes to use 85 per cent of total EDF funds on the peasant agricultural sector and with special attention to improving grain supplies. From the production side, this is to include improving input-supply, agricultural research and extension, rural roads and storage facilities, and rural credit. On the marketing side, it includes partial privatization so that 'grain marketing within and between regions by private operators shall ensure the smooth flow of grain from surplus to deficit areas'. Apart from this, EDF assistance to drought-afflicted areas will focus on 'soil and water conservation, on afforestation and reafforestation as well as on the gradual rehabilitation of the productive agricultural potential' (*Courier* No. 99, Sept.–Oct. 1986:36).

One can only applaud the general change in direction, though the effect will depend very much on *how* the programme is implemented. One may doubt whether private operators will ensure a smooth flow of grain *to* deficit areas, in time of need. But the really important factors lie largely outside the control of EEC policy, however good or bad. The most important will be the state of the different internal wars in the country. The evolution of state settlement policy is another crucial factor.

Ivory Coast – Considering its position as one of the more favoured francophone African countries, the Ivory Coast receives rather modest amounts of aid from the EDF. Rather less than half the amount received by Senegal, for example, though the population is 50 per cent higher. However, Ivory Coast is a major borrower from the EIB and a major recipient of funds from Stabex. The EDF is among its less important aid donors, providing only about 5–8 per cent of total ODA receipts, compared with some 50 per cent from France.

During the period 1981–4, EDF funds went overwhelmingly to agricultural projects, which accounted for 90 per cent of the total. The major items were 'Palm Plan II', to get palm oil production growing again after a period of stagnation, rejuvenation of coffee over an area of 140,000 ha, and to set up a ranch. Most of the remainder went to village water supply. A rather similar pattern shows for earlier years, with EIB loans being used to (part) finance major agro-industrial investments and EDF grant finance being divided between this and investment in smaller-scale production.

Sudan, which has been referred to as the potential 'bread-basket' of Africa, is among the few countries where food production was estimated by FAO to have kept pace with population growth. This has not prevented a major famine, while research by O'Brien (1985) and others relates the rising tide of famine to the very agricultural policies which FAO and other donors have praised for increasing marketed food production. Where do EDF projects fit into this picture?

The EDF programme is certainly strongly concentrated on agriculture, which took over 80 per cent of funds during the 1981–4 period. Almost all of this was allocated to large export-crop projects. The Agricultural Inputs and Rehabilitation Scheme, for the Gezira cotton irrigation scheme, alone took 50 per cent of all funds during the period, largely as fertilizers and other inputs. Much of the remainder has gone to the development of a tea plantation and factory. This part of the EDF clearly fits in with the World Bank and IMF strategy to increase export proceeds.

What concentration on 1981–4 misses out is that, both before and after that period, large sums were allocated to two mountain-area integrated projects, one in Jebel Marra in Darfur province, the other in the Nuba Mountains of southern Kordofan. In order to make any assessment of them, one would need to know something of the specific projects. These have been among the areas worst hit by the recent famine, but I do not know how and where the EDF projects fit in.

Kenya is the largest ACP borrower from the European Investment Bank, but its EDF programme is not especially large. It has, however, been dominated by a number of large agricultural projects, together with an ongoing but lower-level interest in the dairy and livestock sectors.

The most expensive single project has been Bura irrigation scheme on the Tana River, of which the EDF is a part-financer together with others. This large-scale scheme has been taken up by a number of different donors – and dropped by some of them, as it appeared costly and unlikely to deliver the proposed results. It has been apparent for some years that results were likely to be far below initial prognoses, which is hardly uncommon for dam projects. Apart from delays and cost overruns in construction, there have been other technical and social problems. Among the former has been the fact that hydro-electricity generation upstream on the Tana River has led to unpredictable changes in river flow and level, which threaten the viability of irrigation downstream. The social problems are also considerable. The local population are pastoralists and are likely to be most affected by the loss of riverine land and access to dry season grazing and stock watering. They are unlikely to make up the majority of settlers, most of whom are expected to come from the highlands, highlighting further likely problems of adjustment. More recently, other problems have arisen and there are signs that more donors are trying to edge their way out of the project.

The other major use of funds has been the Machakos Integrated Development Programme. As with most other integrated projects, views on this vary from guardedly positive to vitriolic. The anchor crop for the project has been cotton, and much effort has gone into trying to improve the marketing system for this crop.

Other components have been credit, rural infrastructure, food crops, water and health. The project is focussed around the Machakos Co-operative Union and has spent at least some time and funds in trying to help sort out the problems arising from a burst of donor (but not EEC) funded credit in the late 1970s. But, as is the case with other integrated projects, the focus in Machakos tends to shift with changes in government policy, the general economic environment and project personnel. What is true of one period can be completely wrong about another.

While the above section cannot say very much about the content and operations of projects, certain points do stand out. By far the majority of funds spent in the agricultural sector have gone to large projects, most of them focussed on either export crops or irrigation or both. The projects themselves, irrigation and settlement, crop improvement, input-supply etc., give every indication of being part of the standard donor repertoire. The following section looks at two examples from this repertoire in rather more detail, showing that the disparity between what appears in project documents and what actually happens can often be considerable. Both projects are from Tanzania, one EDF, one World Bank, but are aimed at leading into a more general discussion of aid projects and their problems.

Two Integrated Development Projects

The Iringa Region Integrated Rural Development Project, Tanzania was first funded in early 1977, as a four-year project. Its origin, however, lies some years earlier in a series of regional planning studies done in 1974–6. These proposed a general planning strategy for the rural development of Iringa and included a number of specific projects, which could be sent to donors for possible funding. In such cases, it is unusual for the donors in question to accept the overall planning framework which has taken most of the planners' time. They tend rather to pick out specific projects which fit in with their own strategies or resource availability. Thus it can be assumed that the EEC team, which originally assessed the project for funding, chose some parts of the previous Regional Plan, discarded others and added new features on its own initiative or at the desire of the Tanzania Government.

The project was 'to benefit about 70,000 families. The aim (was) to increase production of maize (by 36,000 tons), sunflowers (by 8,000 tons), pyrethrum (2,600 tons) and wheat (2,000 tons)' (Europe Information/DE 29). According to this source, it received a grant of 6.5 million ECU, but others give 6.0 mn. The contract for the project was given to a German firm, Agrar und Hydrotechnik, as was that for the second phase. This started in 1982, with a further 19.3 ECUm. Although 70,000 families may sound a lot, it is only a third of the regional population of about one million people (say 200,000 families). But, in any case, this sort of statement normally only means that there are 70,000 families in the area which the project will cover.

An interim report on the progress of the first phase up to mid-1981 give some idea of the official results. No mention is made of maize production achieved. This is understandable in view of the fact that maize is the major staple crop and grown by almost every household in the region. There would be very little way of telling if 36,000 tons extra had been produced, especially since by far the major portion of what is sold goes outside the official channels. If it could be shown that maize production had increased, there would be no way of telling whether it derived from the project, from the National Agricultural and Food Credit Programme (TRDB/World Bank) which was also operating in the region, or from some other cause – for example changes in official and/or black-market prices. From alternative sources, it would appear that the project may have contributed to increased maize production. Its transport fleet was by far the largest in the Region and was prevailed upon for some years by the regional administration to assist in the physical distribution of fertilizer – most of it used on maize. But this has ceased with project reorganization and seems always to have been a bone of contention between project and administration, rather than intentional co-ordination with local strategy.

For other aspects of the programme there are figures. It had been planned to plant 90 ha of pyrethrum nurseries and this was 93 per cent achieved (i.e. 84 ha). It had been planned to plant an extra 4000 ha pyrethrum (according to this report), but only 1,110 ha had been planted (28 per cent). On the other hand, the target to provide 5 pyrethrum driers had been surpassed by 60 per cent (8 had been delivered). The report does not refer to pyrethrum production in Tanzania during the period, but according to other sources (MDB 1981 – 1982–3 Price Review), it declined steeply due to declining prices, which may explain the shortfall on the planting target. It does nothing to explain why more driers than planned were delivered. According to other reports, there were problems around the payment for pyrethrum drying, which limited the usefulness of the driers.

For sunflowerseed, 88 per cent of the 7,000 ha planting target was achieved and MDB figures show that production did increase significantly during the first two years of the project's life. But Odegaard (1985) shows that taking prices for 1965/6 as the basis, the price index for sunflower was 400 in 1980/81, compared with only 180 for pyrethrum. This would seem to explain much of the difference in 'project success' between the two crops. The project may have had some impact, but with no more than bare figures, there is no way of telling.

Another major concern of the project had been ox-training. All nine of a reduced (from 24) target for Ox-Training Centres were constructed or rehabilitated and 107, rather than the planned 100, Village Ox-Training Units set up. Four times as many farmers were trained as planned, and nearly 6,000 instead of the planned 5,000 pairs of oxen. Unfortunately, instead of the intended 2,500 sets of ox-equipment which it had been intended to distribute, only 614 ox-ploughs were given out. Nor were these extra to normal supplies. 500 of them had been distributed in one year when the project had been quick enough to pre-empt a large proportion of the supply available from Tanzania's UFI factory. According to reports from the area, the need for ox-ploughs was so crying that within days all 500 had been sold for cash. One wonders how many of the farmers trained in fact knew perfectly well how to plough, but thought that entry to the course might qualify them for ox-equipment.[25]

Much of the remainder of the project was infrastructural in nature; to improve and maintain specified lengths of feeder road (25 per cent fulfilled), reafforest 1,000 ha (1,131 done), build 30 stores (28 completed), 56 staff houses (47), 37 dips (35) and set up 5 dairy units (5), provide 1 village with water supply (0), and build 90 buildings for ox-training centres or units (all 90 completed). No figure was given for the number or proportion of target families 'benefitted'.

These figures are cited, not so much because they tell one anything about what the project actually achieved, but because they give some of the flavour of standard project reporting. Physical targets are set and the report indicates to what degree they have been fulfilled. This is both the easiest method and that which corresponds most closely with the work process of a project or government administrator – that is, checking whether work paid for or otherwise set in motion, has been performed. Whether the various facilities were in use or not, is not mentioned. One might imagine, however, that given the general supply situation in Tanzania, the dips might well have been short of acaricide, the dairy units would probably lack several items, while the ox-training centres would be likely to compete with the local population for the grossly inadequate supply of ox-equipment.[26] From an interview with one manager (below), it would appear that many of the hectares reafforested were in the wrong places and contributed little to erosion control.

From other sources, one finds a wide variety of comments on the project. Most seem agreed that the first manager was a disaster, and the project too, while he was in charge. He spoke no Swahili and poor English, severely limiting communication. His replacement spoke both English and Swahili and was reportedly more effective. None the less, a more recent evaluation has arrived at far more negative conclusions about project effectivity, management, co-ordination, 'beneficiary involvement' and 'institution-building', and has resulted in a complete change of structure and personnel. In accordance with a new plan, efforts are now to concentrate on 7 (in other versions 11) chosen villages, which are evidently to be 'focal points'. They are to receive (according to plan) plentiful supplies of inputs and are to be allowed to disperse their settlement pattern (as other villages in Tanzania are not), in order to expand area and production.

While one can accept that this evaluation mission was probably right in its criticisms, it is far clear that such a change will be for the better. If integration with local authorities and 'institution-building' are aims, then it seems curious to select 11 arbitrary (or, more likely, better-off than usual) villages when there are probably more than 700 in the Region. One can understand that it may give more job satisfaction to the foreigners working on the project if they can concentrate their efforts on an area small enough to see the results, but it is far from clear that this is best for the local population. The number of expected beneficiaries must have been reduced to about one thirtieth of the original and there is no question of the favoured villages acting as a demonstration which others can emulate. Quite the reverse; if all the inadequate supplies of farm inputs and implements are to be concentrated on a few villages, then the situation will be even worse for others. One can claim that this allows project experts to test out new ideas, but one can question how useful this is when most of the rural population cannot even get inputs and implements to operate techniques which they 'adopted' years ago.

Co-ordination with local authorities is almost invariably a problem in such projects, since the project administration sits as a foreign body in the local administrative system. In Tanzania, planning and development are the specific responsibility of a government officer, responsible to the Regional Commissioner (the political-administrative chief executive) and to the central government (Prime Minister's Office). This implies a responsibility to follow changes in national and regional policy, including ad hoc 'directives' from above – not to mention such directives as the officer him/herself may think up. The project, however, arrives with a separate 'plan', which has no formal status within the structure, but is backed by foreign currency, one of the most pressing shortages facing government officials. A number of the problems which arose can be cited from an interview with the second manager.[27]

Several of the 'ten factors which can make a project fail', which he cites, are predictable enough. *Logistics* are a perennial problem, and so is *lack of counterpart personnel. Lack of basic data* is hardly uncommon, nor is the fact that *budgets are invariably overtaken by inflation* before they can be put into practice. The *size and heterogeneity of Iringa Region* sounds specific, except that this is generally true of the sorts of area over which integrated projects are defined. His other four reasons are worth citing in more detail:

i). Projects tend to be planned on the assumption that a significant proportion of the work will be done by 'self-help' (i.e. unpaid labour). This self-help component has a tendency not to materialize, because the villagers have not been told what the project is or why they should contribute. In comment on this, it is often the opinion of aid-project workers that it is the responsibility of the government to perform this mediating function – and this is often an effective means of ensuring that nothing gets done. By contrast, a Danida-funded rural water-supply programme in Iringa Region has used a fair proportion of project funds and personnel ensuring that this vital link is made – and apparently with considerable success to date.

ii). Integration with other donors in the Region is poor. Each wishes to start up its own projects. Thus the region has four different dairy projects, cutting across one another, several different farm-input supply programmes (sometimes competing for the same inadequate supplies) and a number of reafforestation projects. In comment on this, there is no reason why donors should co-ordinate with one another *outside* the national planning mechanism – except that the latter scarcely works. It is, however, an obvious problem that different donors duplicate each other's efforts – and one unlikely to be solved by local authorities so long as foreign exchange is brought in by projects and so long as it is in desperately short supply.

iii). Pressures from the Tanzania Government to undertake non-project activities, like the transport of fertilizer. Since the project had more transport and better mainte-nance facilities than anyone else in the Region, it is not surprising that they should have been called upon to help out in times of need. It is true that this may delay project implementation, but there is still the question whether project objectives really are so overridingly important that they must necessarily exclude other activities. One local viewpoint was that transport of fertilizer was the main achievement of the project – and its cessation with the new 'favoured village' structure a disaster.

iv). Perhaps the most interesting is what the manager calls 'the need for immediate action'. Projects are normally started after a relatively lengthy period of planning and discussion, after which it tends to be assumed that the main problems have been identified and are understood. When, as is often the case, it turns out later that this was not so, it is difficult to admit the mistake and change course. One major reason for this is that all concerned are in a hurry to get started on physical implementation. As he put it:

> Who would want a project to be in a region for two years, with nine planners . . . spending money on consultants and all the things they need and there's no output out of it, no outcome, no visible things coming out of it. They don't want that. They think you can start right away . . . and the EEC, would also like their money to be spent right away, because they can't tell people we are sitting and planning in Tanzania. They want an action programme. And the Region wants an action programme. So they quickly determine, in a very short study, that there is a need for reafforestation, for farm implements, for dairy units, for fertilizers and so on and so on. But the real background . . . and problems . . . (get ignored).

Moreover, what the Region expected to get out of the project was rather different from what the project itself had planned. Firstly, the project had brought funds and activity to the Region. Development was being implemented. But the main effect came from project *inputs* of labour, materials and cash. With 1,000 people employed, it was the largest employer in the Region other than the state, and paid out some Sh12 million in wages. Perhaps as important was the Sh10 million spent on building materials and the contribution to local availability of transport and fuel. There was some tendency to lose sight of project goals and output, among all the problems of organizing supplies and labour. Indeed, for many the main benefit of the project derived from the inputs, especially transport. Trucks and spares were then in desperately short supply and project drivers making a small fortune in doing favours for a price.[28]

A major problem was the *need to spend too much money too quickly* and with too little study of local conditions, requirements and problems. To those who consider aid from the macro-perspective and note how small the funds are in relation to needs and indebtedness, this may seem surprising. To anyone with experience (even second-hand) at local or project level, it is not at all surprising. Many aid agencies experience great difficulties in getting rid of funds (especially at the end of the financial year). Because of this, one of the most common ill-effects of projects is to blast (relatively) huge amounts of money into a given area, social system, community or organization with enormous distorting effects on all social and economic relations within it.

The above is undoubtedly an oversimplified account of the Iringa Project, missing out on a number of important dimensions and issues. But it does point to a number of relevant conclusions.

The most important relates to the pressures from all sides to get moving and start spending the money. These come from the EEC (which has to face complaints that it is 'slow to disburse' in comparison with other aid agencies), from the parent company (under pressure from the EEC), from the local administration (whose main interest in the project is the money and materials it brings) and from visitors and teams of consultants who pass through. With the (probable) .exception of the local administrators,[29] none of these knows very much at all about the region, and they tend to assume that certain standard project modules can be transferred from, say, Indonesia, Swaziland or Costa Rica, where they have previous experience.[30] By the time they have found out that this is not the case, if indeed they do at all, it is generally too late. The project structures have been imposed, the materials bought, the targets drawn up and minds closed.

In the case of reafforestation, it was assumed by those planning the project that the problems and answers were known. The problems were excessive tree cutting and erosion, the answer, to plant trees. Little doubt that there has been excessive tree cutting around many villages in Iringa and Tanzania. People have been nucleated in villages of at least 250 families. They continue to use firewood and/or charcoal for cooking and domestic heating. So areas in the vicinity of villages are steadily denuded of tree cover, increasing both erosion and the distance women must walk out and stagger home with loads of wood. But when the project distributed seedlings to villages, they were found not to have been planted where they would combat erosion. The manager assumed that this was because the local peasants did not know how to do this. An alternative explanation would be that they did not even know that combating erosion was the purpose of the project – that they were simply told by the village leaders to plant the trees (possibly in straight lines for decoration) – and without any explanation why. The project then found out that there were not enough extension officers to tell the people where to plant the trees – which prompts the question why the seedlings had been prepared in large numbers *before* anyone had found this out. What it indicates with crystal clarity is that there was no proper phase of preparation in which the whole idea of reafforestation was presented to and discussed with local people.

Lack of co-ordination with local administrators is not just a matter of haste but also of the completely different expectations between them and the project planners. Foreign experts seldom (have time to) co-ordinate with the local authorities and still more seldom wait long enough to hear the opinions of the latter. But even when they do, critical ideas are often not forthcoming. Local administrators tend to see 'landing the project' (and the foreign exchange and materials) as the most important factor – and this not necessarily for corrupt purposes, though it may be. Characteristically (and especially with current budgetary shortages) local administrators, extension personnel and others having to do with projects are desperately short of funds, transport and other materials upon which the performance of their jobs depends. Landing a foreign aid project is sometimes the only way in which they can gain access to vehicles, fuel, or even paper and typewriter ribbons. Under such circumstances, they are not going to look a gift- (or loan-) horse too closely in the mouth. This relates to the enormous economic imbalance, in which even quite a small donor project can have more money at its disposal than a quite substantial administrative structure.

And all this says nothing about the 'beneficiaries', those upon whom the project is to be implemented, but characteristically those with whom it is least of all discussed. Even if they are approached, this can often be at two or three removes from project personnel, through regional and local governmental structures and then through village political leaders, whose interests may not be those of the majority. But apart from this, the definition of project goals normally excludes them, and concentrates on measurable targets like the number of pyrethrum driers constructed.

The World Bank's *Kigoma Rural Development Project (KRDP)*[31] was the first of a number of regional planning exercises, covering most of the Regions of Tanzania in the period 1973–5. The government had invited different donors to plan each separate Region, with the presumed intention of casting the aid net as widely as possible. With fifteen different teams from various organisations and parts of the world operating on very different assumptions, it would not have been easy to achieve co-ordination, but little attempt was made to do so. One guide, members of other teams were told, was to emulate the World Bank plan for Kigoma.[32] This was not very practical advice. Like most government papers at the time, the Kigoma RDP was confidential and unobtainable.

The plan itself sought to increase agricultural production, by identifying and dealing with a variety of different constraints. Planned clearing for new villages would provide sufficient land. Measures to decrease the time taken to collect water, together with improved health care provision, would solve labour constraints. Intensity of cultivation would be increased with a technical 'package', with credit for its purchase and extension to increase receptivity. Marketing services would be improved. All this would (so it was claimed) lead to a doubling of the income of 250,000 people living in villages to be covered by the project. This is only a skeletal outline of a plan seeking to make so many connections and relieve so many constraints that a schematic diagram by Loft *et al.* (1982) contains enough lines and arrows to make the observer dizzy.

The Programme was hit by a major problem at an early stage. It had been planned for the Kigoma Co-operative Union to be the co-ordinating agency, but, together with all other co-operative unions in Tanzania, it was closed down before implementation of the project started. The National Milling Corporation, which took over its crop marketing functions, was clearly unsuited to co-ordinating a regional development plan, and this job was assigned to a 'project management executive' within the office of the Regional Development Director, that is, within the regional administration. Here a number of problems arose between the planners and the administration – as indeed, they almost certainly would have done with the co-operative union.

Another major problem was that KRDP had reckoned on planning the siting of new villages to ensure their suitability for the task in hand (increasing maize and cotton production). But in the major burst of enforced villagization of 1974, the regional administration had jumped the gun (from the World Bank point of view) and by the time the project started, the people were already in villages. The project was left to select 'viable' project villages among them, normally on the basis of sufficient land for expansion.

Loft *et al.* (1982) conclude from their study of KRDP that 'the Kigoma Rural Development Project was a disaster – and contributed significantly to the development of the Region' (p. 4). It must be said though, that their report provides more evidence for the first part of this statement than for the second. It appears that measured incomes did increase significantly during the project period, but mainly from sources entirely unconnected with the project. It had been planned that maize and cotton, with a fertilizer and insecticide 'package' should form the core of the productive part of the programme. Cotton production did increase temporarily, under the stimulus of coercion, while maize production increased under the stimulus of increased prices and improved markets (by NMC, not the project). But in neither case was there much use of the 'packages'. The really significant production increases were of beans and pigeon-peas (purchased by the NMC for the first time and at high prices) and these were not even part of the project. Tsetse-clearing increased the availability of cultivable land, but was organized by the Region independently of the project (with project funds paid, some years after, for the work done). Movement to villages along the more important roads did improve access to markets, but this was specifically the Region's plan, implemented before the project started.

The intention of the project had been to increase and intensify production of maize and cotton through adoption of a seed, fertilizer and insecticide package. Although it was originally envisaged these would be sold for cash, a credit 'component' was inserted. The combination of packages and credit was the most disastrous aspect of the programme.

The package was suitable for neither crop, for separate but related reasons, which Loft et al. discuss in detail (127–35). In the case of maize, the planners had mistakenly concluded from the small size of existing maize fields that there was a labour con-straint in clearing and cultivation, which could be overcome by increasing yields through fertilizer use. In fact, field sizes were limited primarily by the non-availability of markets for surplus produce. What labour constraint existed had been largely the result of out-migration for labour, which declined after villagization. The second problem was that the set of husbandry practices which the local people had evolved to meet the combined and complex requirement of producing the required variety of subsistence products at the right time and with reasonable security, tended to work against the effective use of fertilizers. For example, local peasants inter-planted dif-ferent crops and staggered the planting of maize, to take account of variability in the onset of the rains and the length of the season. In the event, peasants expanded production by increasing acreage, exactly as they would have done if the project had never existed.

In the case of cotton, it was the crop, rather than the package, which presented the major problem. Prior to 1973, when the project was worked out, prevailing prices made the return to labour in cotton production slightly higher than for maize or beans. But in the aftermath of the major food shortage of 1973/4, and with major price increases for food crops, this was never true during the life of the project. Had this been accepted and the project changed, the damage would have been minimized. In fact, the regional authorities, flushed with the 'mobilization', which had gone into compulsory villagization and its 'success' (the people were in villages), decided to push ahead by making it compulsory for all households in villages to grow cotton. This did increase cotton production to a maximum of 3,000 tons in 1976/7, though it fell thereafter. But most of this cotton was produced in a minority of villages where the conditions were especially favourable, and where it had been grown before. In other cases, it took village leaders several years of representation and bargaining to get permission not to grow it.

The problems with the credit programme were two. Firstly, it was tied too closely to the 'package', so that one could only take the whole lot, while many did not want the fertilizer. Secondly, 'supply of fertilizer got off to a flying start, overshooting the target for the first year by over thirteen times, and then ground to a complete halt in the third year because of non-repayment of loans' (p. 54). Large amounts of unexpected and unwanted fertilizers were 'dumped on the first lot of 21 unsuspecting project villages', which had not even been told that they were on credit and would have to be paid for.[33] Further confusion arose over whether the debts should be collected by the Tanzania Cotton Authority (deducting at point of sale), or the Tanzania Rural Development Bank, which had lent the World Bank funds. In the event, both collected about Sh1 mn worth of debts from different villages, and TRDB refused to give further credit to villages which had not repaid. So from having been showered with fertilizer which they did not want, the villagers were punished by the withdrawal of both credit for and physical access to items which they did want.[34] The double collection of debts meant that they had been effectively fined up to one million shillings.

There were also problems with the infrastructural components of the project. The water component appears on paper to have increased the number of piped-water schemes significantly, but this had very little effect on the quality of water supply, because so many were poorly designed (some requiring water to flow up-hill for example), out of order or lacked fuel. Similar problems plagued many of the other components, including one of the more interesting and original ideas – the development of adaptive research (that is research aimed at adapting to local conditions and from local techniques).

While, as Loft *et al.* point out, it might have been a good idea to start the adaptive research *before* specifying the input packages, initiatives of this sort are so rare that one should not complain too much. There were apparently problems with experimental design, but research was done and 'highlighted the complexities and variations in the Region's ecology' so that 'the officers concerned have come to the conclusion that one doesn't really know much' (p. 52). Given general standards of extension and of unwarranted confidence in it, this is a breakthrough of no mean order – or would be if it looked like having any permanent effect or broader impact.[35] Unfortunately the adaptive research institute provided no promotion ladder and, in order to survive, it had to be integrated into the normal research station system, at which point most adaptive research ceased. Evidently it had little impact on the World Bank, for their National Maize Programme repeated some of the same errors at national level, seeking to increase maize productivity with one fertilizer dosage for the whole country.

DISCUSSION

What can one conclude from such studies? That the World Bank is 'worse' than the EEC? Hardly. Loft's report involved several full-time researchers and more research assistants over a two-year period. My information on the Iringa Project comes from two one-hour interviews with managers (about something else), one with the EEC Representative in Dar es Salaam, a few uninformative scraps of 'project documentation', and informal discussion with a few people who were in Iringa at the time. More detailed study would certainly have revealed further problems and complexities. But while the above two examples do not prove anything by themselves, taken together with various other projects which have been discussed at various points in the book, they do allow one to consider some of the standard problems which seem to befall agricultural production projects.

i). *Lack of detailed knowledge.* Time and again one finds that large-scale and expensive projects are initiated on the basis of entirely inadequate study of the areas in which they are to be implemented. Sometimes this results from sheer lack of preparation, borne of the over-confidence that seeds, fertilizer and pesticides must be beneficial. In other cases projects are preceded by up to several years of studies, but the results are still the same. Why?

Part of this relates to the nature of such studies, which are often rather narrow in scope. More important is the fact that what appears in the reports is usually only a fraction of what even the most limited studies actually find out. Moreover, it is often precisely the most relevant details which are left out, either because they are not thought to be relevant (or fit topics for an academic report), or because they refer to aspects of the local political and/or power structure which it is considered embarrassing to include. Apart from this, some of the most useful 'information' is more a matter of having a feel for what is going on rather than anything easily reduced to columns of figures. The problem is compounded when, as so often, the project manager is a 'practical' man with specific concrete expertise in, say, irrigation, crop agronomy or forestry, who finds studies of peasant production or culture irrelevant and incomprehensible.[36]

But this does not refer to incorrect fertilizer or other input recommendations, which are also widespread. Here there are a number of problems. To test input responses takes several years, if one is to find out how responses are affected by climate and other variant factors. It is also very difficult to do this on peasant plots, since these lack the controlled conditions of research stations. By the same token, results from research stations tend to exaggerate the likely response on peasant plots – yet they are almost invariably used for the purpose, even if the research station is several hundred miles away. Still worse than this, there is a tendency, among both consultants and local technical personnel to look 'optimistically' at research station results, ignoring the cautions, provisos and evidence from bad years within which the results are couched.[37] Finally, research on one single crop usually fails to reveal if the local people have other crops or activities which are more important to them – or if their overall pattern of activities allows them time to undertake project innovations. One might suppose that this could be corrected from the local knowledge of field-level extension personnel, but this is only rarely true. In the first place high-level experts, foreign or local, do not often ask the opinion of such persons. Secondly, given the high degree of hierarchy within Ministries of Agriculture and similar organizations, lower-level personnel are unlikely to raise objections even if asked.[38] They are generally far more likely to refer to the backwardness of the peasants.

ii). *More haste less speed.* This is closely bound up with the pressure on all concerned to get into the field and get things moving. It is understandable that aid agencies and governments, seeing large amounts of money being spent on projects, would like tangible evidence of something being done. But it does very clearly add to the problems by increasing the number of ill-thought-out components in the project. But the ill-effects of this are enormously increased by another factor.

iii). *Project definition.* Especially in recent years, it is common for the preamble to any project document to contain a number of references to expected beneficiaries, environmental conservation, health, the situation of women and other desirable goals. The problem is that these are often difficult to 'operationalize' and their achievement still more difficult to measure. In most cases, such aims disappear in the concrete part of the project plan, to be replaced with more tangible goals of the hectares to be irrigated, planted with rice, or fertilized and the number of dairy units, cattle-dips, pyrethrum driers or whatever, to be constructed. Moreover, once this has been done, any objections relating to distributional factors, long-term environmental factors or public health, are likely to be seen by the technical experts as a most unwelcome intrusion.[39] Thus it is that one can get projects which claim to aim at 'the poorest' which nevertheless distribute inputs mainly to richer than usual 'progressive farmers', since this is claimed to increase production more rapidly.

Apart from such obvious effects, this sort of project planning leads to a *time-bounded perspective*, in which all that matters is what can be achieved within the two, four, or however many years of the project's life. It leads to the project being seen as something in itself rather than as an intervention in a far more complex historical process, whose aim should be to make certain lasting changes in areas where they can help. If the project has any useful innovations to offer, it is far better that a number of local people should have learned that they are useful enough to spend their own money on, than to cover the countryside with free or subsidized inputs for a few years, followed by a retreat to the situation before the project started.

iv). *Project subsidies.* In recent years, most aid donors, under encouragement from the IMF, have turned away from heavy subsidization of inputs – at least in theory. All the same, there is a continual tendency for them to creep back in one or other form. Official donor projects commonly have no shortage of grant funds, and they have specific targets to achieve. If fertilizer or other inputs are moving slowly, ways to speed this up include reducing the price or offering them on credit (often in full knowledge that it will not be repaid). But as soon as this is done, one can be sure that those who are better placed politically and economically will get the largest share. Moreover, this can lead to use of too much fertilizer (or whatever) in relation to its real (unsubsidized) economics. Still worse is the practice of giving out inputs 'free' and then charging all peasants through deductions from their crop proceeds, whether or not they have received any inputs.

v). *Input fixation.* Discussion of projects so far has followed one of their major biases in focussing almost solely on purchased inputs, although for many purposes innovations in labour practices are every bit as important. Some of the reasons for this bias are already obvious. Modern agriculture is high-input agriculture and this is what the experts know. Few know much about the details of peasant manual agricultural labour. Aid agencies are set up to provide materials and donor-country firms often have an interest in this. Officials, and decision-makers in the recipient country are also interested in flows of materials, and in many cases this goes for the peasants too. What they are interested in from projects and aid (supposing they are interested at all) is basically inputs to substitute for labour or make it produce more, not advice on how to work harder and possibly for little tangible benefit. Most of them have had enough of that from colonial and post-colonial governments. This can be a particularly difficult problem when it comes to erosion control, since this commonly involves a considerable labour input, often without increasing production significantly in the short run. It also raises difficult problems about what to do about peasants who are unwilling to, say, terrace their plots, when this may spoil a scheme for many others. All this argues for a rather careful process of discussion with the peasants involved, together with grass-roots mobilization. No-one supposes that this is an easy process, especially given the forms which mobilization has often taken.[40] But in very many cases, no effort is made to do it at all, or else it is left to the 'competent political authorities' to pursue with their usual lack of finesse.

This raises one final point – that there is no such thing as a 'right' way to proceed, or a trouble-free project. If a project succeeds without problems, it was probably unnecessary in the first place. One can follow all the right methods, take time to find out the local conditions, and encourage participation, and the result can still be chaos. All one can say is that the chances of doing more good than harm can be increased.[41]

Conclusions

This chapter started out at the most general level, looking at overall financial flows, and gradually narrowing the scope through the EEC aid programme, to country programmes and individual projects. Its purpose was to move closer to what 'aid' really means. Taking up some of the points covered in different order may help to clarify the argument.

Overall financial transfers to the Third World have fallen in real terms since about 1979/80, the main reduction having occurred in non-concessional transfers (that is, loans on market terms) and especially lending from the private sector. Concessional transfers (ODA) have also stagnated since the early 1980s. Since indebtedness and debt service have grown rapidly over the same period, the overall direction of net transfers is currently *away* from the Third World (including Africa). In spite of this, Third World and African indebtedness continues to grow, and since prospects on world markets are poor, this negative balance seems likely to worsen. There is indeed a clear need for an increased transfer of funds to the Third World and notably to Africa. But since it seems unlikely to happen, and since commercial lending at present interest rates would probably worsen the long-term problem, it seems clear that, one way or other, most African countries are going to have to continue to adjust their economies downwards, without this in any way assuring a brighter long-term future. In some cases this may be masked in the short run by new releases of funds and minor adjustments of debts. But even in these favoured cases, the long-term outlook is no better.

This throws the emphasis squarely on to the *quality* of aid and on its ability to have some positive impact *in a situation likely to be plagued by continuing shortages relating to indebtedness and macroeconomic imbalance.* The importance of this proviso is that projects or programmes which are import-intensive and dependent on the functions of sophisticated infrastructural, administrative, marketing and information systems are likely to worsen rather than improve matters, even in economic terms. Since projects of this nature tend also to be large, capital-intensive, and concentrated (where agriculture is concerned) in the richer and climatically most favoured areas, their impact on the livelihoods of poor and vulnerable people in those (and other) areas is likely to be still more negative. Smaller, less capital-intensive and system-dependent projects thus seem preferable in both (short-term) economic and long-term (social, economic, ecological) terms.

But while at the level of rhetoric some or other version of this propostiion is widely accepted, there is little evidence that it will be sufficient to change practices which are not just ingrained with habit but built into the aid process.

One of the most important of these is the relationship between aid and export promotion, which reaches its most serious proportions in the sorts of project funded with 'blended credit'. Here aid money is used to soften the terms for investments whose prime beneficiary is often a donor-country firm, seeking to sell either plant and equipment or a 'system' including such plant and equipment. It is not simply a matter of the more complex and sophisticated systems and equipment earning higher profits. In many cases, this is all the companies in question have to sell, and their main selling point is precisely their modernity and technical efficiency. Moreover, the technical personnel involved in assessing feasibility are inclined simply to take for granted the operation of an infrastructure capable of bearing such systems. An alternative is to complain loudly and at length about its failures (especially in the clubs and bars where expatriates meet) without taking any notice of the implications.

With or without direct material interests at stake, many of those involved in making decisions about projects have a highly technocratic view of development, added in some cases to an almost zealous faith in the particular technical branch which they represent. As an expatriate dairy economist said to me once, regarding Tanzania's

World Bank-funded dairy farms, 'If I wasn't a dairy man, I would say that this whole thing was a colossal waste of time and money'. The implicit premise is that a 'dairy man' will naturally tend to defend dairy projects, while tractor men, fertilizer men, maize men and veterinary men will all tend to favour their own specialities. Indeed, some of them define development solely in terms of numbers of dairies, hectares of hybrid maize planted or of fertilizer applied, number of cattle dipped, loans distributed or repaid, with little regard to the effect on people. But the number of hectares planted, bags of fertilizer distributed, trees planted or ox-training units erected is not the real issue. None of these will be worth anything unless they have contributed in some way to changing people's own practices for the better, even when the project is not there. Even then, they will not help if the only practices improved are those of the rich, and especially if they further increase already existing differentiation. But it is much easier to measure physical quantities, and it is these in any case with which technical experts are more comfortable. Both experts and their employers are thus concerned to find measurable goals.

This is in fact, however, a complete misconception of what a project is and what it should aim at. It is not a small piece of a developed country dumped down in the middle of rural Africa. However 'integrated' and all-encompassing it may attempt to be, it will only affect certain aspects of some people's lives. Its 'anchor crop' may not be the most important for most growers, even if it is the main official cash crop. Its credit programme and input supply will only compose a small fraction of total trade and credit in the area, sometimes even in the very inputs it concerns. A project is in reality a (usually short-term) intervention in one or a few aspects of a complex process of production, reproduction and change. This will have involved changes in agricultural production, in the distribution and composition of resources available, in social structure and dominant power group and many other factors, since long before the project arrived. It will probably also have taken in other projects on the way. It is hard enough to figure out what would be most helpful even if one does take all this into account. Without doing so, it is well-nigh impossible.

It may help to sort out the different sorts of pressure if one distinguishes the different sorts of transfer. In terms of funds, there are three main types of grant aid transfer; grants to projects of several hundred different sorts; grant-aid as a subsidy to otherwise commercial loan finance – or grants to specific aspects of loan-financed projects; food and other commodity aid. Aid funds are used for a wide variety of other purposes, like supply of technical assistance (i.e. personnel), training courses, scholarships, seminars, export promotion (for recipients), assistance to local non-governmental groups, and many others. But these account for only a small proportion of total funds spent (though probably a higher proportion of any positive impact). One other major cost upon the aid budget is administration. This varies widely between agencies, due as much to differences in definition and accounting practice as anything else. Direct comparisons of administrative costs as a proportion of total transfers are thus largely meaningless. Efforts to reduce them, in line with current notions of cost-effectiveness, tend, as so often, to cut into vital learning and preparation processes, leaving the stacked hierarchies of committees and decision procedures untouched.

Of the three types of transfer, those which subsidize loans are the most problematic, since this directly subordinates aid to investments whose purposes are quite clearly commercial and thus involves it in funding ventures selected on (donor-firm) profit-ability criteria rather than criteria related to recipient-country needs. Related to this are projects whose purposes are to support loan-financed developments. For example, the donor may have provided interest subsidies and guarantees for the construction of dairy processing plants in an area where the milk supply is insufficient to keep the factory running at capacity. Under such circumstances it is common to find the grant-aid arm of the agency undertaking dairy development projects in the catchment area. These may or may not be well-designed and successful. The point is that they are only there at all because of the loan project.

The problem with commodity aid (of which food aid is one form) is that it invariably concerns commodities which the donor has in surplus and that the reasons for providing it usually have as much to do with the need to off-load this surplus as with the requirements of the recipient. For similar reasons, it is usually forthcoming mainly at times when the commercial market is slack, and is likely to disappear precisely when most needed. In most cases, it also involves overvaluation, since the fact that it is in surplus often stems from donor protection which maintains prices above world market level. Thus donors exaggerate the value of their transfers while recipients accept grants whose face-value is double what they would have had to pay commercially. Again, commodity aid can be used to sweeten commercial transactions – with wheat for a mill, SMP and butter oil for a dairy, or fertilizers for a large-scale project (or to make up for the shortcomings of a fertilizer factory).

Commodity aid is an extreme form of 'end-use tying' of aid, that is, stipulating that funds shall only be used for one or more specified purpose. It also almost invariably involves 'procurement tying', that is, specifying that the commodities in question shall come from the donor. But both of these can be imposed (and often are) on other forms of aid. Projects are normally 'end-use tied' in any case, to the extent that the personnel and materials to be transferred are largely specified in advance, but it may also be that they are tied to procurement of the materials from the donor. Even when this is not the case, much aid is informally tied through the fact that materials not available in the recipient country will tend to be purchased through the procurement agency of the donor aid agency or some firm appointed for the purpose. Other things being equal, both will tend to favour their own country's materials.

In all this, however, it seems to me that the danger is not so much that normally mentioned – that goods are purchased other than from the cheapest source. Certainly this leads to some waste. Far more important is that it throws such a large proportion of the decision-making about aid projects into hands which have clear material interests at stake – interests moreover which seem clearly to increase the size and wastefulness of projects and so the indebtedness of recipient countries.

The mere fact that official aid is a relationship between governments has another series of effects on its operation, apart from almost necessarily giving support to the regime in power. On the one hand, this relationship pushes policy ineluctably towards large-scale investments and projects with considerable state involvment. Large structured bureaucracies are involved on both sides, together with the usual penumbra of interest groups, offering the special skills of their firms or institutions. If a given amount of money is broken into large project chunks, this reduces the administrative labour for all (donor agency personnel) concerned; it is quite common to hear aid agency officials state that they *cannot afford* small projects, the administrative costs are too high! On the other hand, the conditions for transfer of aid often impose considerable costs on both the recipient state and its citizens. That is, the state will often be called upon to provide certain services or conditions to the project or investment, whether public or private. These may include tax concessions; undertakings to provide infrastructure, transport services, personnel, housing, other local currency costs and even the primary emphasis of whole government departments; alienation of land from its existing uses and users; provision of wage/food-for-work/'self-reliance' labour for work on projects; and many other things. Even the countable costs to the state are often considerable. Those which are not counted at all (or grossly undercounted), like losses of land or pre-emption of labour and farm decision-making, are often far greater.

In some cases, these sorts of costs are not considered at all. Land to be taken is classified as 'unused' or, for example, 'sparsely populated by nomadic herdsmen'. It is quite common in such cases, simply to ignore 'unofficial' populations – who may have migrated recently because of insufficient land in quite significant numbers.[42] Even if such factors are considered seriously, they will seldom be considered sufficiently important to 'retard progress'. Even in countries with severe land-shortage problems (Zimbabwe and parts of Nigeria, for example), projects and private investments continue to be made which involve the alienation of thousands, even tens of thousands, of hectares of cultivable or grazing land for large-scale projects, state farms and private ventures, both foreign and local.

In many parts of Africa, control over land is one of the most important determinants of whether the food situation can be improved for those most in need, or whether future decades will see a further slide into horrific levels of poverty and hunger. Since there is nowhere for the dispossessed to go, except to peri-urban shanty towns, and since there will certainly not be any rapid expansion of urban employment (except in the informal sector, and there through dividing jobs and incomes ever more thinly), it is crucial that policy should aim towards maintaining as many as possible of the peasant population on the land. It is also crucial to achieve sufficient security of tenure to provide the only incentive for long-term intensification of farming methods – that the farm in question provides the continued and reasonably secure livelihood of the family, seen to include succeeding generations.[43] Every time an aid agency or export-credit bank funds a large-scale project, involving the alienation of large tracts of land (or funds large-scale tractor-based cultivation, whether on a shifting basis, as in the Sudan, or more permanently), it probably more than undoes all the work of any projects it may have which are concerned with helping the poorer members of the rural population to hold their own.

One of the major justifications for this is the need to increase aggregate food production in the face of existing stagnation and rapid population growth. In some cases, like the wheat schemes in Nigeria cited in Chapter 3, these do not even produce much food, for all the land and money used. In other cases they do succeed, as in Zimbabwe, which has recovered from the drought of the early 1980s so strongly that, as of early 1987, it had a maize surplus of over 2 million tons.[44] Zimbabwe has increased aggregate production with incentives to both large farmers and to the better-off peasants in the more fertile areas. Over the same period, the resettlement programme has been much reduced in extent, below the original targets – and in part because of advice that this might reduce the level of aggregate marketed production of food. Given that white large farmers still produce some 40 per cent of marketed maize, there is probably some truth in this. A more extensive, rapid and thoroughgoing process of resettlement would have 'reduced confidence' among the white farmers, leading some to leave the country and others to reduce production. But while increasing aggregate production of cereals may increase 'national food security' through allowing the building of stocks for future famine relief, it does so at considerable monetary cost (that of exporting surpluses at a loss or of storing them, above strategic requirements). At the same time, it actually hinders solution of the long-term and more serious problem of land hunger among a very significant proportion of the rural population.

This points to a more general problem, already mentioned, and to be taken up again in the final chapter. There do seem to be serious conflicts between the macroeconomic and national-level implications of aid and possible poverty-oriented emphases. At the national level, policies tend to be concerned, understandably enough, with coping with problems of indebtedness and (where food security is concerned) with maintaining the flow of food to the towns. The first, together with many other pressures, leads to a strong tendency to accept most offers of foreign exchange, however tied. The concern with aggregate agricultural production (whether of food or export crops) tends to generate an emphasis on the 'best' farmers in the most favoured areas. Since large farmers compose a more reliable source of official supplies than small peasants, there is a tendency to emphasize them, which is further intensified during project implementation by a variety of pressures for 'trickle-up'. These tendencies are by no means confined to official aid agencies or private investors, though they are certainly present there. They also mark the decision-making of most African states, posing the question, to be considered in the final chapter, of the relation between the state, the mass of the population and emerging class forces.

EEC DEVELOPMENT AID

In this context, we can return to the EEC development programme and re-assess some of the points made above.

The EEC operates a medium-sized aid programme. But since EEC aid is subtracted from the other aid transfers of its member countries, the size of the EEC programme is only of relevance in so far as it allocates funds differently from the way the members would have done, or do so more or less efficiently. There are a number of obvious differences. The EEC concentrates its aid to a greater extent on Africa, and transfers more of its total aid in the form of food aid. Where efficiency is concerned, the EEC is widely rated as slow, bureaucratic and cumbersome, though its direct administrative costs are below average. On the other hand, some claim that the Lomé framework offers special advantages.

The concentration on *Africa* has been criticized, with some justice, as avoidance of the huge problems of the large and poor countries of Asia. It is, moreover, clearly in part a geo-political decision, marking out Africa as an EEC sphere of interest in continuation of the colonial tradition. But this question lies outside the scope of this book.

The high proportion of food aid has the obvious effect that EEC aid is, on this score more overvalued than that of other donors, especially since the 'face-value' of EEC food aid is further above its actual market price than is the case for other donors. EEC food aid, with its high proportion of dairy products, is more than usually open to criticism. While this preponderance of food aid clearly relates to the CAP and surplus generation, the significance of comparing EEC food aid with that of the member countries is reduced by the fact that EEC members find it more convenient (and cheaper) to give food aid through the EEC than directly. In addition, there are the more general points about food aid considered in Chapter 8.

One reason for the apparently low administrative costs of the EEC is that the Commission maintains a skeletal aid bureaucracy and undertakes little or no direct implementation or supervision of projects.[45] This is almost invariably done by private firms, one effect of which is to include a larger proportion of administrative expenses within project budgets. It therefore makes little sense to compare these with the costs of agencies which perform more of the supervision directly.

On the other hand, there are significant costs involved in the use of private firms. The 'bottom line', and condition for survival, for private firms is profit. Large projects are more profitable, since revenues are related to the size of deliveries, total costs or turnover, and administrative costs are subject (at least in theory) to economies of scale. Put the other way round, small projects often do not provide sufficient revenue to cover a company's overhead costs. But, at the same time, almost any project is usually preferred to no project, so that there are strong pressures to arrive at 'positive' conclusions about proposed projects. It is not just that the firm itself often stands to gain from a positive decision. The donor country (especially where represented by its export-promotion department) and national firms, like equipment suppliers, may be depending on the contract. Moreover, the recipient government and its decision-makers will normally have a variety of reasons for wishing a project to go ahead, even if it is demonstrably unviable. It would be naive to suppose that official donor-agency structures and personnel are not subject to the same pressures. Official agencies also prefer large projects to allow them to distribute the specified amounts of materials without increasing staffing.[46] All the same, it is probably easier to present negative views of proposed projects in official agencies than private firms.[47]

One reason for the cumbersomeness of the EEC's aid operations is that many decisions have to be agreed by the Council of Ministers, which often means lengthy political bargaining. EEC Commission officials also complain that the involvement of the European Parliament in aid matters (which is far more active than in most national parliaments) also leads to delays. This might not seem to apply to the Lomé Convention, for which funding is secured over a five-year period, but it does, at least to the extent that all project expenditures have to be approved (or at least rubber-stamped) by the European Development Fund Committee, composed of member representatives. But while this is widely mentioned as being a major problem of EEC aid, it is not clear that it is either significantly slower than other donors,[48] or more importantly, that this is the most relevant criterion. If the time taken was used fruitfully to improve the quality of projects, it would be well worth the delay. The problem is that most of the time is spent in bureaucratic paper-pushing and in waiting for meetings of committees which, when they eventually meet, will rush through fifteen assorted projects in one day with no time to discuss any of them properly.

The Lomé Conventions represent some innovative features in the EEC aid programme, though initially grossly oversold as a way towards a 'new international economic order'. There is first just the idea of the conventions themselves, in which trade and aid are negotiated for five-year periods between the EEC as a whole and the ACP countries en bloc. How much practical difference this has made is another matter. The negotiations for the second and third Lomé Conventions have shown clearly that negotiation en bloc certainly does not mean easy concessions to the ACP. The funds have declined in real terms per capita at both renewals, while the trade concessions started and have remained very limited. Supporters, like former aid Commissioner Edgard Pisani, stress the opportunities for steady improvement through learning as one convention follows another, and the fact that, with indicative programmes set for a five-year period, EEC aid is politically neutral i.e. unlikely to be cut during the five years. He also stresses that both the contractual form of the conventions and the current emphasis on food strategies emphasize and contribute to the national sovereignty of the recipients (Interview in Cimade (Paris), Dec 1986:10–13). The first of these seems to me purely formal. The second, in defining the food problem as primarily a national problem, risks diverting attention from the problem of hunger.

The EEC does seem to use political leverage to a considerably lesser extent than some donors (notably the USA). To tell whether any real learning process has taken place, it will be necessary to find out whether the changes proposed under Pisani's directorship become reality. One must also recognize one negative aspect of the development of the ACP as a bloc in the process of the Lomé negotiations. This is that most of the gains which it is possible for the ACP to achieve, can only be made at the expense of other Third World countries. This is obviously true of trade preferences, though they have not so far been important enough to make much difference, and is in principle true of efforts to increase aid flows under the Lomé Convention.

EDF aid during the period 1981–4 was clearly dominated by large projects. Projects costing over 10 million ECU took over 40 per cent of all funds, both in aggregate and to the agricultural sector. Projects over 3 ECUm. took almost 80 per cent of the total. For the non-agricultural sector, these were primarily transport and other infrastructural projects. Within agriculture, they were mostly composed of large dam and irrigation schemes, cash-crop development or rejuvenation projects and 'integrated projects'. There was not much sign at this level of the vaunted concentration on food crops, though, to be fair, the policy changes would not have affected expenditures much during 1981–4. In one respect, the above computation underestimated the overall average size of EDF projects, since it ignored the EIB loan projects to which interest subsidies were given. Nearly all EIB projects are over a 3 ECUm. and the majority of them over 10 ECUm.

The different country programmes seem generally to reflect recipient policies and choices more than any standard 'EEC pattern'. Given that the EEC is only one (and usually not the largest) of the donor agencies working in any one country, there is no reason to expect EEC aid to form any specific pattern. At the level of one-sentence project descriptions, most of the projects (large dams excepted) at least sound potentially useful. But this, unfortunately, does not tell one very much, for whatever the content of a project, one can assume that its sponsors will give it a title which gives that impression.

Where aid is concerned, the prime issue is not whether it comes from the EEC or other source. There are, of course differences of style and some of substance between different aid agencies, but the similarities are far more important than the differences. Nor is the issue simply one of developed countries exploiting tropical African countries. One of the more depressing aspects of the whole process is that African states do not protect the interests of their citizens. Indeed, their decision-makers share many of the attitudes which generate projects which are more part of the problem than its solution. Lying behind these, on both sides are class interests and processes of accumulation which are related to the way aid policy is formulated, and an ideology which defines development and projects in conformity with these interests. This is not to say that useful aid projects do not exist. But this chapter has pointed to very severe problems with the overall structure within which aid is transferred and projects thought up and implemented. This discussion is continued in Chapter 10.

NOTES

1. The OECD definitions of 'aid' (ODA) and 'grant element' can be found in OECD (1985a:171 ff). To give an example, a loan at 3% interest, with 28 years maturity (time to pay it off) and an eight years' grace period, would have had a grant element of 60% at the time. It was not stated whether the interest rate was real or nominal, but presumably the former.
2. The distinction between 'aid' (ODA) and investment, or commercial lending, is not quite the

same as that between concessional and non-concessional funding. Not all lending from the World Bank, European Investment Bank or other state and international 'development credit' agencies is concessional. On the other hand, state export-credit agencies, whose primary purpose is export promotion, do often subsidize their loans. The whole issue of 'blended credits' (loans with the interest rate subsidized down with grant funds) has recently been under discussion. The USA has objected strongly to France's use of then as 'unfair competition' – and then responded with counter-use of blended credits to dislodge France from specific deals and contracts.

3. A figure based on the same table, in the OECD Annual itself (1985a:Chart III–2) manages to show the development of ODA as rapid and continuous, almost exponential, growth. One reason for this is scale – their chart is set the other way round on the paper. Another is that they use four-year moving averages, which conceals the decline since 1981. Another point is that my Figure 9.1 is not taken directly from Table VI–2, but adjusts in line with recommenda-tions in note 'a' to the table. This points out that much of the private lending shown in recent years was in fact rescheduling of previous debts. Even so, the Table may still overestimate the real value of net flows to developing countries in recent years. For one thing, as noted in Chapter 3, the size of private debt and debt service is not fully known, and is thought to be higher than in any statistical source. For another, the OECD 'aid deflator' (the figures used for deflating current flows to 'real' terms – see OECD, 1985a:Annex Table 27) has for the past few years consistently been an *inflator*, implying (quite falsely) a negative rate of infla-tion. This is probably the unintended result of the methods used, during a period when the dollar was high, but does not alter the fact that it overstates the real growth of aid and other financial flows. For example, their Table 25 shows the 'real' growth in ODA from 1970 to 1984, to be 14% *higher* than in current terms.

4. If one computes from OECD, 1986 (which uses the concept 'Official Development Finance' (ODF) instead of 'ODA', and gives slightly different figures), total financial flows to the developing countries were $80 bn in 1985 (ODF was $49 bn). This compares with total indebtedness in that year of $991 bn and debt service of $133 bn. In this case, debt service exceeds flows to the developed countries by 66% and is three (2.8) times ODF. In absolute terms, this implies a net flow to the developed countries of more than $53 bn.

5. OECD, 1986:Tables V.13 and V.14. For debt and debt service, I have taken the average of 1982 and 1983, and compared with ODA for 1982/3.

6. Since OECD (1985a) provides no figures beyond 1982/3, I have used those from the World Bank *Development Report 1986* (Annex Table 21) to compute ODA flows. While OECD is given as the source for these data, the computation is probably different, and they are not fully comparable.

7. UN Special Report cited in Kenya *Daily Nation* 16 October 1987. This source finds inflows of development assistance to be slightly above debt service, but computes a negative overall flow by calculating a loss of $19 bn from falling export commodity prices.

8. Computed from OECD (1985a:Annex Table 1). There is no separate row for the EEC, whose ODA is included in that of the member states.

9. Multilateral agencies as a whole provide about twice as much non-concessional as conces-sional funds. The World Bank gives three times as much non-concessional (IBRD) as conces-sional (IDA) finance ($9.3 bn as against $3.2 bn in 1984 – OECD, 1985a:298).

10. But a high proportion of US aid takes the form of 'military aid', though if this is included in the figures cited it is very carefully hidden.

11. Loan subsidies to large capital investment projects can often be considered at least as much as 'export promotion' and thus subsidies to developed country exporters, as 'development aid' and subsidies to the recipients. In struggles between different countries (see note 2), one's own soft loans are generally seen as 'development aid,' those of others as 'unfair competition'. In OECD and World Bank statistics all are considered to be development aid.

12. In recent years, it has become fashionable to talk of 'development co-operation' rather than 'development aid', since this stresses the two-sided nature of the relationship and the partici-pation of the recipient. Ironically, this has occurred during a period when tight funds and ever more stringent conditionality have made the reality yet more one-sided.

13. This reflected in part the undoubted talent for public relations of the then EEC Development

Commissioner, M. Claude Cheysson.

14. OECD (1985a:Annex Table 12) provides comparative but incomplete lists of *major* recipients for all OECD/DAC donors plus the EEC, from which one can deduce the approximate proportions of total bilateral ODA going to tropical Africa. This shows Belgium (75%), Italy (64%) and Denmark (54%) directing more to Africa than the EEC (53%), while the UK (40%), France and Netherlands (28%) and Germany (18%) allocate less. The figures for France are misleading, in that about 60% of French ODA goes to 'Overseas Departments and Territories', a group of small and generally rich islands still belonging to France. If these are excluded, 65% of French bilateral ODA goes to Africa. By comparison, of other major donors, Sweden directs 58% to tropical Africa, compared with 7% for the USA and 4% for Japan.

15. I have used two sources for this, the EEC's *Courier* and *Telex Africa*, since neither provides a full listing. The *Courier* lists only decisions passed, while *Telex Africa* lists projects to be discussed at forthcoming meetings, thus allowing comparison of the two. Both provide brief (and only sometimes informative) notes on what the project is supposed to do.

16. While there is certainly discussion and disagreement over certain projects (for example a pesticide-heavy tsetse-clearing project for southern African countries), this seldom seems to be reflected in decisions against projects. The two persons I know who have attended EDF Committee meetings (as delegates of national development agencies) describe them as a complete waste of time. Since up to ten, fifteen or even more projects have to be discussed at one sitting, there is considerable pressure not to delay proceedings by going into detailed discussion. This problem is by no means confined to the EEC or EDF.

17. It is not always possible to distinguish, since some projects and authorizations refer, for example, to 'all ACP' or 'all ACP and OCT' (Overseas Countries and Territories).

18. This also includes projects which are 'infrastructural' in nature, but bound up with 'directly productive' projects (an access road to a mine, for example). Here the 'directly productive' project is often funded with an EIB or other soft loan.

19. Part of this may arise from linguistic confusion. In French, 'plantation' can simply mean 'planting', but has often been translated as 'plantation' (large company-run farm).

20. This terminology derives from the 'diffusion of innovations' school of extension theory, which is very influential in donor and African government circles. It assumes that 'adoption of innovations' is always beneficial and that it derives from personal characteristics of the farmer (thus 'progressive', 'laggard' etc). Thus the theory excludes, virtually by definition, the possibility of misconceived innovations, innovations relevant only to larger farmers, or difficulties in getting access to inputs.

21. That, at any rate, seemed to be the case in the early 1970s, when I worked there and studied both research station findings and farmer responses. Fertilizers are probably economic for the small proportion of peasant farms on which coffee is grown in pure-stand. (Raikes 1976)

22. This may have changed since the devaluations of the shilling, starting in 1986, since these would increase the relative price of coffee.

23. This is a quotation from M. Pisani, the then Development Commissioner.

24. Mike Speirs, personal communication.

25. It seems a common misconception among aid personnel that ox-training (by foreign experts) is always a crucial component of the expansion of ox-cultivation. All experience indicates that factors like availability (and price) of oxen and equipment, grazing land and dry-season fodder, are often far more important. On a consultancy visit in 1983, colleagues and I visited a village in Tanzania which had been chosen as the site for a large FAO integrated project – though apparently without informing the village leaders, still less the others. Ox-training was a major component of this. When we asked the village leaders their reaction, they burst out laughing saying that even children knew how to use ox-ploughs. If we really wanted to help, could we not help them get hold of spare plough-shares and bolts, which had been unobtainable for some years. A visit to a nearby farm machinery factory – recently set up with donor finance – revealed that they would only produce complete ploughs, no parts, even though their plough was so poorly designed that, in spite of shortage, no-one would buy it.

26. Kjaerby, personal communication, rated the ox-training programme highly. But given the

severe shortage of equipment, I find it hard to believe that the money would not have been better spent on supplying it.

27. Taped interview by Carola Boehnk, 1985.

28. Esbern Friis-Hansen, personal communication.

29. In most ex-British colonies, there is a tradition of posting civil servants away from their home areas. While they will be far better informed about the locality than foreign experts, they will often be partly 'foreigners' themselves, not speaking the local language, for example.

30. This point about standard 'project modules' was raised by Jan Bahnson Jensen in the course of the consultancy mentioned in note 25 – and in specific relation to the proposed project being visited.

31. This section comes mainly from Loft et al. (1982).

32. I was, at the time, a member of one of the other teams whose plan was, to the best of my knowledge, never implemented at all. A team member visiting the area some two years afterwards called on the Regional Planning Officer and saw on the floor of his office a large cardboard box containing 48 copies of the plan – still unopened and with the seal unbroken.

33. This is not the only example of this criminal 'error'. The same was done in parts of western Kenya under the World Bank funded Integrated Agricultural Development Programme (IADP), whose effect on the co-operatives in the area was little short of catastrophic. Started to assist 'weak' co-operatives, it left many so deeply in debt as to be virtually non-operational. In some cases peasants dare not sell their produce to the co-operative, for fear that the proceeds would simply be attached as debt. In one area, so I was informed by a Kenyan extension officer involved, peasants were initially unwilling to take loans. They had, so they said, had problems with credit before. On returning to the officer in charge of the programme, they were told that the money had to be distributed, and within a deadline, so peasants should be told that they were receiving grants!

34. Supply conditions in rural Tanzania at the time were such that many inputs were available *only* through credit programmes.

35. I understand, however, that the adaptive research unit did produce a mass of detailed and interesting material, most of it never published, which could be of great use if and when anyone studies it.

36. Nor are they always to be blamed for this. Apart from the tendency of many academics to write only for those with Ph.D.s in their own subject, it is not uncommon for studies to focus on whatever controversy is fashionable within the discipline rather than to bring out points of relevance to a project.

37. For example, in Kagera Region, Tanzania, a wheat scheme was started in Karagwe District, on the supposed basis of colonial, trials which 'showed that ten bags per acre could be achieved'. What the trials really showed was an average yield of 3 bags/acre, ten being the best single result. The project, which did not last very long, seldom achieved more than 2 bags/acre, with costs estimated at 8–10 bags/acre.

38. For example, there are numerous examples of extension agents claiming steadfastly that the recommended practices are 'the best', even when it emerges that they do not use them themselves, and for good reasons.

39. During a consultancy in Zimbabwe, a team of which I was member visited a half-completed irrigation project financed by another donor. The donor had taken the care to have public health specialists assist in designing a system of channels and intersections which would avoid standing water in which bilharzia-carrying snails could live and breed. But the contractor had done a sloppy job in this respect, so that both standing water and snails were already present. The senior local official seemed to think that questions as to whether this might not detract from the quality of life (and even production) were largely irrelevant; after all the purpose of the project was irrigated production, not public health.

40. Although one tends to think of mobilization as a process of persuasion and building up enthusiasm, the reality is as often to get the peasants moving with a boot in the backside, a bayonet in the back or the threat of a fine or jail. To put it mildly, this does not assist the task of subsequent mobilizers.

41. One NGO project I know of has spent a considerable amount of time considering different aspects of tree-planting for increased woodfuel. One of the difficulties now faced is that if the

project is to have any widespread effect, it will need to work through one or other government agency, since its own personnel are far too limited. The problem is that neither the agricultural extension service nor the forestry service is willing or even capable of discussing the alternative merits of different types of small woodlot at length and in depth with peasants. The former has many years of experience of giving 'advice' in the form of brief commands. The latter generally regards humans as a regrettable obstruction to the growth of forests, and to be excluded from them wherever possible.

42. Thus in Kagera region Tanzania, in 1974, one large area of land had been planned for three separate 'modern' uses (ranch expansion, large-scale mechanized food-crop production, dam and irrigation for rice production). Not only did these plans take no notice of one another, they also ignored the large number of peasants who had migrated into the area and were growing coffee and bananas. Not only did this latter offer a livelihood to a far larger number of people; even in straight economic terms, it seemed the most viable use of the land.

43. As will be discussed in the following chapter, monetized goals, especially when interest rates are high, do not provide an incentive to the development of sustainable agriculture.

44. This is not without its problems, since Zimbabwean internal prices are well above those on the world market, while neighbouring Mozambique, which desperately needs maize, can less well afford to buy it than Zimbabwe could to sell it cheaply.

45. Another, until recently, was that part of the administration of food aid came under the Directorate for Agriculture and thus the CAP, rather than the aid budget.

46. Here as in other aspects of neo-conservative thinking on the improvement of efficiency in state bureaucracies, the emphasis is heavily on reducing labour costs rather than improving services.

47. In a seminar in Copenhagen in Autumn 1986, professionals working for consultancy firms spoke to researchers and students about the conditions of work. One woman recounted how she had produced a series of figures for the internal rate of return on a project which showed it to be unviable. She was called to the manager's office and told to change them. It is perhaps a sign of the times that only one or two questions to the panel focussed on this type of issue. Most concerned pay, conditions of work and the possibility of getting jobs.

48. I know of several cases involving other agencies where it has taken over five years from initial discussion until the first funds were transferred – and others which were cancelled after three or four years' discussion.

10

Perspectives on the African Food Crisis

Introduction

The purpose of this chapter will be to tie together some of the many loose ends from other chapters, consider some points inadequately covered and discuss their implications. It starts by covering some of the material presented in a different order.

Income and food security

To start at the beginning again, hunger arises from lack of income, not from the physical absence of food. This is true even in the case of famine, the most dramatic and gruesome expression of hunger. But far more millions are mal- or undernourished than are affected by outright famine. Even if one substantially reduces the World Bank estimate, cited in Chapter 4, that 50 per cent of Africa's population is malnourished and 25 per cent seriously so, one still has tens of millions of malnourished people. These are the poorest urban famlies, poor peasants and landless rural labourers.

In terms of numbers, the largest group are poor peasants, with some land but not enough (or not good enough) to provide subsistence without additions from wages or remittances from a wage-worker. While this group includes a whole range of situations, one could take, as a sort of mid-point, a family which can just get by on farm produce in a good harvest year, must struggle and do wage labour in an average one, and encounters really serious problems in a poor one.

Clearly climatic patterns enter here. People go short and hungry at particular seasons of stress, typically the growing season prior to the main or only harvest. At this time, food stocks are lowest, labour requirements tend to peak, and in many cases rainy weather means increased incidence of illness. But there is no doubting the socio-economic determination of hunger. Poor peasants suffer most seasonal stress because they have least reserves and so are forced into casual labour or borrowing at an earlier stage. Because they face higher risks, they must undertake risk-avoiding production strategies which tend to reduce output and income. In addition, they are often more bound by the traditions and customs which underpin kinship and other security networks, another form of risk reduction which may take time, funds and attention from farming.

This, in turn, means that periods of seasonal stress, especially in poor harvest years, can be the occasion for a further turn of the downard ratchet, with loss of on-farm labour, and so production, to casual wage work, increased indebtedness from borrowing to get by, and finally sale of land when all else fails. For others who have the resources and cash to spare, this is not a season of stress but rather of opportunity. Food prices rise and so the profits from selling it. One can lend money or food and gain labour, clientelage or interest, or buy a plot of a neighbour's land to 'help him over the crisis'. There is good reason to believe that with the penetration of market relations, this sort of response is increasingly replacing previous redistributive patterns. The former were doubtless far from perfect and certainly not without their exploitative elements. But there are significant differences. Traditional leaders and patrons were (and some still are) concerned to keep the subject or client as an operating member of the rural community. The accumulator of land has no such interest. If land is scarce, labour most probably will not be. This is not to say that kinship and other social networks have disappeared, and patronage certainly has not. What this implies is that, apart from economic vulnerability, there are some who are more vulnerable than others on social grounds – those least integrated into existing social networks for one or other reason. These will include people from other areas, clans or tribes, (unless they have their own networks), often widows and grass-widows, and a variety of outcasts, ranging from beggars to witches. Lone women are often especially vulnerable, not only because they tend to have smaller plots and more limited access to labour and credit, but because in many societies they are regarded as the food providers, not to be provided for. In short, while seasonal stress clearly has a climatic component, it hits specific vulnerable sections of rural communities and is part of a process of accumulation and differentiation.

Even famine and exceptional food shortage are not so closely and directly related to climate as is sometimes thought. Sen demonstrates that there is no clear relationship between the size of the harvest shortfall and the seriousness of the famine. The effect of any given climatic pattern will depend on soil, topography and vegetation, and will clearly be affected by human practices. Deforestation and other erosion-accelerating factors reduce both convectional rainfall and the amounts of moisture which enter and are stored in the soil – as do husbandry methods.

The mechanisms which turn a serious harvest shortfall into a famine are socio-economic. The *expectation* of a bad harvest can be as devastating as the reality, if it sets off a spiral of hoarding, price increase, panic-buying, further price increase etc. The better markets are functioning, the more effectively they suck produce from the poorer areas and social groups towards those which have 'effective demand'. The more delayed, inadequate, inflexible or plain perverse the response from the state or international agencies, the worse the situation will be. Again, one can outline specially vulnerable groups in Africa, partly in terms of location but also in terms of socio-economic status:

a. Small peasants living at high density in relation to soil fertility and unable, by reason of poverty, to undertake husbandry improvements to conserve the soil, to fallow the soil for long enough, and to avoid further denuding forest cover in their search for firewood. Harvests, and thus incomes, vary increasingly with rainfall variations, while poverty also means a lack of reserves.

b. Small peasants who are pushed into increasingly marginal rainfall areas, where land pressure, low income and high risk inhibit improvement, and the adjustment of cropping methods developed under conditions of higher and more reliable rainfall. Here cultivation practices and fuel-wood cutting can serve to increase the variability of already highly irregular rainfall and harvest patterns. If savings are held largely in the form of livestock, these lose their value whenever food is in short supply or the need for cash pressing.[1]

c. Pastoralists, often pushed in front of the above, whose physical livelihood is weakened by the loss of dry-season grazing, while their security in times of shortage (livestock for sale) lose much if not most of their value just when most needed.

d. Rural or small-town landless labourers, who are affected by rising food prices and falling levels of economic activity in the aftermath of a bad harvest.

e. Dependants of any of the above.

f. Widows, labour-migration grass-widows and other women living alone with children on small plots.

g. People living in areas ravaged by war, civil war, armed incursion or police action against the same.

Another group are significant in the generation of famine conditions but do not suffer from it themselves:

> Farmers or businessmen with tractors, farming marginal and vulnerable areas, with low-cost, short-term techniques which are a form of mechanized shifting-cultivation. These people face little or no personal risk – indeed they are likely to profit from famine as holders of stocks or transporters. They do, however, increase the risks of other groups, notably peasants in marginal areas and pastoralists.

Clearly levels of production and productivity are significant factors lying behind famine and food shortage, but they are only part of the story. Moreover, not all ways of increasing aggregate production necessarily help, even if they are successful. This shows clearly in the previous paragraph, since this sort of farming both pushes peasants and pastoralists off their land *and* reduces its fertility.

Hunger and the crisis

If hunger and food shortage derive primarily from lack of income, then they are clearly related to the current crisis in Africa. This is *not* just because of the claimed decline in per capita food production, which has been much exaggerated in any case. More important has been the impact on the level and security of incomes for those at the bottom of the pile.

There is no need to go through the figures of the crisis once again. The situation is one of crippling indebtedness, which is steadily increasing. Many countries can scarcely even cover debt service from total exports, and thus have to borrow more simply to survive. Even with further borrowing, most countries have to accept drastic reductions in imports, since loans are increasingly difficult to get and deterioration in terms of trade means declining export proceeds. At present the flow of funds actually seems to be away from Africa, and would be more so if more countries were able to maintain their debt-service payments. Beyond that, indebtedness has forced most of the countries of Africa to submit to IMF conditionality, implying currency devaluation, reduction of subsidies, reduced spending on social services and lowered tariff barriers. Given the situation, some of this seems necessary and even beneficial (reduction of subsidies, for example, and some devaluation), since shortages and black markets have often in any case denied their benefits to those most in need. But the underlying premise of IMF and World Bank conditionality – that it is possible to export one's way out of debt – seems unfounded. Reduction in social service spending may reduce waste, but often cuts essential services, leaving the waste to continue.[2] Reduction in tariff barriers may get rid of some wasteful industrial ventures, but also hits others and obstructs the development of any industrial sector at all.

All this has eliminated most, if not all, of the income gains of the 1960s and 1970s, leaving poor peasants worse off than before. Population increase and accumulation have reduced the availability of land and usually increased its price far more rapidly than wage levels or product prices have risen. Indebtedness, conditionality and short-age of credit have reduced levels of non-agricultural activity, both through breakdown for lack of spares and through declining demand. Conditionality regarding tariff bar-riers threatens more industrial activity, all of this reducing wages, availability of jobs, and demand. If reported per capita incomes are down to the levels of the early 1960s, as cited in Chapter 3, then real levels of income will be significantly lower for the poor. Part of the increase derived from the monetization of subsistence income, invariably increasing its supposed value in the process, while subsistence values lost (minor crops, wild fruits and foods) were usually ignored. Apart from this, the period of income growth involved a significant process of accumulation, class formation and differentiation, which has not been reversed during the period of declining incomes. Part of this income decline relates to the volume of agricultural production; far more of its does not, but is rather a matter of worsening terms of trade both internationally and internally. Taken in conjunction with the factors mentioned above, it is clear how this increases the pressures on the poorest groups and renders them more vulnerable to seasonal and exceptional stress.

Processes generating hunger and the crisis

If one looks at the historical processes leading up to the current crisis, it can be seen that in many cases they interrelate with processes generating hunger in its current form.

Colonialism implied a major break with the previous organization of life in a number of important respects. The war and disruption which accompanied colonial conquest led in many cases to famines, epidemics and social chaos, which reduced the capacity of local peoples to withstand drought, in many cases for years on end. The imposition of colonial boundaries blocked off one response to drought by pastoralists and others, while the imposition of flat-rate taxes increased risk during the worst years. It also led to other changes. The major purpose of taxation was to force African peasants into wage labour or export-crop production, in both of which it was successful. Since this primarily affected men, it left an increasing proportion of farm work to women. This, together with the imposition of boundaries and (later) increased population, also changed the nature of farm work, reducing fallow lengths and thus the amount of clearing labour (previously mainly a male task), while increasing the burden of weeding (mainly a female job). Where improved techniques and spare labour were not available to make up for the loss of naturally regenerated fertility, this meant reduced yields and variability of diets (as low-yielding but nutritious and tasty minor crops were dropped to reduce labour inputs and increase production of basic staples). In other cases, it meant replacing cereals with root crops, which give greater yields for given area and field labour, but (as in the case of bitter cassava) may imply greatly increased time spent in processing (by women). Another effect was increased female labour burdens which sometimes lead to decreased frequency of infant-feeding, low-ering the nutritive value of any given level of food intake.

As against this, wage labour and cash-crop production led to increased cash incomes and, for some, a shift into higher income brackets. At a certain level, this clearly improves family income and food security, but the degree to which (and the income level at which) it does so depends very much on the disposition by males of 'their' incomes. While some of this is invested in increased production (in or out of agriculture), improved housing and education, it is less common for it to be used for family subsistence. While nutritional studies show varying trends, there are at least a significant number which show declining levels of female and child nutrition with adoption of export or other cash crops. Where security against bad harvest is concerned, it seems likely that a certain degree of cash-crop production improves security, especially if cash crops have different growing periods and/or are more resistant to drought than food crops.

The post-colonial period. If the colonial period saw the beginnings of a process of commoditization and integration of African peasants into both national and international markets, the process accelerated after independence and took on the forms which would lead to the current acute crisis.

To some extent this can be explained in terms of the sequence of events since independence. Colonial regimes had maintained fiscal and foreign-exchange balance by dint of extreme parsimony, including gross underinvestment in infrastructure and social services (except where there were white settlers and then only for them). It was obviously politically impossible for newly independent states to continue with this policy, and hardly to be expected that they would do so. This introduced an expansionary dynamic into state spending which, under almost any circumstances, would have placed pressure on budgets and foreign-exchange resources, though much aggravated by factors to be considered below. As if this was not enough the first strains of this process in the early 1970s were 'relieved', only to be aggravated later, by the interest-rate whiplash of the 1970s and 1980s. Low interest rates and easy money encouraged further spending, most of it tied to development projects, which much increased the level of indebtedness just in time for the increased interest rates and tight money of the 1980s.

If, however, this was all there was to the debt crisis, it would be very much less severe than it is now. Behind it lie two sets of powerful forces and the ideologies which support and justify them.

International capital

The underlying premise of 'development' seen from the viewpoint of international capitalism and its development institutions is further integration of Third World countries and their peasantries within the world market. This implies emphasis on increasing exports to gain access to the requirements for development, themselves seen to derive largely from imports in one or other form. While there is far more talk of 'conditionality' now than in the 1960s, it has been there all the time. Apart from their own problems (see below), African regimes which have attempted to 'go socialists' or cut links with the world market (often more rhetorically than in reality) have consistently been punished with lack of credit or investment. By comparison, those which are grossly corrupt suffer no such punishment, so long as they keep their economies open to foreign firms. In short, there have been strong general pressures towards development strategies increasing international integration.

But capitalism is not just about general directions. It is composed of firms whose concern is to maximize profits by whatever means available, and a variety of factors have combined to make 'development' a very profitable business. Where sub-Saharan Africa is concerned, very little of this has been 'investment', if by that is implied taking part in the equity and risk of developments. By far the major proportion of external capitalist involvement in Africa has been not only loan-financed, but financed under loans that are guaranteed by both donor and recipient states and often subsidized by the former. Foreign firms *supply* plant and equipment, factories and management contracts for payment. Their own interest, after the initial profit on setting them up, is largely confined to continuing sales of expertise and spare parts. Not only are rates of profit often very high indeed on such transactions, but overall profits are related to the quantity, sophistication and price of items or systems supplied, and not to the profitability of the factory or other unit constructed. Few if any of the firms involved have any interest in reducing the level of Third World development imports; on the contrary, this is what they profit from. This tendency is aggravated by certain aspects of the aid relationship, notably aid-tying, credit subsidies and the ways in which projects are formulated and defined. All these have the effect of increasing the export-promotion aspects of aid and thus the imports and indebtedness of the recipient country. Excessive sophistication and system dependence reduce the productive impact, while encouraging concentration in areas where infrastructure is better than usual. In agricultural terms, this also tends to imply concentration on areas of higher and more reliable rainfall, unless irrigation is a feature of the project – which in the most expensive cases it not infrequently is.

One might suppose that foreign firms would be ambiguous about import-substitution projects, since these would reduce their future markets, but for several reasons this does not happen. In the first place, the firms supplying the equipment may not be those whose supplies would be substituted for. Even if they are, the sales of plant and equipment are likely to be both larger and far more profitable, since aid contracts, with their tying conditions and subsidies, make overcharging quite easy, while markets for complex equipment are far less transparent than those for final products.[3] In many cases, import substitution actually increases overall levels of imports, since imports of raw and semi-processed materials, spare parts, services and repairs outweigh reductions in imports of final products. Finally, with IMF conditionality pressuring for lowered tariffs, firms may be able to have it both ways, profiting from building large and expensive plants meant to operate under protection and then regaining the markets when its reduction bankrupts them.

In agricultural development, similar tendencies prevail. The tendency has been for official aid agencies to focus on export crops, though increasingly in recent years on food-crop production for the market. In either case they tend to emphasize increased imports of inputs and equipment to increase production. Here again, the tendency has been to move into whole 'systems' of production, with state or project farms, often associated with processing factories. These have as often increased imports as self-sufficiency.

Class-formation in tropical Africa

If capital-intensive development would have been impossible without the encouragement and funding of foreign aid donors and financial institutions, it certainly does not mean that it has been forced upon unwilling African governments. This has to do both with the ideology of modernization, which they share with aid-donor personnel and with a state-centred process of class-formation.

In most African countries, the state stands at the interface between the country and the outside world, and between the urban and rural areas. This is symbolized by, but not confined to, the control over foreign exchange and monopoly food marketing. The control is no more complete in the one case than in the other – and for similar reasons. In both cases, the state is concerned to extract revenue or increase its control of resources, and in both cases, traders, peasants, its own employees and leaders undermine that control. But, whether or not effective in achieving its formal aims, state control over these interfaces is extremely important as a means for accumulation and class-formation. This is not at all to propose that the state as such is a class. Comprising not only its civil and military employees and the structures within which they operate, but aspects like the monetary and legal systems, the state is structured differently from a class and more akin to the arena in which class-formation takes place. It is not in any sense a neutral arena, however, including not only the ring and the sand but the emperor, and the lions.

With very few exceptions, like Ethiopia's previous feudal regime, the states of Africa came to independence in an economic and political power vacuum. Previous ruling groups had been smashed or subordinated under colonialism and were, in any case, mostly insignificant in size and power in relation to ex-colonial countries, let alone the international capitalist system. In this situation, the state was forced to take a leading role in both 'normal' state functions and the setting up of industries and agricultural ventures which would usually (under capitalism) be left to private sectors. In turn, this put servants of the state in a position to dispose of considerable amounts of resources, much of this together with their counterparts from aid agencies. Moreover, recruitment to state bureaucracies, and thus to positions of power, lay primarily through formal education, an education which, both during and after the colonial periods, laid great stress on the superiority and modernity of things Western and on the task of development being to adopt such features.

The development of state-classes is sometimes depicted as a process of pure 'kleptocracy', unbridled theft from the public purse, and the reasoning or ideology lying behind it as simple fraud and propaganda. This seems to me both to distort history and to involve a dangerous oversimplification – implying, for example, that if one had honest leaders the problems would be solved.

The reality is that the sorts of processes observed can quite easily be (and have been) set in motion by honest leaders whose genuine concern is development and even equitable development. Indeed, given the situation at independence in most cases, it is hard to see what else political leaders could have tried. Except where there were exploitable minerals, overwhelmingly the most important sector of the economy was peasant agriculture. To the extent that a local capitalist class existed, it was often dominated by foreigners and was in any case too small and weak to withstand foreign competition without help from the state. Moreover, given that the peasantry was not, for the most part, subjected to landlords or heavily in thrall to merchants, only the state was available to tap its surplus labour and produce for development. Many if not most of the controls and revenue-raising devices operated by post-independent sub-Saharan African governments had already been imposed during colonial periods and were accepted, especially by civil servants, as natural facts of life. Their very operations and the efforts of peasants and others to evade them tended to give rise to problems which pointed the way to more controls and regulations to solve them. This process also dated from the colonial period, but both before and after independence, once the underlying ideology of modernization had been accepted, every conflict with peasants in the course of implementing unpopular policies merely tended to confirm its initial assumptions – that peasants were 'resistant to change' and needed regulations, control and force to make them accept it.

The urgency of this was aggravated by the fact that, where there were no minerals, virtually all revenue for development purposes would have to come from (or be repaid from) the peasant sector, as the only substantial non-state sector – since tax holidays and subsidies made the industrial sector a user rather than a source of revenue. Peasants are at the best of times not an easy class for states to deal with and extract revenue from. Income taxes are virtually impossible to collect from small, semi-subsistence producers, who usually keep no written records and certainly would not show them to tax officials. Flat-rate poll or hut taxes yield a limited revenue (what can be squeezed from the poorest), do not increase with the level of output, and are almost always intensely unpopular. (A significant part of the mobilization for independence had often been related to their abolition.) This leaves taxation on produce sold and on items purchased by peasants, both of them bearing directly and negatively on the real returns to agricultural production and thus its level. All of this would apply even without a process of class-formation, if the state really was the neutral, development-oriented entity assumed in so many official reports. But there can be no doubt that the effects have been aggravated by a concurrent process of class emergence and formation. The primary characteristic of this is that it is, to a very large extent, a struggle for rent,[4] for privileged access to free or subsidized resources through the state, which can then be disposed of at a profit because the very processes which give privileged access to some generate shortages for others.

The most common method of (individual) accumulation in most African states has been through 'straddling', that is, using funds accruing through state employment or connections for investment in private enterprises both agricultural and other. The ways in which these funds are achieved in the first place range from saving and investing monthly wages,[5] through taking advantage of privileged access through the state, to outright corruption, some of it on a massive scale. Not all of those involved are state employees. Some of the most significant may be politicians and private business-men, but the source of the funds is the peasantry, through state controls. Thus grain monopolies pay low prices to peasants and use the 'surplus' derived therefrom to fund programmes of input distribution or subsidized credit for the well-placed. More directly, they supply cereals to the urban public at subsidized rates, to political cronies for even less, and lower the producer price to cover this. In general, licensing statutory monopolies on different forms of trading, subsidized credit, rations and subsidized distribution systems all provide opportunities for the well-placed to gain privileged (sometimes free) access to resources controlled by the state but originally extracted from the peasantry.

This refers not simply to material resources but also to conditions of production. It is not simply that successful straddlers manage to milk the funds and materials to set up their own businesses. They are further assisted by licensing monopolies which limit competition, access to imports at the official rate, and credit from cheap official sources. While this does result in the setting-up of 'productive' enterprises, the overall effect is not to enhance production and certainly not economically viable production. Investment in a given sector by leaders or the powerful may result in the closing down of large numbers of other enterprises in the same field. The profit in such ventures comes more from monopoly restrictions and subsidies than from productive efficiency. This tends to imply a concentration of investment in the transport and trading sector, where monopoly profits can more easily and rapidly be made and circulated. The problem with this is not that these are 'unproductive' sectors as such, but that the profits on them relate to monopoly and corruption rather than the function of distribution – they derive from worsening, not improving, the supply situation. Finally, this process requires a substantial degree of unproductive reinvestment of funds in the purchase of influence, often accompanied by luxury consumption, not to mention outright bribery. While the process contains a number of superficial likenesses to capitalist growth and is clearly a relation of sorts, there are a number of very significant differences. The most important of these is that surplus is accumulated as a form of rent to privileged access, even when it takes the formal shape of profit. For this reason and because its reproduction is of privileged access rather than capital stock, socially it involves no accumulation.

Because it is basically unproductive, concentration of resources means a direct reduction in the living standards of others, notably peasants. But it also means an increasingly fierce struggle over control of a stagnant or declining material resource-base. One effect of this is political instability and a rotation of elites through coups d'état. And this very instability, together with the licensing and controls which underlie the system, limits the possibilities for any productive investment: encouraging those with funds to diversify into a variety of fields, hold funds in foreign accounts and avoid the long-term tying-up of funds which is essential for investment in expanding production. Even where production does form part of the pattern, its efficiency is reduced by excessive diversification and consequent dilution of management control. A further factor in this respect is mentioned by Sara Berry (1984). With the power of the state both extensive and weak, the importance of kinship, tribal and other patronage relationships increases. But while such persons can more easily be trusted with secret dealings and cash transactions than outsiders, it is often difficult, if not impossible, to exert the sort of labour discipline upon which an efficient productive unit depends.

Among other effects is a weakening of the state qua state, which has led some to minimize its importance as a means for individual accumulation. In an interesting discussion of the 'second economy' (corruption and the informal economy in general) in Zaïre, Janet MacGaffey (1983) refers to this as being 'autonomous from the state'[6] and suggests that 'the use of the state is merely an incidental resource, rather than the primary factor in class-formation'. But to state that fortunes are made by evading state regulations, by no means implies that the regulations are unimportant. If they were not there, there would be no fortunes to be made from privileged evasion, and she makes it clear that political power is a major condition for this. The other side of the matter is that state control most certainly is an important means for getting cheap produce from the peasantry – and by no means least in Zaïre, where Schoepf (1985) and others have shown how forced cultivation is still used to extract commodity production at very low prices. In cases like Zaïre and Amin's Uganda, the state may cease to perform any useful functions, but its control of force remains the backing for surplus extraction from the peasantry.

This mode of accumulation also has great significance for forms of development spending. Even if funds come in the first instance from foreign grants and loans, there are also local costs associated with development projects, while the loans will have to be repaid. This is not private but state money, and the access of the state to surplus extracted from peasants does not (at least in the short run) relate to levels of production, let alone surplus production. It depends on political power, the capacity of the state to squeeze revenue out. As discussed in Chapter 3, the form of the extraction *does* eventually set its own limits, since exactions on the prices of crops sold or products bought lowers both incentives and capacity to produce, reducing the revenue base. But it often takes time for this to show, especially when credit is involved, while the modernization ideology which underlies this form of accumulation points to other reasons for declining production, ranging from peasant conservatism to technical details.[7] In some cases it only shows after several years have been spent in trying to force peasants to increase production with rules and regulations, fines and jail sentences. But during the 1970s, with loans easy to get, development spending was not even limited by a falling revenue base, since it consisted of spending the future revenue to be extracted from peasants. Given this, given the absence of a tight circuit of reproduction, and given the interests and ideology of decision-makers, it is scarcely surprising that local states should have matched foreign firms in their penchant for the expensive, sophisticated and grandiose in development projects.

The ideology of development

It was emphasized earlier that the motivations which underlie state-class formation are not simply private profit from the public purse. Indeed, the cynicism which pervades the public sectors of many sub-Saharan African countries is as much a result as a cause of the problem. Much the same can be said of decision-makers within official aid agencies. While there are cynics and no shortage of self-interested pressures from donor country firms, there still are plenty of people who believe in what they are doing and feel that aid has a major contribution to make to the development process. And among these, there are plenty who involve themselves in just the sorts of project which have aggravated the situation in Africa.

It is thus important to consider the ideology of development (and notably agricultural development) which is held in common by a large proportion of development decision-makers in both developed and Third World countries. It is not just that narrow self-interest is an oversimplification, which fails to explain some aspects of projects which seem to benefit no-one at all. It is also important to understand because, like any effective ideology, it has a number of self-fulfilling aspects, while certain of its most dubious propositions seem so obviously true that they provide a very effective shield against criticism.

At the base of the ideology lies a sharp distinction between 'traditional' society, community, way of doing things etc. and a 'modern', scientific, rational, order of society and pattern of production. This disctinction has its roots in the work of Weber and other sociologists of development and is certainly not a completely invalid distinction. There are a number of differences between what can be called in other terminology pre-capitalist and capitalist modes of production and they do affect the behaviour of those within them. The fault lies rather in making the distinction far too simple and rigid. It is quite true that peasants are motivated by a variety of factors other than 'profit maximization' – and given the importance of various forms of risk avoidance, very sensibly so. But to assume from this that peasants only react from conservatism and 'resistance to change' is sheer nonsense, as the rapid growth of cash production whenever it is profitable makes clear. Indeed, the failure of peasants to adopt innovations from the extension service often relates directly back to state-peasant conflict. State agencies are primarily concerned to increase forms of production which are channelled through official institutions, to yield revenue and foreign exchange for the state. Peasants are concerned so far as possible to direct their energies where such exactions are minimized.

The other side of the ideology is a great faith in the power of modern techniques and scientific methods to solve all problems – though not in the open discussion and application of logical argument, which many would take to be the hallmark of scientific method. For example, almost without exception agricultural extension involves passing knowledge from the top down through, say, ministry of agriculture hierarchies, being progressively simplified on the way and with little or no opportunity for those at lower levels to question (except purely for information) or disagree. The final result, to be passed from the lowest level extension personnel to the peasants, normally has all traces of reasoning removed from it in the process of reducing it to simple commands. Even the agents themselves often have little idea why the advice is being given. Since advice of this nature is often incorrect (there cannot be one correct planting date if the onset of the rains changes every year), or takes no account of other demands on peasant time or land, it is often not accepted with enthusiasm by the peasants (even the extension personnel often ignore it on their own farms). Since the notion of knowledge is hierarchical rather than questioning, this means that the designers of such policies have no means of finding out why their advice is rejected, and reinforces the view that traditionalism and resistance to change are the root of the matter. This again increases pressures for further simplification of the message and the use of force.[8]

The ridiculous aspect of the whole thing is that the 'modern' is supposed to represent the 'rational', as opposed to the 'ascriptive', hierarchical structure of traditional society, yet it is precisely the hierarchical control over modern agricultural knowledge and the preference for orders over explanation and reason which account for the appalling quality of so much extension advice. As can be seen from a variety of the examples of projects in this book, modernization is shot through with absurdities of this sort. This, in turn, relates to what is considered to be 'knowledge'. 'Scientific' knowledge, for example of soil responses to fertilizer, is 'knowledge', even if wrong. Knowing that a particular patch of land is unsuitable for growing groundnuts – possibly without being able to say why – is not.[9]

But there is a real core to modernization. One can hardly doubt the immense power of capitalism as an engine of production in comparison with other means of social organization. Again, science and technology have contributed enormously to this. 'Northern' agriculture has achieved the most staggering increases in production and productivity over the past century and especially the last forty years. Therefore if one wishes to increase agricultural production, this seems the obvious pattern to emulate – or does it?

No-one doubts the achievements in terms of production – after all the surpluses it has generated are the cause of huge problems on the world market. But it has expelled staggering numbers of people, taking agricultural populations from over half the total down to a few per cent in not much more than a century – and the most rapid advance has occurred after most of them had already left. It has also radically altered energy balances, making agriculture a large net user of energy rather than a positive contributor. At the same time, agriculture has become one of the most highly capital-intensive sectors in many developed countries. Moreover, much of this has been achieved under heavy subsidies, without which many of the innovations would not have been profitable and certainly not in their present form.

So there are very serious doubts whether current modern agriculture is a useful model for emulation in Africa, and these doubts do not stem from any romantic attachement to the 'traditional'. But they only begin to arise when one puts the various aspects of the system together, and unfortunately the last places one would expect this to be done are agricultural colleges, universities, research institutes and ministries. Certainly there are likely to be courses in the first two, attacking the CAP and protection in general (this goes without saying if the staff includes economists). But there tends to be very little on the inappropriateness of methods developed under heavy protection, high wages and well-developed infrastructure, to places with heavy taxation, low wages and poor infrastructure. The problem is that 'technical' subjects are taught as if they were solely technical. Even if social scientists, including economists, are allowed a few critical remarks, these do not penetrate the 'mainstream'. The same is still more true of ministries of agriculture and research institutes, where price relations, if considered at all, are generally taken as given and not related to policies. In all cases a further tendency is to take 'background costs' (like those of expanding the extension service or improving communications) for granted, since they will usually not be assigned to the budget of the innovation being considered. None of this is confined to agricultural specialists, and the above are simply examples of how a widespread process operates in that sector. At least part of the problem is the atomization of knowledge which derives from specialization, though the refusal to listen to alternative views indicates other factors as well – notably the sectoral loyalty referred to in the previous chapter.

This whole issue is much easier to discuss at a more concrete level, in relation to aid projects, for example. But it may be useful to summarize some of the main points. The notion of development through modernization involves the implantation of innovations and modern techniques, structures and patterns from the outside, through agents of change at the interface between the modern world and the traditional. It is envisaged that the superior power of these is such that it will simply erase the traditional patterns of production, culture etc. It is seldom considered that the latter have anything to offer themselves. Even where social scientists are employed to find out about the existing system, this is more often with a view to eliminating 'barriers to change' than considering if there is anything useful in the existing system which could be incorporated, let alone change the conclusions.

The scientific knowledge involved is a very *partial* knowledge, very much bound up with technical experiments in laboratories and research stations and heavily focused on purchased inputs and equipment. True enough these imply changes in patterns of labour input (though these are often not considered), but it is much more rare to find programmes which consider the whole farming (let alone socio-cultural) system with a view to considering what might be most useful *in that specific case*.

But here one returns to the interest component of the ideology. Aid is in part a form of export promotion, and the materials of which it is composed do form an important aspect of state-class accumulation. From both donor and recipient sides funds are being taken from taxpayers and spent on subsidizing development projects which benefit specific groups at either end to the cost of many more.

Aid projects as a paradigm of perverse development

Official aid projects involve the combination of funds, materials and expertise, transferred from developed to developing countries via a variety of official bureaucracies (and sometimes private, firms) for the achievement of certain desired 'developmental' ends. In most cases this involves trying to increase production of something or other, either directly or through training and education, disease control, or provision of other infrastructure. A high proportion of total aid expenditures go on large projects, for dams and hydro-electricity, roads and railways, mining, factories and large-scale farms. To some extent this is inevitable. Railways and tarred roads are expensive whichever way one looks at it, and there is little point in building only a part of one. *If* it is worthwhile to build large dams and hydro-electric schemes the same applies, as often to factories and mines. But by no means all is inevitable or helpful. There are plenty of cases of roads being up-graded at huge expense, when the traffic density warrants nothing of the sort, or of the redoing of several hundred kilometre stretches, when the real problems concern a few kilometres of trouble-spots. Similarly there are more and more examples of large dams which, even if they do ever fill, are very wasteful and divert funds which could far better have been used on improving rainfed agriculture.[10] The number of factories which would never have been built if there had been a realistic (and honest) economic assessment probably runs into hundreds, if not thousands, and the same applies to large-scale farms of the state and project type. Precisely what proportion of total aid expenditures has gone on white elephants and 'cathedrals in the desert' is not possible to say. But that it is substantial and has added significantly to the African debt burden is beyond doubt.

One of the reasons it is not possible to measure precisely (apart from disagreement over what constitutes a white elephant) is that pressures and processes generating waste and lowered effectiveness do not apply only to the most grandiose projects. They are also built into a large number of other 'curate's eggs', which have useful and helpful features. Since the mid-1970s, increasing numbers of donors have been including in their project statements references to the fulfillment of basic needs, assisting the poor, women and other disadvantaged groups, preserving the environment and other laudable aims. And in a (smaller) number of cases, agencies and their personnel have made real efforts to achieve such aims. But those who have made the most genuine efforts would be the first to admit their inadequacy and to point to aspects of the aid relationship which make the task harder.

One of the most insidious and difficult to deal with is 'production-orientation'. In the first place, most agricultural projects are initially suggested by agencies or interests concerned with increasing production of one or other crop or livestock product. Much of the initial survey thus focuses upon the means for achieving that purpose, normally based on information from ministry of agriculture or livestock research institutions. This will almost invariably focus upon larger than usual 'progressive' farmers, because they are easier to work with (speak English or French, for example), have the capacity to purchase the required inputs, equipment and livestock, and are primarily 'economic operators' (i.e. their subsistence needs do not take much time, space and other resources or 'get in the way of' improvement). This is perfectly natural for a ministry of agriculture institution, which is concerned with increasing production from the most efficient units – and if there is a ministry of agriculture in Africa which does not focus on this, I have yet to come across it.[11] In some cases, social scientists may have been involved in the initial process of drawing up the project, though usually in a subordinate capacity or off to one side. If they make too much noise about things like basic needs and concentrating on the poor, they will rapidly make themselves very unpopular.[12] Even remarks on the significance of social relations and structures are often considered irrelevant to what is seen as a technical problem of reducing tick-infestation, increasing rice yields, dealing with maize stalk-borer or introducing a new crop variety, regardless of the fact that social organization is highly relevant to every one of these and that taking it into account can mean the difference between success and failure. In other instances, the project has been almost entirely developed by technical personnel, in which case the sole reference to the various laudable aims is likely to be in the overall preamble to the project. By the time one comes to the practical aims and methods of achievement, this will have been replaced by output targets.

At some point in the proceedings, the project will be assessed by a 'mission' from the donor agency, whose main task is to come up with clear recommendations (go ahead, go ahead with clearly specified changes, do not proceed) in a period of, say, three weeks. Of this, at least one week will be spent on courtesy calls and discussions with government or aid-agency officials in the capital city, another week on writing up the report, and the third week in visiting the 'project area', much of which will be taken up with further courtesy calls and discussions with regional authorities. In some cases, it will include arranged visits to groups of 'intended beneficiaries' who have been primed to say how much they are looking forward to the project, or how well it is working, if already in operation.[13] Even if the consultants are concerned to give the project a poverty-, women- or environmental-orientation (and this is not always), there are limits to what they can do in the time available, given that most of the details have already been worked out. There certainly are ways in which 'development tourists', as Robert Chambers calls them, can significantly improve their work and impact, most especially on disadvantaged groups, and a number of them are well set out in his book *Putting the Last First* (1983). But even to get this far requires some willingness on the part of donor and recipient bureaucracies, and this can certainly not always be taken for granted. More often consultants are forced into 'damage-limitation', trying to get rid of the worst aspects of a project to avoid its becoming a total (and publicized) disaster. One problem with this is that in trying to solve technical and economic problems, those connected to distribution usually fly out of the window. In any case, the environment is often unconducive to their mention.[14] Even if a project should be turned down, there will be a line of other donors waiting to fund it.[15]

This relates to another, less-mentioned problem of official aid, namely, that it is often difficult to get rid of the money within the budget period, and this also gives a bias towards wasteful and grandiose spending. It also leads to the 'year-end' or 'Christmas present' phenomenon, in which, as the end of the financial year approaches, virtually any spending is approved, since it is necessary to spend all of the funds by that time. If the funds are not spent, not only has the branch failed to fulfill its job – and can be criticized from higher-up for this – but next year's allocation of funds is likely to be cut. One might argue that if the funds cannot be spent, they might as well be cut, but from the agency (and recipient country) point of view this is out of the question. A certain transfer of funds has been agreed and must be maintained. If headquarters does cut the budget next year, one can be quite sure that they will cut in the wrong place, leading to shortages under some headings. Since calling for more funds during the financial period is frowned upon (bad forward planning), it makes sense (from this viewpoint) to have a bit to spare – and then find something to spend it on. For the recipient a certain transfer of funds or materials has been agreed upon, and should be transferred regardless of when. This conflicts with the donor setting of a 'rate of disbursal', requiring that funds be spent within the financial year. There may be perfectly genuine reasons why a fixed rate of disbursal does not suit the recipient agencies (though there may also be lengthy bureaucratic delays). But whatever the reasons, this does imply pressure on donor agencies to get the funds spent within a given financial period. The results include a preference for projects with a high proportion of material transfers in relation to organization and planning, hasty administration to get the funds out in time and, if that fails to clear the budget, spending on projects which would otherwise not be accepted.

The above refers primarily not to the largest projects, but to medium-scale or even small ones. It also refers to projects where the material supply element is not enormous, and where direct pressures from commercial interests are largely absent. But even in these cases, prior project definition often makes it very difficult even to see how basic needs, the specific needs of women, or environmental factors can be included. How does one 'improve the situation of women' in a project set up to give credit and other assistance to male-dominated co-operatives? How can one 'take environmental factors into account', when the project is designed around provision of fertilizers, to monocropped hybrid maize? How can one 'improve basic needs' through agricultural improvement when the extension service is only concerned with 'progressive farmers'? The answer is the same in each case – that one can only do it through completely different sorts of project.

But if rigid production-oriented definition and technicization are the actual processes involved, they clearly relate back to the pressures and ideologies mentioned earlier.

There is no avoiding the fact that aid is partly an exercise in export promotion by donors. All the signs are that with increasing strain on world markets and increased protection all round, the importance of this aspect is increasing, as is the incidence of aid-tying. Most pernicious is the combination of aid-tying and subsidized credit. Not only does this give considerable scope for overcharging, as cited above, it also gives donor firms an enormous advantage over those from the recipient country. Coughlin *et al.* (1987) give a number of examples from Kenya, where manufactured goods were supplied on aid contracts by donor firms, when local firms could have supplied them competitively. What the local firms could not compete with was heavily subsidized long-term credit. In a number of cases, so important was the project market that donor preferences threatened to bankrupt the local firms.[16] The combination of aid-tying with grants and subsidized loans generates three problems: it gives enormous scope for overcharging (especially on plant and equipment for which market prices are difficult to assess); it gives huge preference to donor firms, even in relation to local producers; and it strengthens the tendency to increase the import- and capital-intensity of projects. One might suppose that the IMF and World Bank, with their well-known aversion to protection when imposed by Third World states, might have objected to this sort of practice. If so, one would be wrong.[17]

Again, however, one must stress that this is not a matter of the Western countries exploiting the Third World. It is rather a matter of powerful interests within both combining to exploit the rest. It is true that labour unions in the West can sometimes be bought into uncritical support for aid as export promotion, because it holds out the promise of jobs, but this does not mean that this represents the interests of more than a small proportion of workers. It is also probably true that a large proportion of the African population thinks of aid as a 'good thing' – at least when confronted with the decision whether to accept a given grant or not. Still more emphatically, this does not mean that aid of the present sort is in their long-term interest. But in fact the majority of the population has no more say in the definition of aid projects on the one side than the other. In both cases most of the information is 'confidential' and kept from the public. Moreover, because of its daunting complexity and the deadly tedium of its style, few have the stamina to delve deeply into the official material, while many of the most dubious assumptions are hidden in minute footnotes to sentences in the middle of, say, page 382 or Annex XVII.[18]

To summarize, one can encapsulate the aid relationship as both central to and indicative of the processes which have generated increased indebtedness, poverty and hunger in Africa.

The following section provides relatively brief notes on a number of issues which do not fit within the framework of the chapter up to now, many of which pose unresolved problems.

Economics versus poverty-orientation

The nations of Africa are heavily, and some inextricably, indebted. Apart from the negative effects on the population, this means that in many cases states are breaking down under the dual strain of this and unproductive accumulation. Whatever the nature of the particular state, it is obvious enough that a good proportion of its policy will be directed towards trying to stave off this breakdown, whether militarily or through attempts to maintain flows of revenue and the services provided from them. This itself is likely to be a highly ambiguous process. Short-term economic survival depends on access to further credit and the key to this is acceptance of IMF conditionality, with its emphasis on devaluation, exports, privatization and lowering trade barriers. Since all of these are policies which tend to reduce the control of the state, they are likely to be resisted and evaded wherever possible. In neither case will the well-being of the poorest sections of the population be a major issue, and this applies especially to the rural population which is out of sight and out of mind and poorly organized politically. All the same, there are issues at stake which do affect their long-term interests.

IMF conditionality. Throughout this book, there has been a certain ambiguity towards IMF conditionality. On the one hand, it does seem necessary that some of the over-spending of the 1970s should have been curbed. But there are a number of disturbing features about the way it has been done.

In the first place, the IMF has acted throughout as agent for the creditors, seeing its main job as protecting their 'investments', whatever the cost to be imposed on the debtors. This is natural enough for a banking institution, especially given the fears of a collapse of the international financial system. Considered as a 'development institution', it makes little sense. Moreover, it contrasts strongly with its treatment of developed countries, notably the USA, where there has been no strong presssure to devalue a heavily overvalued currency. Again, the rigid macro-conditionality applied by the Fund takes no account of the particular situations of different countries, ignoring factors which may lead it to make things worse, even in its own terms. Its bias towards export orientation, and optimism as to the possibility of exporting one's way out of debt, makes the whole strategy dubious in economic terms, though predictable enough given the interests represented.

Where subsidies are concerned, however, it is hard to be so critical. Food subsidies have sometimes had a major and positive impact on the incomes of lower-income urban dwellers. Sometimes they have mainly served the better-placed and formally employed, leaving the poorest to purchase on black-markets at prices further inflated by subsidy-derived shortages. To the extent that subsidies to consumers have come from reduced producer prices, this has aggravated the above. It may be that in some cases getting rid of subsidies is actually helpful to the urban poor. On the other hand, since the livelihoods of large numbers of people are at stake, this requires consideration of each particular case rather than rigid across-the-board demands. On subsidies to agricultural inputs, I am heartily in agreement with the IMF, since these lead to excessive use by (or speculative profits to) those who have the influence to get hold of materials at the official price, reducing their availability to others or forcing them to purchase on black markets at higher prices. My only reservation here is whether this will in fact suffice to eliminate the indirect subsidies and cross-subsidies which creep into donor programmes when the materials transferred are moving more slowly than intended. On the more general issue of devaluation (which can be seen as eliminating a foreign-exchange subsidy to those with access at the official rate), the important issue is less the rights and wrongs in principle of a given exchange rate than the pragmatic issue of what happens with failure to devalue. Controlled exchange rates and rationing of foreign exchange usually lead to flourishing black markets and severe shortages of goods, partly because of the preference given to 'development imports' but partly due to corruption in allocation of foreign exchange. It is true that

black-market rates often reflect factors other than simple demand for commodities –
movement of capital out of the country by minority groups, for example. But some
degree of devaluation is required in many cases, simply to get the official market functioning again. Other points of ambiguity are protection, and monopoly grain marketing,
which are considered at slightly greater length below, followed by more general points
about economics and food security.

Protection and industry. Any policy to maintain, let alone improve, the livelihood of the
most vulnerable sections of the African population must centre around keeping as
many as possible on the land, while providing improved conditions for their survival
and reproduction. But this certainly does not imply dropping industrial and non-
agricultural development Whatever happens, huge numbers of people are likely to be
squeezed out of agriculture in the coming decades. It is therefore imperative to
develop jobs for them. Whatever the particular strategy for this, it would seem to
imply some degree of import substitution, focused on lower levels of capital intensity
than previously, and this clearly involves some degree of protection. Part of this is
simply to counter the very effective protection embodied in aid projects and other
developed-country export subsidies. But apart from that, industries in small African
countries, producing for limited markets, with low levels of infrastructure and limited
numbers of trained and skilled personnel, are almost inevitably going to be higher-cost
and produce rougher quality products than those of the developed countries or the
major south-east Asian producers. No-one denies that this involves costs to the
remainder of the society, and there is clearly a problem of weighing-up what level of
increased costs should be borne to achieve what results. On the one hand, to develop
an effective industrial sector is a learning process which takes time. On the other
hand, IMF free-traders have a point when they say that excessive protection relieves
its beneficiaries of the need to go through that learning process. The problem is
further complicated by the fact that 'policy decisions' over such matters are often not
simply a matter of social optimization. If, for example, important leaders have interests in import-substitution industries, pressures for both tariff protection against
imports and licensing protection against local competitors are likely to be strong. If, in
addition, such leaders find it profitable to increase the import content of production,[19]
then the country gets the worst of all worlds. The situation does not have to be as crude
as the above for political interests of various sorts to muddy the waters. All told,
however, it seems clear that some degree of protection is a necessity if African industrial sectors are to be maintained, even on a slimmed-down basis.

Cereal monopoly marketing. Throughout most of this book, I have been highly critical
of state monopoly grain marketing agencies, and I do think they have caused considerable harm. But it is simply unrealistic to expect African states to relinquish control of
food marketing more than they are actually forced to. In the first place, staple food is
simply the most important and most political commodity traded in most African
countries. It may not enter the national income statistics as such, but it accounts for
well over half the total consumption of the majority of the population and is, for them,
quite literally a matter of life and death. Any government which fails to provide at
least minimally for urban food needs is not likely to last in power long, and certainly
not unless it can be seen to be doing its best to provide. At the same time, the power of
the state, in the sense of control over revenue surplus, derives from its position at the
interface between the agricultural sector and the towns, and to give this up would
involve significant loss of control.
       The above are reasons why state monopoly marketing in one or other form probably

will continue. But there is also a separate reason why – at least in some form – it *should*. This is because private markets tend to move food *away from* rather than towards areas of shortage when there is no purchasing power there. Given that the danger of local famines and severe food shortages is considerable in most African countries, it is clear that agencies are needed which will move food into such areas in time of need. One could argue that this should more properly be some form of 'relief and rehabilitation agency' than an overall grain marketing monopoly, but unless such an agency is to be heavily dependent on food aid, it needs some means of access to locally produced cereals. There are at least two good reasons for not getting too heavily dependent on food aid. The first can be illustrated by Ethiopia in 1984 – the risk that international agencies will mess around for months on end while people starve. The second is that it makes little sense to import grain and reduce local prices if supplies can be purchased locally. It is true that to accumulate sufficient reserves for such situations does not necessarily require an overall monopoly. But it does require a state agency with the funds to purchase considerable amounts of grain. It also requires that storage facilities and a transport fleet are available. These are all costly activities and there is every likelihood that their budgetary cost would actually increase if divorced from monopoly marketing, since apart from idle capacity (trucks and stores), there would be no 'normal marketing' from which to subsidize them.[20] Given the nature of most such agencies, and the fact that they tend to perform such famine relief as they undertake poorly, there is no easy answer to this question. But that includes the apparently easy answer of the IMF and World Bank – privatization.

Land tenure is a particularly difficult issue. On the one hand, security of tenure is of vital importance if peasant families are to be encouraged to undertake long-term oriented intensification of production methods. It is far from clear that individual private tenure actually provides such assurance to individuals, though it *is* among the better assurances against mass movements of peasants, as in villagization and settlement programmes.

One of the more common reasons given for preferring individual private tenure is that makes it easier to provide credit for farm improvement. This seems to me a highly dubious argument. Historically one of the major processes by which peasants have been expelled from their land is foreclosure on mortgages. This is not widespread in those parts of Africa where I have worked, as yet, but this is precisely because in most cases land is not used as security, even where there is formal private tenure. The result of this has been a large number of bankrupt credit projects and co-operatives, but, weighing the two problems, this seems preferable to large-scale dispossession of peasants from their land.[21] While state and co-operative credit projects may not have had this effect, however, there do seem to be cases where peasants have sold privatized land to repay private debts, or even to maintain consumption standards.[22]

Another related problem with the registration of land titles is that it usually involves losses of rights to women. Traditional tenure systems often allowed for multiple rights in land, so that while a husband or patriarch retained overall control, certain parts of the land were allocated to women. With the transition to private property, these rights are usually lost in the assignation of the whole plot to the male 'head of household'. Not only does this reduce women's capacity to decide on the allocation of land between crops, it also means that a husband can pledge (and perhaps lose) land which would previously have been regarded as his wife's. It also means that women have little or no access to credit when it is related to property mortgages.

My own reason for favouring secure tenure is somewhat different. If African peasantries are to intensify cultivation in ways which allow permanent increases in land productivity without draining the soil, this will require considerable labour input into the development of husbandry systems. Much of this work will not yield short-term returns, and will certainly not compete in market terms with short-term improvements from application of fertilizers and other inputs. It is therefore most unlikely to be undertaken as part of a market strategy. The only motivation which I can see as having any chance of generating such change is sufficient security of tenure so that the land can be seen not simply as the living of the present generation but as the means of continuing the family in succeeding generations. The market-oriented would doubtless throw up their hands in horror at the implications, which are that land tenure should be secure but not disposable by sale, since this prevents the allocation of land to the most efficient users through time-honoured capitalist methods. But this in turn prompts the question 'efficient for what?'. Does the concentration of land in the hands of those with the resources and commercial outlook to develop it generate the best solution? Considered from the viewpont of food security, the answer seems definitely to be no.

Economic optimization versus food security. This poses an even harder question, raised initially at the beginning of the book. How does one combine economic decision-making with a concern for the food (or other) security of the most vulnerable, when they seem often to point in opposite directions? Economic optimization is concerned with aggregates without regard to internal distribution, and even when specific distri-butional 'constraints' are included in economic plans, they have a tendency to fall out along the way. One can see how this operates in project definition, where concern for the smaller poorer farmers can be inserted in project aims and yet still be brushed aside as hampering the maximization of production (though this itself may not be economic). Similarly, one can see how this has directed EEC agricultural policy into support only for 'viable' farms (though the CAP is definitely not economic). Given the considerable degree of waste in development spending, there is clearly some need for economic control, but the sort that arises seems often both to exclude the poor and to involve subsidies and other wasteful elements. Even when one has become used to it, there is still something amazing about the sleight of hand with which economic analy-sis is used to justify grossly wasteful development projects and programmes like the CAP, but to brush off any suggestion of using funds to benefit the poor on the grounds that it is 'uneconomic' and 'non-productive'.

This relates to another aspect of economic analysis as currently applied to African agriculture (or the CAP again), a strong *bias towards the quantifiable*. Sacks of wheat, maize, coffee and pyrethrum can all be counted and valued (though in the EEC case, that valuation is inflated by large amounts of subsidy). Quality of diet, security of income and food supply, fitness and health are much more difficult to quantify and attempts to do so often distort and under-rate them. Often they are left unquantified and thus disregarded in economic calculations. This is not just a matter of economic method and technicalities, it reflects the underlying values of capitalism as expressed through the discipline closest to its heart (or pocket). So one sees increased production by rich farmers hailed as a solution to the hunger of the poor, without considering where they will get the income to buy it. Similarly one finds self-congratulation on the CAP's 'success' in increasing productivity, even though this costs EEC citizens vast amounts in high prices and subsidies to get rid of the surpluses. If the irrigation project cited in the previous chapter increases both rice and the incidence of bilharzia, the former will be attributed to the project as a success. The latter will not even be connected to it in calculations of return or 'social cost-benefit'. And even if by some chance it were to be, this would refer only to the days' work (and thus rice production) lost, or the costs of curative treatment, without mention of the misery of feeling sick and weak much of the time. Even to mention the latter sort of thing is to invite accusations of sentimentality and soft-headedness in the quarters where the 'hard' decisions are taken.

One cannot deny the immense power of such values, however, and the enormous difficulty of pursuing any policy which runs against the stream of short-term individual self-interest. And this concerns not just decision-makers and the rich and powerful, but most of mankind when in the role of producers and consumers of commodities, that is, when operating as individuals facing the market. One can expect different behaviour from people operating in groups, as kinsfolk, friends or sharers of political views. But these activities involve the minority of people's time, and a diminishing share with the penetration of commodity production. This seems fairly clear in relation to atomized individuals in the consumer cultures of the West. But also where African peasants are concerned, it would appear that redistributive mechanisms which previously provided some food security have to a considerable extent broken down.

This seems as good a place as any to insert brief conclusions on the EEC.

EEC policies and the African crisis

In the course of this chapter, the EEC has largely dropped out of sight as a causative factor. In part, this is because, with the focus on Africa, the EEC recedes into being just one of many external factors. But in part it also reflects the fact that, considered from Africa, there really is no great difference between the EEC and other external agencies. It is true that the EEC has a greater effect on world cereal market prices than other producers, not because of the size of its surplus production, which is far smaller than that of the USA, but because of the means of subsidizing its exports down to whatever the world market price is, thus further lowering it. But although EEC grain surpluses and food aid are important factors, the really important factors generating hunger seem to lie outside the food sector as such, in the patterns of aid, export promotion, development expenditure and class formation which have contributed to the general crisis in Africa and to the dual process of concentration and marginalization in agriculture. CAP agriculture, with its very high subsidies, provides a particularly inappropriate model for emulation. But there is no clear evidence that other donors (non-EEC members) are ahead in terms of supplying appropriate technology.

In this respect there seems little to choose between the EEC and other Western powers and donors. Given the enormous pressures towards high cost, capital intensity and system dependence in almost all aid programmes, the fact that the EEC is (perhaps) slower to deliver than some others loses much of its significance. EDF aid is no more and no less tied than that of most other donors, and like the others the real extent of tying is higher than formally. EDF aid may have resulted in some unusually gross white elephants, like the Diama and Manantali dams, but during the Pisani period, it also had one of the most serious and questioning directors of any Western aid programme, who tried to press for a significant change in direction. Doubtless many of the decisions on large projects are to a considerable extent ruled by cynicism and national rivalry, but then so are many of those in bilateral agencies concerned with national interests. At the other end of the scale, some of the better aid and projects, from both the EEC and other donors come from the 'miniproject' or equivalent funding, primarily through NGOs. For a variety of reasons, moreover, the EEC has been among the donors which have channelled most funds through NGOs. The major problems for Africa lie with the system of which the EEC is a part, not with the EEC as such. For EEC citizens, the problems of the CAP are considerably more central, but that is another story.

What are the possibilities for change?

It would be pleasant to finish this book with an uplifting conclusion full of certainties about the way ahead. But then it would have been a very different book (and by a different author). My main optimism lies in the uncertainty that things are as bad as they seem to be and the hope that the possibilities may be better than they appear. Thus, while this section discusses a number of proposals which have been made, this is more to indicate their limitations than to endorse any particular line.

Further integration into the international capitalist system and exporting out of debt does not seem a very hopeful solution. There is no sign of any massive expansion in world markets for products exportable by Africa; rather the reverse, given the trends in terms of trade during the 1980s. The long-term prospects seem, if anything, worse, with bio-technology and other forms of developed country import substitution likely to be more effective (and more heavily protected) than anything Africa can offer. Various authors have been calling for years for 'massive transfers' of funds to Africa and the Third World, though the consistent lack of response to that cry has somewhat dampened their enthusiasm. While every few months sees the announcement of another plan for the World Bank, IMF or other agency to step in and solve the problem, the amounts involved are normally trivial in comparison with the size of the debt. In any case, since one of the major purposes of the exercise is to maintain Africa's capacity to import, most of such funds would be as closely tied to materials from the West as those which generated the problem. The chances for debt renegotiation seem rather better, if for no other reason than that many African countries have no hope whatever of keeping up their debt-service payments. So long as the international financial system survives this and other much larger Third World (and US) debt intact, one can expect the renegotiation to take forms which bind the debtors quite closely into the system and change rather little. If the whole system breaks down, the consequences are far beyond my capacity to predict.

What then are the chances for African countries breaking out of the system and 'going it alone'? Rather thin, to put it plainly, and most unlikely to be attempted other than in rhetoric. The countries of Africa are certainly no less dependent on the West economically than they were at independence – and in many ways far more so. They may produce rather more consumer goods. But demand has also increased enormously, and there has been a massive increase in imports of capital and intermediate goods for use in import-substitution industries. Without continued flows of these a large proportion of local production ceases, as has indeed happened already in a number of cases.

Moreover, to talk of the economic possibilities for breaking away ignores the underlying political factors. Those in charge of most African states may be willing to tighten the belts of others under IMF conditionality. There is no sign that they are willing to tighten their own belts very much more, for the uncertainties and privations of economic autarchy. I have not used the term 'comprador bourgeoisie' to describe the 'state-class' because it seems to me inaccurate and an oversimplification in several respects. But there is no doubt of the international linkages and dependency of the ruling groups in most of sub-Saharan Africa. The likelihood of their leading any major breakaway from the international system is, to put it mildly, slim.

What then of socialism and a breakaway under new leaders, either to independence or to dependence on the USSR? Until he was murdered in Autumn 1987, Sankara's military socialist regime in Burkina Faso was the most recent leading contestant for independent socialism as the answer. While there seemed to be attractive and sensible features of the regime, I am not qualified to assess its potentialities. Clearly it was vulnerable to military rivalries and would seem likely to continue vulnerable to rigidities inherent in the military outlook. Otherwise, socialism seems rather on the retreat as Tanzania and Mozambique unwillingly accept conditionality for lack of funds.

In any case, these forms of socialism did not appear to provide a solution. Tanzania's food situation is probably better than that of most African countries, though this has as much to do with relatively low population densities and varied climate as with anything else. The socialist element in villagization dropped out in the early 1970s, leaving a policy of nucleation, which had generally negative effects on agricultural production and incomes (assisted in the latter by high marketing board margins). Certainly Tanzania was not successful in supplying the towns with foodstuffs or the mass of the population with basic consumer goods. The more 'Marxist' forms of socialism, in Ethiopia, Mozambique and Angola, seem, if anything, to have fared worse. In the latter two cases the situation has been made impossible by destabilization and direct attacks from South Africa, while Ethiopia has also been fairly constantly involved in civil wars. But, in addition to this, soviet-style planning seems to have transferred the Soviet genius for producing enormous agricultural deficits. Even if more regimes of this sort do arise, they do not seem to be a solution to the problem of food security.

Any socialist regime in Africa, as elsewhere, has to run the gauntlet of US, IMF, and other external disapproval, and withdrawal of funds and credit. In many cases, this is also accompanied by destabilization, forcing the regime to allocate more funds to the military, thus not only diverting funds from productive purposes but accentuating any authoritarian tendencies it may have. But even without this, socialist states in Africa, and especially those closest to the USSR, have adopted grandiose modernization schemes to rival the worst thought up by Western-oriented regimes and their aid donors. Moreover, the more simplistic forms of Marxism generate a dismissive attitude to peasants which leads to schemes for their forceful mobilization and intended transformation into modern agricultural co-operators, villagers and the like. Most of these schemes have had negligible or negative effects on agricultural production, while doing much both to waste state energies and funds and to generate opposition. In doing so, they have often squandered large reserves of goodwill, enthusiasm and willingness to work with which they started out.

Peasant-based politics. If most of those vulnerable to food shortage are peasants, is not one possibility a peasant-based politics? And if so, of what would it consist? I have seen no clearly articulated version of what this would imply, but one can guess at it from the writings of peasant sympathizers. One aspect would be not unlike the IMF scenario, with freeing of markets from state monopoly and a general reduction in exactions and regulations from the state. Another would be a shift of agricultural policy and aid projects away from rich peasants and capital intensity to appropriate technology and concern for the disadvantaged and the environment. One could add features, but these will suffice to show the problems. Not only is it difficult to see an effective political basis for either of these policies, but they are to a large extent inconsistent with one another. The first is basically a 'rich peasant' policy of letting the market rip and the devil take the hindmost. Undoubtedly it would find support from many peasants, but it is hard to see what sort of ruling class in charge of the state would adopt such a policy. It is also hard to see how it would solve problems of food security. The second is a more utopian, if more attractive, 'basic needs' policy, likely to find more adherents among NGO radicals and some poor peasants than anywhere else. Its likelihood of achieving an effective political base is less than slim.

There are distinct limitations to the notion of peasant politics in any case. The term 'peasant' spans a wide variety of small-scale mostly agricultural producers, having in common that they produce both for subsistence and for sale and that they are politically subordinate to other classes. Otherwise, they range from well-to-do rich peasants, hiring labour, possibly owning agricultural machinery and selling a substantial amount of produce, down to small peasants scratching a bare living or less from inadequate plots and supplementing this with casual labour as and when they can get it. Although middle and poor peasants form the large majority, this does not imply any effective political division between them and rich peasants. Peasant political structures and social networks in much of Africa are characterized by kinship, patron-client and other vertical relationships which hold particular sections of the poor peasants closer to their kin or patrons among the rich than to other poor peasants. The likelihood of mass peasant movements is still further reduced by differences of culture and language between different local peasantries.

During the 1970s, there was a tendency among Marxist observers to characterize African peasants as a 'concealed proletariat'. This related to an analysis in which subsistence production was seen as subsidizing the wages of migrant labourers, but also involved characterization of peasant income from sale of cash crops as a concealed wage. Except in a small minority of cases where peasants on highly controlled export-crop settlement schemes really did receive incomes in a form closely resembling wages,[23] this argument has always seemed unconvincing to me, since there are significant differences between the forms of both producer income and surplus extraction. Moreover, it glides over one crucial point in the Marxist analysis of capitalist production, that the proletarian is 'doubly free', that is, not forced to work by direct political means and at the same time dispossessed of the means of producing subsistence other than by engaging in wage labour. Neither of these is true of the peasant in Africa, who both retains at least some land and is not infrequently subject to forced labour (for 'nation-building', 'self-reliance' etc.) or to the enforcement of agricultural practices and minimum acreages of crops. This argument was related to a conception of 'worker-peasant alliance' which would form the basis for socialism. Given the not inconsiderable differences of interest over factors like urban food subsidies and peasant producer prices, this seemed even less convincing.[24]

In more recent Marxist analyses peasants are more commonly referred to as 'petty commodity producers' or even 'petty bourgeois', which continues the tradition of trying to find some more 'analytical' term than 'peasant'. While many of the analyses are interesting and useful, the search for a term which fits better with Marx's analysis of capitalism seems to me misplaced. 'Peasant' may be an imprecise descriptive category, but this is because the reality to which it refers is also diverse and hard to place within the analysis of capitalism, at least in the traditional categories. The above two terms are preferable to 'concealed proletarian' in that they refer to sale of commodities rather than labour and allow for the political ambiguity of peasantries. But neither captures the political subordination which seems to me an essential element of peasantry. 'Petty bourgeois' suffers from the further problems that it derives from words meaning 'small townsman' and that it is also the category to which many Marxists assign the state-class. A class analysis in which the vast majority of predominantly rural societies, including their ruling groups, are referred to as 'petty bourgeois' seems not to take one very far.

In one respect, however, it supports the conclusion which I would draw from the political subordination of peasantries and from the nature of their political structures – that the peasantry as such is unlikely to form the basis for a major change in political structure. While most peasants would prefer a reduction in state revenue extraction and controls, they follow the leadership of larger farmers and patrons in real political manoeuvring, which implies an ambiguous policy of pressing for a general reduction in controls and taxes, combined with efforts to secure privileged access to resources for their own patron-client network as against others. In short, they are drawn into political processes which are really for the benefit of more powerful others, rather in the way that holes in developed country tax laws allow wage-earners and small business operators the *feeling* that they are getting something for nothing, while in reality they are paying more to make up for the really substantial tax avoidance of huge corporations.

The above is a summary account of peasant 'civil' or non-revolutionary politics. Quite different patterns can emerge in extremes of subordination and desperation. But even here, apart from localized insurrections and jacqueries, peasants seeking to overthrow an existing political system or leadership normally remain subordinate to other groups which are more effectively integrated at national level. In summary, this implies that the chances of peasants assuming political power are slim, while the most likely form of state to be based on a peasant political base would be not unlike what already exists, with a combination of populist rhetoric and manipulation of networks for sectional interests.

To return to external factors, if the chances of changing official aid are limited, do not NGOs offer a better hope? In some cases, yes, they do – but not always. The term NGO spans an enormous spread of opinions and approaches, the division into religious and secular being one of the less important. One finds a range from thoroughgoing reactionaries (religious and secular) to some of the most useful and positive initiatives for increasing long-term food security (again both religious and secular). Thus one cannot be certain that increased channelling of aid through NGOs will support the more progressive, especially if this implies channelling more official funds through them, as has been a tendency in recent years. In the first place, this tends to favour the least critical and outspoken, but at the same time it can place pressures on others to tone down their approach to make it more agreeable to the official mind. Apart from this, simply running the gauntlet of the daunting bureaucratic procedure of official aid agencies can waste enormous amounts of time and lead to funds being used in ways which fit in with official categories even though they reduce project effectiveness. In short, NGOs which accept official aid funds suffer the same sorts of dependency as do recipient states. At the same time, channelling aid through NGOs does offer the chance of going outside official recipient-state bureaucracies and getting closer to the grass roots.

Women and food security. Women suffer more from food shortage than men, while in most cases working considerably longer hours. Their access to sources of income, from farming, wage labour and petty trading is poorer, as is their access to means of production and credit. They have very much weaker, in some cases virtually non-existent, rights in land, most especially where private property in land has been registered. This all relates to their generally subordinate social situation and to a specifically subordinate relation to men as fathers and husbands.

This situation contributes to insecurity of food supplies, not only for women themselves, but for the whole family. Women's days are already full, they often cannot take decisions or borrow money – and yet they are largely in charge of providing food. In some cases they are also assigned much of the work in cash-crop production, which can have negative effects on family cash income. If the woman is forced to do this in addition to all her other work, and especially if she receives none of the proceeds, as may well be the case, then she may be unwilling (or unable) to exert herself.

It is possible to think of technical changes which could reduce women's work, but although these can help, they do not attack the real problem – which is the male definition of what is 'women's work' and of how it shall be valued. That is, women tend to be assigned the longest, hardest and most tedious tasks and their definition as 'women's work' involves a downgrading of status and value. Not only do men often feel it demeaning to undertake such work, but it is assigned no value and thus not rewarded, so that any transfer from husband to wife is considered to be a 'gift' rather than an exchange or obligation. Since women's work is primarily concerned with household reproduction, it does not result in the production of a measurable amount of some commodity, with a market price and thus value. Of course, some part of the work does result in the production of goods which could be commodities (food crops, for example). But the fact that they are not sold, but provided to the husband and family (with further preparation and cooking labour added) as his right in the marriage relationship, deprives them of imputed value. The other services provided as part of household reproduction are even more invisible, so that husbands and national income statisticians can refer to wives as 'dependants' where, in terms of labour and production of use-values, it is more often the other way round. At the same time as being a specific gender problem, this is also an example of the phenomenon noted above, that non-marketed goods and services (direct use-values) are invariably downgraded, if not ignored entirely in comparison with commodities.

Again, it is not hard to propose changes. The question is rather from where the political basis for their achievement should come, since change involves a radical transformation of the whole structure of peasant (and non-peasant) societies. There is little question that socialist regimes have been more sensitive to this issue than others, though actual achievements have been relatively sparse, and there has been a tendency there too, simply to ignore the existence of women's work, in calling for them to be more 'integrated into the economy'. Certainly 'peasant politics' does not seem to be the solution to this problem.

The problem of women's status relates to another aspect of peasant social structure – age hierarchies and the respect accorded to (certain) old men. The absurdity of basing extension advice on the hierarchical control and dispensation of 'knowledge' was noted above, but that should not be taken to imply an absence of such hierarchy in peasant societies. There is a very real sense in which social control by old men, many of them relatively privileged by reason of both gender and social status, retards adjustment to changing circumstances, since these are the people who have insulated themselves from the problems most effectively. This is especially problematic in cases and areas where respect for seniors inhibits disagreement with them, and particularly so because it applies with more than normal force to public occasions on which important decisions are taken.

Thus the factors generating hunger can be seen to range from international capitalism to male dominance and gerontocracy in peasant communities, underpinned from one end to the other by ideologies which take privileged status for granted and assign value primarily, if not solely, to commodities having a market price. This does not, to my mind, point towards any 'grand strategy', rather it points to the limitations of such an approach.

For the general reader, the purpose of this book is not to draw general conclusions but to stimulate thought and discussion which can help people deepen and broaden their own opinions on the topic. It was my aim, when I started out on the research for the book, to draw a coherent set of conclusions, but they have continually eluded me. One of the reasons for this is that conclusions drawn from one area of the study constantly came up against analysis from other parts. Another is that it has led to a continual broadening of the scope of the book, rather than the narrowing down required for drawing conclusions. One can think of all sorts of conclusions about how development policy and aid projects should be redefined and reorganized. But once one comes to analyse the forces within and working upon official aid agencies, one has to conclude that the opportunities for changing things are limited. They exist, and they should be taken, but for specific strategies as to how, the reader is advised to read Robert Chambers' *Putting the Last First* (1983) or a similar work.

One general problem with covering topics as broad as that treated here is that one continually finds gaps which need filling and areas where the analysis is too superficial – not to mention loose ends which need tying. But considered as a basis for critical reading and discussion, this may be no bad thing, since it provides opportunities for the reader to interpose (and counterpose) his or her own ideas and analysis.

The African reader will approach this book with a far more detailed knowledge of the politics of his or her country than I have. For such a reader, I would suggest that the usefulness of the book is the same: as a basis for thought and discussion. Do the economic and political features resemble those described here? Or are there points of similarity which can be illuminated by the analysis here? If not, what are the differences and how do they alter the implicit conclusions of the book? It is only by digging more deeply into the underlying causes of hunger and insecurity of food supply that adequate solutions can be found. It is to that debate, and to the more difficult task of actually changing the situation, that the book is directed.

NOTES

1. A variant is that cattle prices dip when school fees are due, though this is not necessarily related to other seasonal stress periods.
2. This is a standard process where cuts in government expenditure are concerned. To put it crudely, those out doing the work are less well-placed to protect their jobs than those who administer and sit on the endless committees set up to supervise 'improved efficiency'.
3. Coughlin et al. (1987) give examples from Kenya involving overcharging, in one case by 166% and in several others by between 50 and 100%, also citing similar cases from elsewhere. In general, it is far easier to check whether prices are competitive for, say, hoes, fertilizers or biscuits than for pieces of equipment to be used in factories for making them.
4. The significance of referring to rent rather than profit is that the former is defined as a payment due to the owner or controller of a resource which is in fixed supply. It does not imply any participation by the rent-recipient in the production process or its organization. Nor does it depend on reinvestment in increased production or productivity. On this point, Ricardian (Western) and Marxist theories of rent are in agreement.
5. This was historically important as a start to the process, but in many African countries state-sector wages are currently barely enough to live on and certainly not the source of any significant accumulation. Moreover, land prices have usually risen much faster than wages, even during the early years after independence.
6. Here she cites Kasfir (1982) on 'magendo' (Swahili for smuggling, but more generally applied to all black marketing and corruption) in Uganda. The notion of autonomy *from* the state seems to derive from (or be opposed to) the Marxist notion of the autonomy *of* the state from the dominating class.
7. Thus in Tanzania, when cashewnut production began to fall, a wide variety of technical factors were raised and discussed earnestly. But while they may have hastened the decline,

there seems little doubt that villagization was the major problem. Why, one wonders, should insects and plant diseases suddenly take hold just at the time when the population was villagized away from their cashewnuts?

8. There is a close relationship between simplification of extension messages and the use of force, since only a very simplified and direct command can be made the subject of controls. One can check whether peasants have planted by Dec. 12th. One cannot check whether they have planted 'as soon as possible after the onset of the rains', and not with the certainty required for imposing a fine or jail sentence.

9. This example is not merely hypothetical. It underlies the grandfather of all development white elephants, the Tanganyika Groundnut Scheme.

10. That is, the local expenditures could be better used on it. The capital funds would probably not be available for such purposes, since tied to the particular end-use.

11. This is an example of one of the apparently obvious but actually dubious bases for modernization thinking. If development means maximizing the increase in aggregate production, then it is obvious that one should concentrate resources where they achieve the highest return.

12. In the Danish development consultancy world, sociologists and anthropologists are referred to derisively as 'bloede dyr' (soft animals). Their presence is increasingly recognized as necessary, to iron out problems and (especially) to write the general project goals about attention to basic needs, women etc. But they should be careful not to 'spoil' projects by over-insistence on such matters.

13. One of the worst of these at which I have seen concerned a 'women's project', which was in fact merely the key to unlocking much larger sums of money for a male-controlled co-operative society. We were greeted by women lined up to sing songs of welcome and gratitude (or perhaps not, since we could not understand the language). Then, after a visit to their 'demonstration plot', we were given a meal cooked by the women. A secondary purpose of the latter was perhaps to erase the memory that, because of late delivery of seeds and fertilizer (and possibly lacking enthusiasm), the sunflowers in the demonstration plot were at least a foot shorter and considerably poorer than those in surrounding fields. On another occasion, a World Bank team attended the formal opening of a credit project (which in fact came to nothing). Someone from the audience stood and spoke excitedly for a few minutes, this being translated to the assembled dignitaries as an expression of gratitude. What he had actually said was roughly 'Why can't you cut out the messing around and give us the bloody money?'

14. During a consultancy related to small-scale irrigation, a colleague and I asked whether women were among the present or planned plot-holders. Not except for widows working them for their sons we were told, 'If, women got plots of their own, how would we get them to marry?' Amid gales of laughter which rattled the windows, it seemed pointless to pursue the question or write in a requirement which would surely be ignored.

15. I once turned down a 'small-scale dairy project', based on the 'Keiskammahoek model' (and on a blatantly over-priced purchase of land) for one donor agency, this being on account not only of its distributive effects but of the fact that it seemed both unworkable and grossly uneconomic. Within six months, so I was informed, the same project had been funded by another agency.

16. Examples they cite are hand-pumps for shallow wells, steel reinforcing rods, electrical cables and water-pipes. In the case of cables (aluminium conductors) the local firm had lost half its previous market to grant-aid donor supplies. For water-pipes, local industries were running at 35% (plastic) and 13% (steel) capacity, while large quantities were being imported. The article also includes an interesting section on negotiating strategies to offset some of these problems.

17. Doubtless one can find critical writings somewhere from within these organizations, and even 'expressions of deep concern', but there has never been any major attempt to tackle this problem.

18. The World Bank which, if nothing else, is a master in the art of writing reports, has a particular penchant for this. One finds that tractors are expected to last ten years, or that maize is expected to yield 5 tons per hectare, in notes whose type is so small that one almost needs a magnifying glass to read them.

19. Reasons for this could include profiting from overvaluation in the form of bribes or hidden rebates, and the accumulation of hard-currency reserves.

20. This would also be the case under existing monopoly powers, if various hidden subsidies and periodic debt write-offs were entered in formal budgets.

21. State credit agencies seldom dispossess small peasant debtors, partly to avoid the political odium involved, but also partly because small peasant plots are not much use to large state agencies.

22. Three examples from Kenya. In Maasailand, the privatization of group ranches has led to some selling-off of land by pastoralists. The same is reported to be happening among the hunter-gatherer Okiek in Narok District, with Kipsigis and Gusii farmers moving in and paying derisory prices (Cory Kratz, personal communication). In Kisii District some small peasants sell their land to repay debts and maintain consumption (Orvis, 1985b, 1986 and personal observation). In all three cases, it appears that drunkenness on the part of male sellers of land is a significant factor, but this does nothing to help their wives and families who are dispossessed.

23. Even in many of these cases, the analogy was strained by the fact that many of the 'concealed proletarians' in question turned out to depend largely on hired labour.

24. This was in fact pointed out by many of those who used the 'concealed proletarian' formulation.

References

ABC/IDS (1982) – An Evaluation of the EEC Food Aid Programme, by the Africa Bureau Cologne in association with the Institute of Development Studies, Univ. of Sussex, Brussels/Brighton/Cologne, June.

Ali, m. (1984) – 'Women in Famine: The Paradox of Status in India' in Currey and Hugo (eds).

Alvares, C. (1985) – Another Revolution Fails, Ajanta Publications. New Delhi.

Anderson, K. and R. Tyers (1983) – 'European Community's grain and meat policies and US retaliation: effects on international prices, trade and welfare'. Research School of Pacific Studies, National University, Canberra, Australia, Mimeo (cited in BAE 1985).

Andrae, G. and B. Beckmann (1985) – The Wheat Trap: Bread and Underdevelopment in Nigeria. London. Zed Books.

Andrews, D., M. Mitchell, & A. Weber (1979) – The Development of Agriculture in Germany and the U.K.: Comparative Time-Series, 1870–1975. Centre for European Agricultural Studies, Wye College, Ashford.

Arhin, K., P. Hesp and L.v.d. Laan (eds) (1985) – Marketing Boards in Tropical Africa, London, KPI.

Ba, T. Aliou and B. Crousse (1985) – 'Food Production Systems in the Middle Valley of the Senegal River', International Social Science Journal 37(3), 389–400.

BAE (1985) – Agricultural Policies in the European Community: Their Origins, Nature and Effects on Production & Trade. Policy Monograph No. 2. Bureau of Agricultural Economics, Canberra, Australia. (Distributed by Australian Government Publishing Service).

Bairoch, P. (1973) – 'Agriculture and the Industrial Revolution', in C.M. Cipolla (ed.) The Fontana Economic History of Europe, Vol. 3. London, Fontana.

Bantje, H. (1985) – 'Time Related Aspects of Food Security'. Paper to Workshop on Alternative Approaches to food security, Mikumi, Tanzania, May. Mimeo.

Bard, R.L. (1972) – Food Aid and International Agricultural Trade., Heath & Co. Lexington, USA.

Barnes, C. (1976) – An Experiment with African Coffee Growing in Kenya: the Gusii, 1933–50 Ph.D Thesis, Michigan State University.

Barnet, R.J. and R.E. Mueller (1974) – Global Reach, Simon and Schuster, New York.

Bates, R.H. (1981a) – Markets and States in Tropical Africa. Berkeley, USA. Univ. of Calif. Press.

Bates, R.H. (1981b) – Food Policy in Africa: Political Causes and Social Effects. Food Policy, August: 147–57.

Beckman, B. (1986) – 'Peasants and Democratic Struggles in Nigeria'. Paper to AKUT Conference on Labour & Democracy, Uppsala, Sweden, Sept. and to Review of African Political Economy Conference, Univ. of Liverpool, Sept. Mimeo.

Belshawe, D. (1978) – 'Taking Indigenous Technology Seriously: the case of inter-cropping systems in East Africa'. Workshop on the Uses of Indigenous Technical Knowledge. Inst. of Development Studies, Univ. of Sussex, Mimeo, April.

Berry, S. (1984) – 'The Food Crisis and Agrarian Change in Africa: A Review Essay', African Studies Review, 27(2), June. 59–111.

Bryceson, D.F. (1978) – 'Peasant Food Production and Food Supply in relation to the Historical Development of Commodity Production in Pre-Colonial and Colonial Tanganyika'. BRALUP

Discussion Paper, Bureau of Resource Assessment and Land-Use Planning (now Institute of Resource Assessment), University of Dar es Salaam, Mimeo.

Bryceson, D.F. (1982) – 'Tanzanian Grain Supply: Peasant Production and State Pollicy', – *Food Policy* May: 113–124.

Bryceson, D.F. (1984) – 'Nutrition and the Commoditization of Food Systems in Sub-Saharan Africa', Paper to Conference on 'Political Economy of Health and Disease in Africa and Latin America'. Mexico City. Mimeo.

Bryceson, D.F. (1985) – 'Peasasnt Household Food Consumption'. Paper to Workshop on Alternative Approaches to Food Security, Mikumi, Tanzania, Mimeo, May.

Buch-Hansen M. and H. Secher-Marcussen (1982) – 'Contract Farming & the Peasantry: Case-Studies from Kenya.' *Review of African Political Economy* No. 23, Jan.–April.

Bull, D. (1982) – *A Growing Problem: Pesticides and the Third World Poor*. OXFAM, Oxford.

Burch, D. (1987) – *Overseas Aid and the Transfer of Technology*. Avebury, Aldershot.

Burniaux, J-M and J. Waelbroeck (1985) – 'Agricultural Protection in Europe – Its Impact on Developing Countries', in *EEC and the Third World: a Survey*, No. 5, London, Hodder & Stoughton for ODI and IDS.

Bush, R. (1985) – 'Drought and Famines', *Rev. of Afr. Pol. Econ.* No. 33, August: 59–63.

Caspari, C. (1983) – *The Common Agricultural Policy, the Direction of Change*. Economist Intelligence Unit, Special Report No. 159, London.

Cathie, J. (1985) – 'US and EEC Agricultural Trade Policies', *Food Policy*, February: 14–28.

Chambers, R. (1983) – *Rural Development: Putting the Last First*. London, Longman.

Chambers, R., R. Longhurst R. and A. Pacey (eds.) (1981) – *Seasonal Dimensions to Rural Poverty*. London, Pinter.

Chambliss, M. (1982) – 'US food aid develops cash markets', *Foreign Agriculture*, July: 6–8, USDA. Washington.

Chapin, G. and Wasserstrom R. (1981) – 'Agricultural Production and Malaria Resurgence in Central America and India', *Nature* 293(17).

Chayanov, A.V. (1966) – *The Theory of Peasant Economy* Thorner, B. Kerblay and R. Smith (eds) Homewood. Illinois, USA.

Clairemonte, F. and J. Cavanagh (1981) – *The World in their Web: the Dynamics of Textile Multinationals*. London, Zed Press.

Clay, E.J. (1985) – 'The 1974 and 1984 Floods in Bangladesh; from Famine to Food Crisis Management', *Food Policy*, August. pp. 202–6.

Clay, E.J. and H.W. Singer (1982) – 'Food Aid and Development: The Impact and Effectiveness of Bilateral PL 480 Title I – Type Assistance', AID Programme Evaluation Discussion Paper No. 15, Washington D.C., USAID, December.

Cliffe, L. & R. Moorsom (1979) – 'Rural Class-Formation and Ecological Collapse in Botswana.' *Review of African Political Economy*, No. 15/16: 35–52.

Clough, P. and G. Williams (n.d.) – 'Decoding Berg; The World Bank in Rural Northern Nigeria', in Watts M. (ed) (1987:168–201).

Collinson, M.P. (n.d.) – 'A Farm Economic, Survey of a part of Usmao Chiefdom, Kwimba District, Tanganyika', Western Research Station, Ukiriguru, Tanzania, Mimeo.

Collinson, M.P. (n.d.) – 'Farm Management Survey Report No. 3, Luguru Ginnery Zone: Maswa Disrict, Shinyanga Region, Tanganyika', W.R.C. Ukiriguru, Tanzania, Mimeo.

Comm. (1985a) – *The Agricultural Situation in the Community, 1984 Report*. Commission of the European Communities, Brussels.

Comm. (1985b) – *Perspectives for the Common Agricultural Policy*. Commission of the European Communities, COM(85)333, Final, Brussels.

Coughlin, P. (1987) – 'Tied Aid, Industrial Development, and New Tactics for Negotiations: Observations from Kenya.' paper to the Kenyan Economic Association discussion meeting, 17/12/1987, Mimeo, Nairobi.

Coulson, A.C. (1977) – 'Agricultural Policies in mainland Tanzania'. *Review of African Political Economy* No. 10. Sept.–Dec.

Crotty, R. (1980) – *Cattle, Economics and Development*. Farnham, Commonwealth Agricultural Bureaux.

Currey, B. and G.T. Hugo (eds) (1984) – *Famine as a Geographical Phenomenon*. Reidel. Dordrecht.

Cutler, P. (1985) – 'Detecting Food Emergencies: Lessons from the 1979 Bangladesh Crisis', *Food Policy* 10(3), August: 207–24.

Dinham, B. and C. Hines (1983) – *Agribusiness in Africa*, Earth Resources Research, London.

Djurfeldt, G. (1981) – 'What Happened to the Agrarian Bourgeoisie and Rural Proletariat under Monopoly Capitalism?', *Acta Sociologica* (24), 3:167–91.

Doering, O.C. III (1982) – 'The Effects of Rising Energy Prices on Agricultural Production Systems', in W. Lockeretz (ed.) *Agro-culture as Consumer and Producer of Energy*. Westview, Boulder, Colorado. pp. 9–24.

Doornbos, M. & K.N. Nair – 'Operation Flood Re-Examined; Report of a Workshop.' *Economic and Political Weekly*, 14.2.87: 266–68.

D'Souza, F. and Shoham (1985) – 'The Spread of Famine in Africa.' *Third World Quarterly* 7(3), July 1985:515–531.

Dutkiewicz, P. and G. Williams (1986) – 'All the king's horses and all the king's men couldn't put Humpty Dumpty together again': A prolegomena to any future proposals for African development'? IDS Bulletin 18(3) July 1987: 39–44.

Ellis, F. (1982) – 'Agricultural Price Policy in Tanzania', *World Development*. 110(4):263–83.

Ellis, F. (1983) – 'Agricultural Marketing and Peasant-state Transfers in Tanzania'. *Journal of Peasant Studies*. 10(4).

EC Court of Auditors (1980) – *Special Report on Community Food Aid*, Brussels.

Fagen, (1975) – 'The United States and Chile: Roots and Branches.' *Foreign Affairs*, 33(304).

FAOa (annual) – *FAO Production Yearbook*. Rome. (year of printing, i.e FAOa 1982–1983 Yearbook)

FAOb (annual) – *FAO Commodity Review and Outlook*. Rome. (FAOb 1985 refers to 1984–85 number)

FAOc (annual) – *FAO Trade Yearbook*, Rome. (year of printing, i.e. FAOc 1982–1983 Yearbook).

FAOd (annual) – *State of Food and Agriculture*. Rome. (year of printing and reference same)

FAOe (annual) – *Food Outlook (annual) Statistical Supplement*. Rome (published year following that of reference).

FAOf (monthly) – *FAO Monthly Bulletin of Statistics*. Rome.

FAOg (annual) – *Food Aid in Figures*, Rome.

FAO 1985 – *Food Situation in African Countries Affected by Emergencies – special report 26.3.1985. Rome.

FAO (1986) – *Food Supply Situation & Crop Prospects in Sub-Saharan Africa – Special Report 21.4.1986. Rome.

Fennell, R. (1979) – *The Common Agricultural Policy of the European Community*, London, Granada.

Ford, J. (1971) – *The Role of the Trypanosomiases in African Ecology*. Oxford, Clarendon Press.

Francois, P. (1982) – 'Class Struggles in Mali', *Review of African Political Economy*, No. 24, pp. 22–38.

Friedmann, H. (1980) – 'Household Production and the National Economy: Concepts for the Analysis of Agrarian Formations', *Jo. of Peasant Studies*, 7(2):158–84.

Froebel, F. (1985) – 'Changing patterns of World Market Integration of Third World Countries'. Paper to conference on *Economic policies and Planning under Crisis Conditions in Developing Countries*, Harare, Zimbabwe, Mimeo.

Froebel, F., J. Heinrichs and O. Kreye (1977) – *Die Neue Internationale Arbeitsteilung*. Reinbek.

Garcia, R. (1981) – *Drought and Man; the 1972 Case History. Vol. I: Nature Pleads Not Guilty*. Oxford, Pergamon.

Garcia, R. (1984) – '*Food Systems and Society: A Conceptual and Methodological Challenge*', Food Systems Monograph, UNRISD, Geneva.

Gartrell, B. (1985) – 'The Roots of Famine in Karamoja', *Rev. of African Pol. Economy* No. 33, Aug 1985:102–110.

George, S. (n.d.) – *Feeding the Few: Corporate Control of Food*. Institute for Policy Studies, Washington D.C./Amsterdam.

George, S. (1985) – 'Rejoinder: on the need for a broader approach' (to Paulino and Mellor 1984), *Food Policy* 10(1), February 75–79.

Gibbon, P. and M. Neocosmos (1985) – 'Some problems in the political economy of "African Socialism" ', in H. Bernstein and B. Campbell (eds) *Contradictions of Accumulation in Africa. Studies in Economy and State,* Beverly Hills, Sage.

Gill, P. (1986) – *A Year in the Death of Africa.* London, Paladin.

Gittinger, J.P., J. Leslie and C. Hoisington (eds) *Food Policy: Integrating supply, distribution and consumption.* Johns Hopkins Univ. Press, for World Bank, Baltimore.

Good, K. (1986) – 'Systemic Agricultural Mismanagement: the 1985 "Bumper" Harvest in Zambia.' *Journal of Modern African Studies* 24(1): 257–84.

Government of Kenya (1979) – *Child nutrition in Rural Kenya Nairobi,* Ministry of Planning & Economic Development, Central Bureau of Statistics.

Greenough, P.R. (1982a) – *Prosperity and Misery in Modern Bengal: the Famine of 1943–44.* OUP, New York.

Greenough, P.R. (1982b) – 'Comments from a South Asian Perspective', *Jo. of Asian Studies.* XLI(4):784–97.

Grigg, D. (1985) – *The World Food Problem 1950–80.* Blackwell, Oxford.

Guyer, J. (1984) – 'Women's Work and Production Systems: A Review of Two Reports on the Agricultural Crisis', *Review of African Political Economy* No. 27/28, pp. 186–91.

Harris, S.A. (1980) – 'EEC Sugar and the World Market', Paper to the *International Sweetener and Alcohol Conference on the Future of Sugar,* 1–3 April 1980, London.

Harris, S.A. (1984) – 'Rejoinder: the Mixed Blessing of Domestic Support Policies', *Food Policy* 9(4), November, pp. 328–30.

Hansen, M., L. Busch, J. Burkhardt and W.B. Lacey (n.d., but 1985) – 'Plant Breeding and Biotechnology', Paper for Kentucky Agricultural Experiment Station Journal. Mimeo. (received from L. Busch).

Heald, S. and Hay A. (1985) – 'Problems of Theory and Research: Comments on Buch-Hansen and Marcussen'. *Review of African Political Economy* No. 34, December, pp. 89–94.

Hedlund, H. (1979) – 'Contradictions in the Peripheralization of a Pastoral Society: the Maasai', *Review of African Political Economy* No. 15/16 May–Dec. pp. 15–34.

Hewitt, A. (1983) – 'Stabex: an Evaluation of the Economic impact over the First Five Years', *World Development* 11(12).

Heyer, J., R. Roberts and G. Williams (eds) (1981) – *Rural Development in Tropical Africa.* London. Macmillan.

Hoffmeyer, B. (1982) – *The EEC's Common Agricultural Policy and the ACP States.* CDR Research Report No. 2, Centre for Development Research, Copenhagen.

Holt, J.F.J. (1983) – 'Ethiopia: Food for Work or Food for Relief', *Food Policy* 8(3), pp. 187–201.

Hyden, G. (1980) – *Beyond Ujamaa in Tanzania.* London, Heinemann.

Iliffe, J (1983) *The Emergence of African Capitalism,* Minneapolis, University of Minnesota Press.

Jackson, T. (with Barbara Eade) (1984) – *Against the Grain.* Oxfam, Oxford.

Jensen, Kr. M. and A. Reenberg (1980) – *Dansk Landbrug: Udvikling i Produktion og Kulturlandskab.,* Geografforlaget, Brenderup, Denmark.

Johnson, D.G. (1975) – *World Food Problems and Prospects.* American Inst. for Public policy Research, Washington D.C.

Jones, W.O. (1961) – 'Economic Man in Africa', *Food Research Institute Studies,* Stanford, California, USA.

Jonsson, U., V. Leach and B. Llungqvist (1985) – 'Household Food Security and Child Survival in Tanzania', Paper to IRA /INUSC *Workshop on Food Security,* Mikumi, Tanzania, May.

Josling, T. and R. Barichello (1984) – 'International Trade and World Food Security: the role of the Developed Countries since the World Food Conference', *Food Policy* 9(4), November, pp. 317–27.

Josserand, H.P. (1984) – 'Farmers' Consumption of an Imported Cereal and the Cash/Foodcrop Decision: An Example from Senegal', *Food Policy* 9(1), February, pp. 27–34.

Junne, G. (1986) – 'New Technologies and Third World Development', *Viertel Jahres Berichte,* No. 103, Friedrich-Ebert Stiftung, Bonn, March, pp. 3–10.

Kasfir, N. (1982) – 'State, Magendo and Class-Formation in Uganda', Paper to American African Studies Association annual meeting, Washington D.C.

Kitching, G. (1980) – *Class and Economic Change in Kenya: The Making of an African Petite-Bourgeoisie* New Haven and London, Yale U.P.

Kitching, G. (1982) – *Development and Underdevelopment in Historical Perspective*. London. Methuen/OUP.

King-Meyers, L. (1979) – 'Nutrition, in Kenya: Problems, Programmes, policies and Recommendations'. USAID Staff Report, Mimeo, Nairobi.

Kjekshus, H. (1977) – *Ecology Control and Economic Development in East African History*. London, Heinemann.

Klaasse Bos, A. (1981) – 'Constraints on Strategies for Reaching Food Self-Sufficiency in Sahel Countries'. Paper to *Polish-Dutch Seminar on 'Barriers and Alternatives to Agricultural Development in the Third World'*. Mimeo, Warsaw, June.

Koester, U. (1982) – *Policy Options for the Grain Economy of the European Community: Implications for Developing Countries*. International Food Policy Research Institute (IFPRI), Research Report No. 35, Washington D.C.

Koester, U. and M.D. Bale (1984) – *The Common Agricultural Policy of the European Community*, World Bank Staff Working Paper No. 630, Washington D.C.

Koester, U. and A. Valdes (1984) – 'Reform of the CAP: Implications for the Third World', *Food Policy* 9(20), pp. 94–8.

Körner, P., G. Maass, T. Siebold and R. Tetzlaff (1984/1986) – *The IMF and the Debt Crisis*. London, Zed Press 1986 (translation of 1984 German original).

Lattimore, R. and S. Weedle (1981) – *The World Price Impact of Multilateral Free Trade in Dairy Products*. Agriculture Canada Working Paper, Ottawa. (cited in BAE 1985).

Leftwich, A. and D. Harvie (1986) – 'The Political Economy of Famine: a preliminary report' Mimeo, Dept. of Politics, Univ. of York.

Lemma, H. (1985) – 'The Politics of Famine in Ethiopia.' *Rev. of African Pol. Economy* No. 33, August:44–58.

Leys, R. (1985/1986) – 'Drought and Drought Relief in Southern Zimbabwe.' *Forsknings-rapport 1985/2*, Inst. of Political Studies, Univ. of Copenhagen. Also in Lawrence (ed, 1986: pp. 258–77).

Loft, M. *et al.* (1982) – 'An Evaluation of the Kigome Rural Development Project', ERB (restricted) Paper, No. 82. Mimeo, Economic Research Bureau, University of Dar es Salaam.

MacGaffey, J. (1983) – 'How to Survive and Become Rich amidst Devastation: the Second Economy in Zaïre'. *African Affairs*, 82(3–28): 351–66.

Maletnlema, T.N. (1980) – 'Food Aid, Views and Experiences of a Third World Country', in *Nordic Food Aid*, Nordic Food Council, Stockholm.

Maletnlema, T.N. (1985) – 'Food and Nutrition Policy as a Tool in Programmes for Food Security in Tanzania', TFNC Report No. 937, mimeo, Dar es Salaam, also presented to workshop on food security, Mikumi, Tanzania, May.

Mamdani, M. (1982) – 'Karamoja: Colonial Roots of Famine', *Rev. of African Pol. Econ.* No. 25, September–December, pp. 66–73.

Mariam, Mesfin Wolde (1984) – *Rural Vulnerability to Famine in Ethiopia, 1955–77*, Delhi, Vikas Publishing House.

Matthews, A. (1985) – *The Common Agricultural Policy and the Less Developed Countries*, Trocaire with Gill & McMillan, Dublin.

Maxwell, S. (1986a) – 'Food Aid: Agricultural Disincentives and Commercial Market Displacement'. Institute of Development Studies, Discussion Paper (IDS/DP) 224, Mimeo, Univ. of Sussex.

Maxwell, S. (1986b) – 'Food Aid to Senegal: Disincentive Effects and Commercial Market Displacement', IDS/DP 225, Mimeo, Univ. of Sussex.

Maxwell, S. (1986c) – 'Food Aid Ethiopia: Disincentive Effects and Commercial Displacement', IDS/DP 226. Mimeo, Univ. of Sussex.

Maxwell, S. and H.W. Singer (1979) – 'Food Aid to Developing countries: A Survey', *World Development* 7(3). pp. 225–46.

Mitchell, M. and C. Stevens (1983) – 'Mauritania – the Cost-effectiveness of EEC Food Aid', *Food Policy* 8(3), August, pp. 202–8.

Mooney, P.R. (1979) – *Seeds of the Earth*, ICDA, London and Toronto.

Mooney, P.R. (1983) – *The Law of the Seed*. Special Issue of *Development Dialogue*, 1–2.

Morgan, D. (1979) – *Merchants of Grain*. London, Weidenfeld and Nicholson.

Moss, J. and J. Ravenhill (1983) – 'Trade between the ACP and the EEC during Lomé I', in *EEC and the Third World, a Survey*, No. 3, London, Hodder & Stoughton for ODI and IDS.

Morrison, T.K. (1984) – 'Cereal imports by developing countries: trends and determinants' *Food Policy* 9(1), February: 13–26.

Narayanan Nair, K. & M.G. Jackson (1985) – 'Alternatives to Operation Flood II Strategy.', in Alvares C. (ed.) (1985).

Newbury, C. (1984) – Unpublished paper in Zaire for Workshop on 'Food Systems in Central and Southern Africa.' Mimeo. School of Oriental and African Studies. London.

Nielsen, G.Aa. (1986) – *Dansk Landbrug den for EF* (Danish Agriculture outside the EEC), Copenhagen, Flkebevaegelsen mod EF.

O'Brien, J. (1981) – 'Sudan: An Arab Breadbasket?' MERIP Reports, September.

O'Brien, J. (1985) – 'Sowing the Seeds of Famine: The Political Economy of Food Deficits in the Sudan', *Review of African Political Economy*. No. 33, August: 23–32.

Oculi, O. (1979) – 'Dependent Food Policy in Nigeria, 1975–79.' *Review of African Political Economy*, No. 15/16: 63–74.

OECD (1984a) – *Development Cooperation, 1984 Review*. Paris 1984.

OECD (1984b) – *External Debt of Developing Countries; 1983 Survey*. Paris 1984.

OECD (1985) – *Twenty-five years of Development Cooperation, a Review*. Paris.

OECD (1986) – *Financing and External Debt of Developing Countries – 1985 Survey*. Paris.

Oedegaard (Ödegaard) K. (1985) – *Cash Crop versus Food Crop Production in Tanzania*. Lund Economic Studies, No. 33, Lund, Sweden.

Orvis, S. (1985a) – 'Men and Women in a Household Economy: Evidence from Kisii'. Institute of Development Studies (IDS) Working Paper No. 432, Mimeo, University of Nairobi, Oct.

Orvis, S. (1985b) – 'A Patriarchy Transformed: Reproducing Labor and the Viability of Smallholder Agriculture in Kisii', IDS Working Paper No. 434, Mimeo, University of Nairobi, December.

Orvis, S. (1986) – 'Men, Women and Agriculture in Kisii, Kenya', Paper to African Studies Association meeting Oct. 29–Nov. 2, Mimeo.

Orvis, S. (1987) – 'State-Peasant Relations in Kenya: the Development Policy Arena' for Midwestern Political Science Association Meetings, Mimeo, Chicago, April.

Oxfam (1984) – *The Sahel: Why the Poor Suffer Most*. Oxfam Public Affairs Unit, Oxford.

Paarlberg, D. (1984) – 'Tarnished Gold: US Farm Commodity Programmes after 50 Years', *Food Policy*, May.

Paarlberg, R.L. (1986) – Responding to the CAP: Alternative strategies for the USA', *Food Policy*, May.

Petit, M. (1985) – *Determinants of Agricultural Policies in the United States & the European Community*. Washington D.C. IFPRI Research Report No. 51, November.

Pinstrup-Andersen, P. (1981) – 'Nutritional Consequences of Agricultural Projects'. World Bank S.W.P. No. 456, Washington.

Pinstrup-Andersen, P. (1983) – 'Export Crop Production and Malnutrition'. Inst. of Nutrition, Univ. of North Carolina, Occasional Paper 11(10), 1983.

Pinstrup-Andersen, P. (1985) – 'The Impact of Export-Crop Production on Human Nutrition' in M. Biswas and P. Pinstrup-Andersen (eds) *Nutrition and Development*, O.U.P. London.

Paulino, L.A. and J.W. Mellor (1984) – 'The food situation in developing countries: two decades in review', *Food Policy* 9(4), November 291–302.

Pottier, J. ed. (1986) – *Food Systems in Central and Southern Africa*. London, School of Oriental and African Studies.

Poluha, E. (1986) – 'The Producers Cooperative as an option for Women – a Case Study from Ethiopia. Paper presented at a seminar on Cooperatives Revisited', Uppsala, November, Mimeo.

Price Gittinger J. *et al* (eds) (1985) – (see under Gittinger).

Raikes, P.L. (1975) – *The Development of Mechanized Commercial Wheat Production in North Iraqw, Tanzania*. Unpublished Ph.D. Thesis, Stanford Univ. Calif, USA.

Raikes, P.L. (1976) – 'The Development of Coffee Production in West Lake Region, Tanzania'. CDR Papers A.76.9, Centre for Development Research, Copenhagen, Mimeo.

Raikes, P.L. (1981) – *Livestock Development and Policy in East Africa*. Scandinavian Institute of African Studies, Uppsala.

Raikes, P.L. (1981b) – 'Seasonality in the Rural Economy', in Chambers, Longhurst and Pacey (eds) (1981) *Seasonal Dimensions to Rural Poverty*, pp. 67–73.

Raikes, P.L. (1982) – 'Djurfeldt's 'What Happened. . . . under Monopoly Capitalism?' A Comment', *Acta Sociologica* (25), 2:159–65.

Raikes, P.L. (1983) – 'State and Agriculture in Tanzania'. Unpublished book-length typescript.

Raikes, P.L. (1984a) – 'Stabex: Offsetting Shortfalls or Distributing Windfalls?' CDR Project paper D.84.6, Mimeo, Copenhagen.

Raikes, P.L. (1984b) – 'Food Policy and Production in Mozambique since Independence', *Rev. of African Pol. Econ.* No. 29. (also in Pottier 1986).

Raikes, P.L. (1984c) – 'EEC Aid in Perspective; the Size and Direction of Flows', CDR Project paper D.84.11. Mimeo, Centre for Development Research, Copenhagen.

Raikes, P.L. (1985a) – 'National Food Balance or Food Security as a Basic Need. Is there a policy option for Tanzania?' Paper to Workshop on Alternative Approaches to Food Security, Mikumi, Tanzania, May. mimeo.

Raikes, P.L. (1985b) – 'The Performance of Agriculture in Africa: Country Case-Study of Tanzania', for Advisory Council for Development Cooperation, Dublin, Ireland.

Raikes, P.L. (1985/6) – 'Socialisme og Landbrug i Tanzania', *Den ny Verden* 19(1), 10–40.

Raikes, P.L. (1986a) – 'Socialism and Agriculture in Tanzania.' Paper to Conference on *Tanzania after Nyerere*, School of Oriental and African Studies, London, Mimeo.

Raikes, P.L. (1986) – 'Eating the Carrot and Wielding the Stick: the Agricultural Sector in Tanzania'. in

Raikes, P.L. (1986c) – 'Flowing with Milk and Money. Agriculture and food production in Africa and the EEC.' in Lawrence P. (ed., 1986: pp. 160–76).

Raikes, P.L. (1988) – 'The Common Agricultural Policy and the African Food Crisis.', *C.A.P. Briefing No. 13, The CAP, African Food Security and International Trade*, London, CIIR (Catholic Institute for International Relations).

Raikes, P.L. and W. Meynen (1972) – 'Dependency, Differentiation and the Diffusion of Innovations; a Critique of Extension Theory and Practice', East African Universities Social Science Council, *Conference*, Nairobi, December, Mimeo.

Rama, R. (1985) – 'Do transnational agribusiness firms encourage the agriculture of developing countries? The Mexican experience', *Int.l Social Science Journal* 37(3), 331–44.

Ravenhill, J. ed. (1986) – *Africa in Economic Crisis*, London, Macmillan.

Redda, A. (1984) – 'The Famine in Northern Ethiopia', *Review of African Political Economy* No. 27/28, February, pp. 157–63.

Richards, P. (1983) – 'Ecological Change and the Politics of African Land Use', *African Studies Review* 26(2).

Richards, P. (1985) – *Indigenous Agricultural Revolution*. London, Hutchinson.

Richards, P. (1986a) – *Coping with Hunger: Hazard and Experiment in an African Rice-Farming System*. London, Allen & Unwin.

Richards, P. (1986b) – 'Coping with Hunger in Tropical Africa; What do we know? What can be done?' Paper for the Conference of the African Studies Association of the UK, Univ. of Kent 17–19 September, Mimeo.

Richards, P. (1986) – *Indigenous African Economics*

Roberts, I.M. (1982) – 'EEC Sugar Support Policies and World Market Prices: a Comparative Static Analysis'. Paper to *Annual Conf. of the Australian Agric. Econ. Society*, Melbourne. (cited in BAE 1985 of which Roberts was principal author).

Rogers, B.L. and M.B. Wallerstein (1985) – *PL 480 Title I: A Discussion of Impact Evaluation Results and Recommendations*. USAID, Washington D.C. February.

Rotberg, R.L. ed. (1983) – *Imperialism, Colonialism, and Hunger: East and Central Africa*. Lexington, Mass. & Toronto, Lexington Books.

Ruivenkamp, G. (1986) – 'The Impact of Biotechnology on International Development: Competition between Sugar and new Sweeteners', *Viertel Jahres Bericht* 103, March, pp. 89–102.

Ruiz, O. (1981) – 'Economic Policies and the Nutritional Status of the Urban Poor in Chile, 1968–76.' in G. Solimano and L. Taylor (eds.) *Food Price Policies and Nutrition in Latin America*. United Nations Univ. *Food and Nutrition Bulletin*, Supplement No. 3, Tokyo.

Sanders, D. (with R. Carver) (1985) – *The Struggle for Health: Medicine and the Politics of Underdevelopment*, London, Macmillan.

Sano, H-O (1983) – *The Political Economy of Food in Nigeria, 1960–1982*. Research Report No. 65, Scandinavian Institute of African Studies, Uppsala, Sweden.

Sano, H-O (1985) – 'Danmarks Foedervarebistand' (Denmark's Food Aid), Mimeo, Copenhagen.

Sarris, A. and J. Freebairn (1983) – 'Endogenous price policies and international wheat prices', *American Journal of Agricultural Economics* 65(2), 214–24.

Schoepf, B.G. (1985) – 'Food Crisis and Class Formation in Shaba', *Review of African Political Economy*, No. 33, August, pp. 33–43.

Sen, A.K. (1981) – *Poverty and Famines*, Oxford, Clarendon Press.

Sexauer, B. (1980) – 'Critical Choices in World Food Aid and Nutrition: Discussion', *Amer. Jo. of Agric. Econ.* December, pp. 993–4.

Shanin, T. (1972) – *The Awkward Class*, London, OUP.

Shenton, R. (1986) – *The Development of Capitalism in Northern Nigeria*. London, James Currey.

Shenton, R. and M. Watts (1979) – 'Capitalism and Hunger in Northern Nigeria'. *Review of African Political Economy* Nos. 15/16, May–December, pp. 53–62.

Shepherd, J. (1975) – *The Politics of Starvation*. Washington, Carnegie Endowment for International Peace.

SIPRI (1984) – *SIPRI Yearbook 1984*. Stockholm International Peace Research Institute. Stockholm.

Southall, A.W. (1981) – 'The Ciskei Scandal: Keiskammahoek, 1981, after Thirty Years'. Paper to *African Sudies Association* Meeting.

Speirs, M. (1987) – 'From food aid to food strategy – the case of Mali'. CAP Briefing No. 3. CIIR, London.

Spencer, I.R.G. (1983) – 'Pastoralism and Colonial Policy in Kenya, 1895–1929', in Rotberg (1983).

Spitz, P. (1985) – 'Food Systems and Society in India: the origins of an interdisciplinary research', *Int.l Social Science Jo.* 37(3), 371–88.

Sporrek, A. (1985) – *Food Marketing and Urban Growth in Dar es Salaam*. Lund Studies in Geography. No. 51.

Stevens, C. ed. (annual) – *EEC and the Third World: A Survey*. London, Hodder & Stoughton in association with ODI and IDS.

Sunkel, O. (1977) – 'The Development of Development', *IDS Bulletin*, March, Sussex.

Svedberg, P. (1987) – 'Undernutrition in Sub-Saharan Africa: A Critical Assessment of the Evidence'. WIDER Working Papers WP15, UN University, Helsinki, Mimeo.

Swift, J.J. (1977) – 'Sahelian Pastoralists; Underdevelopment, Desertification and Famine', *Annual Review of Anthropology* 6, pp. 457–78.

Talbot, R. (1979) – 'The European Community's Food Aid Programme', *Food Policy*, November, pp. 269–84.

Tangermann, S. and W. Krostitz (1982) – *Protectionism in the Livestock Sector, with particular reference to the International Beef Trade*. Study for FAO Intergovernmental Group of Meat (cited in BAE, 1985).

Tarrant, J.R. (1980) – *Food Policies*, Wiley, Chichester & New York.

Teke, T. (1986) – 'The State and Rural Cooperatives in Ethiopia'. Paper presented at Seminar on 'Co-operatives Revisited', Uppsala, Sweden, November, Mimeo.

Telex Africa (fortnightly) – mimeographed newssheet with information on EEC and Africa. Brussels. (also as 'TA').

Terhal, P. and Doornbos (1983) – 'Operation Flood, Development and Commercialization.' *Food Policy*, Aug 1983:235–39.

Thomson, A.M. (1983) – 'Egypt – Food Security and Food Aid', *Food Policy* 8(3), August pp. 176–86.

Thomson, A.M. (1983b) – 'Somalia – Food Aid in a Long-term Emergency', *Food Policy* 8(3), August, pp. 209–19.

Thompson, R.L. (1986) – 'Recent developments in United States agriculture and food policy.' *Jo. of Agricultural Economics (UK)*, 37(3): 311–316.

Tibaijuka, A. (1981) – 'Food Aid and Food Production Programmes in Tanzania'. Uppsala, Sveriges Lantbruksuniversitet, Mimeo.

Tibaijuka, A. (1984) – *An Economic Analysis of Smallholder Banana-Coffee Farms in the Kagera*

Region, Tanzania, Uppsala, Swedish University of Agricultural Sciences, Dept. of Economics and Statistics, Report No. 240. (Published Ph.D. Thesis).

Tickner, V. (1985) – 'Military Attacks, Drought and Hunger in Mozambique', *Review of African Political Economy*, No. 33, August, pp. 89–90.

Timmer, P.C., W.P. Falcon and S.R. Pearson (1983) – *Food Policy Analysis*. Baltimore and London, Johns Hopkins.

United Republic of Tanzania/FAO (1984) – *Tanzania National Food Strategy*. 2 Vols. Rome.

Valdes, A. and J. Zietz (1980) – *Agricultural Protection in OECD Countries: Its Cost to Less Developed Countries*. IFPRI Research Report No. 21, Washington D.C.

Vengroff, R. (1982) – 'Food Aid and Dependency: PL 480 Aid to Black Africa', *Jo. of Modern African Studies* 20(1) pp. 27–43.

Wallace, T. (1985) – 'Refugees and Hunger in Eastern Sudan', *Review of African Political Economy*, No. 33, August, pp. 64–68.

Wallerstein, M.B. (1980) – *Food for War – Food for Peace*. London and Cambridge Mass., M.I.T. Press.

Watts, M. (1983) – *Silent Violence: Food, Famine and Peasantry in Northern Nigeria*. Berkeley & Los Angeles, Univ. of California Press.

Watts, M. (ed.) (1987) – *State, Oil and Agriculture in Nigeria*. Inst. of International Studies, Univ. of California Press, Berkeley.

Weber, A. and N. Gebauer (1986) – 'Fertilizer Use and Grain Yields in World Agriculture: A Cross-sectional Analysis', *Quarterly Journal of International Agriculture*, DLG-Verlag, Frankfurt/Main, pp. 198–216.

Weiner D., S. Moyo, B. Munslow and P. O'Keefe (1985) – 'Land Use and Agricultural Productivity in Zimbabwe', *Journal of Modern African Studies* 23(2). June.

Whitehead, A. (1981) – *A Conceptual Framework for the Analysis of the Effects of Technological Change on Rural Women*. World Employment Research Monograph. ILO.

Wijkman, A. and L. Timberlake (1984) – *Natural Disasters: Acts of god or Acts of Man?*, London, Earthscan.

Williams, G. (1981) – 'The World Bank and the Peasant Problem'. In Heyer, Roberts and Williams, *Rural Development in Tropical Africa*.

Wisner, B. (1983) – 'Energy and Self-Reliant Struggle in African Development'. Draft assessment of Basic Needs Approaches, Beijer Inst. Stockholm, mimeo.

World Bank (1981) – *Accelerated Development in Sub-Saharan Africa*. Washington. D.C.

World Bank (1986) – *World Development Report 1986*, Washington D.C.

Index